ORGANIC NONLINEAR OPTICAL MATERIALS

Advances in Nonlinear Optics
A series edited by Anthony F. Garito, *University of Pennsylvania, USA*
and François Kajzar, *DEIN, CEN de Saclay, France*

Volume 1 Organic Nonlinear Optical Materials
Ch. Bosshard, K. Sutter, Ph. Prêtre, J. Hulliger, M. Flörsheimer, P. Kaatz and P. Günter

This book is part of a series. The publisher will accept continuation orders which may be cancelled at any time and which provide for automatic billing and shipping of each title in the series upon publication. Please write for details.

ORGANIC NONLINEAR OPTICAL MATERIALS

Ch. Bosshard, K. Sutter, Ph. Prêtre, J. Hulliger
M. Flörsheimer, P. Kaatz and P. Günter

Institute of Quantum Electronics
ETH-Hönggerberg HPF
Switzerland

GORDON AND BREACH PUBLISHERS
Australia • Austria • Belgium • China • France • Germany • India • Japan
Malaysia • Netherlands • Russia • Singapore • Switzerland • Thailand
United Kingdom • United States

 Printed in Singapore.

Gordon and Breach Science Publishers SA
Postfach
4004 Basel
Switzerland

British Library Cataloguing in Publication Data

Bosshard, Ch
Organic Nonlinear Optical Materials. –
(Advances in Nonlinear Optics Series,
ISSN 1068-672x; Vol. 1)
I. Title II. Series
621.36

ISBN 2-88124-975-2 (hardcover)
ISBN 2-88449-007-8 (softcover)

CONTENTS

INTRODUCTION TO THE SERIES

Advances in Nonlinear Optics is a series of original monographs and collections of review papers written by leading specialists in quantum electronics. It covers recent developments in different aspects of the subject including fundamentals of nonlinear optics, nonlinear optical materials both organic and inorganic, nonlinear optical phenomena such as phase conjugation, harmonic generation, optical bistability, fast and ultrafast processes, waveguided nonlinear optics, nonlinear magneto-optics and waveguiding integrated devices.

The series will complement the international journal *Nonlinear Optics: Principles, Materials, Phenomena and Devices* and is foreseen as material for teaching graduate and undergraduate students, for people working in the field of nonlinear optics, for device engineers, for people interested in a special area of nonlinear optics and for newcomers.

1. INTRODUCTION

Recent developments in the field of nonlinear optics hold promise for important applications in optical information processing, telecommunication and integrated optics. Electro-optic, nonlinear optical and photorefractive materials are active media which can be used in a variety of devices in which light waves have to be manipulated by electrical and optical fields.

With the improvement of quality and output power of diode lasers, it is expected that instruments and systems employing lasers will be used more and more. Light, especially a laser beam, has a number of beneficial characteristics, thanks to its good coherence, high density, parallel processing ability, high-speed responsivity and diverse wavelength and frequency. All of these characteristics are essential for realizing high-speed, large-capacity information transmission and processing, high-density data recording and storage. In order to utilize the various optical functions of a laser beam fully, the second or third-order nonlinear optical response of a material for an electrical field at optical or radio frequency is employed in one form or another. In particular, the development of electrically and optically controlled devices such as frequency converters, electro-optic modulators or photorefractive nonlinear devices employing the second-order nonlinear optical response, including the first-order electro-optic effect, is actively taking place.

Electro-optic devices provide many basic functions for modulating and deflecting a laser beam. Electro-optic materials which, in addition to the electric field induced changes of indices of refraction, also show photoconductivity have been shown to be efficient nonlinear optical materials. It has been demonstrated, that optically released charge carriers from impurity ions can produce space-charge fields which lead to refractive index changes by the electro-optic effect. Since large photoinduced changes of refractive indices can be induced at low light intensities (mW level) nonlinear optical processing of arrays of pixels (images) can be achieved. Several applications of dynamic image processing using nonlinear optical and photorefractive materials have thus been proposed and investigated in recent years. They include: optical frequency conversion, electro-optic modulation and deflection, dynamic holography, real-time interferometry, optical storage, optical phase conjugation, optical image amplification, spatial light modulation, optically induced beam deflection and optical interconnection and several types of optical information processors, etc. Most of the features of these applications critically depend on the materials used.

It has been known for many years that certain classes of organic materials exhibit extremely large nonlinear optical and electro-optic effects. The electronic nonlinearities in the most efficient organic materials are essentially based on molecular units containing highly delocalized π-electrons and additional electron donor and electron acceptor groups on opposite sides of the molecules. The most highly active molecules thus tend to be highly polarized. The most polar molecules have a tendency to crystallize in a centrosymmetric structure which does not show a linear electro-optic effect. The choice of special molecules and the crystal growth methods used are of main importance for obtaining electro-optically active single crystals of good quality and large size. A main advantage of the organic materials is the possibility of altering the molecular structure for optimizing the electro-optic or nonlinear optical properties. For the future development in the field of nonlinear optical organic materials a multidisciplinary effort combining organic chemistry, crystal growth, materials science, physics and electrical engineering is essential. This topical volume presents the results of the preparation, characterization and application of organic electro-optic materials in this emerging combination of fields including chemistry and physics.

This review deals with second-order nonlinear optical materials. Noncentrosymmetry is the main requirement for obtaining optical frequency doubling, optical sum and difference frequency generation, optical parametric amplification, linear electro-optic effects (Pockels effect) and also, in most cases, photorefractive effects. Different types of acentric nonlinear optical material are described. They include bulk crystals and thin crystalline films, Langmuir-Blodgett films and poled polymers.

In Chapters 2 and 3 of this monograph the basic molecules used for second order nonlinear optics and electro-optics are described and an introduction into the fields of electro-optics and nonlinear optics is given. From the measured properties of molecules the limits of electro-optical and nonlinear optical properties are estimated, assuming optimum arrangement of molecules for these applications. Chapter 4 gives a short review of methods to grow bulk and thin film crystals. The preparation and properties of thin organic nonlinear optical films deposited on such (cheap) amorphous substrates as Langmuir-Blodgett films or poled polymers are discussed in Chapters 5 and 6. The basic experimental techniques of evaluating optical, electro-optic and nonlinear optical properties of organic materials are summarized in Chapter 7. The electro-optic and nonlinear optical properties of several organic nonlinear optical crystals, Langmuir-Blodgett films and polymers are described in Chapter 8. Its properties are critically confronted with the predictions based on molecular properties and available theoretical models. Chapter 9 gives a survey of guided-wave electro-optics and nonlinear optics using organic materials. In Chapter 10 the photorefractive effect in organic materials is discussed. Some photoconductivity parameters measured in organic compounds are reviewed in order to compare the expected photorefractive properties in bulk crystals, amorphous and crystalline polymers. The first known examples of photorefractive organic crystals and polymers are discussed and its properties compared with the ones reported in inorganic materials.

2. ORGANIC ELECTRO-OPTIC COMPOUNDS

In this chapter we discuss the basic molecular units and structures that are essential for electro-optics and nonlinear optics. We briefly introduce the concepts necessary to obtain an efficient electro-optic and nonlinear optical response.

The exact definition of molecular polarizabilities is given by quantum mechanics through time — dependent perturbation theory (*Ward 1965*). These expressions are usually given in terms of sums of transition matrix elements over energy denominators involving the full electronic structure of the molecule. Although accurate, these expressions offer little physical insight into synthetic methods that are suitable for optimizing the molecular nonlinear optical response. For the case of the second-order polarizability, β, the understanding was considerably clarified by the recognition that β is primarily determined by strong, low energy, charge transfer electronic excitations. When this charge transfer (CT) state dominates the perturbation expression for β, one arrives at a two-level approximation between the highest occupied and lowest unoccupied molecular orbitals (see e.g. *Oudar 1977a, Oudar et al. 1982*). The resulting β_{CT} is a function of the energy gap between the two states, the oscillator strength of the CT transition, the dipole moment associated with that transition, and the fundamental laser photon energy (see Chapter 3). The spectroscopic energy gap is related to the frequency (or wavelength) of the UV-VIS absorption spectra of the molecule, and the oscillator strength is related to the extinction coefficient of the absorption (*Atkins 1983*).

A further simplified model (the Equivalent Internal Field, EIF model) of a free electron gas corresponding to the delocalized π electron density of a conjugated system of length L has been derived (*Oudar et al. 1975, Jain et al. 1981*). In this approximation $\beta \propto L^3$, which shows the strong nonlinear dependence of the hyperpolarizability on the length of the conjugated π system. These two models, although somewhat simplistic, have provided the foundation of a rational strategy for designing highly polarizable nonlinear optical chromophores.

An optimized nonlinear optical chromophore can then be expected to have an extended conjugated system (large L), a low energy transition (long wavelength absorption) with a high extinction coefficient, and a large dipole moment between the excited and ground state electronic configurations (charge asymmetry). Charge asymmetry is introduced when different functional groups are substituted into the molecule, as discussed in the next section. Our discussion concentrates on second-order effects. Nevertheless a short presentation of third-order effects and associated material groups is also included in Chapters 3 and 8 in order to complete this review.

2.1 BASIC MOLECULAR UNITS FOR ELECTRO-OPTIC RESPONSE

In order to engineer optimal nonlinear optical organic materials, the origins of the nonlinear optical response must be understood. In consideration of organic crystals, this requires a knowledge of the bonding properties of the atoms in the molecule, and also of the bonding of the molecules in the crystal. It is typically assumed for organic crystals, because of their weak intermolecular bonding interactions, that the molecules function independently of each other and only their net orientations within the crystal lattice are important to the contribution of the macroscopic optical properties (this approach is known as the oriented gas model, see Chapters 3 and 8 for a further discussion).

The wide diversity of the properties of organic compounds is primarily due to the unsurpassed ability of the carbon atom to form a variety of stable bonds with itself and with many other elements. This bonding is primarily of two types, which differ considerably in the localization of electron charge density (*Pauling 1960*). A two electron covalent σ C_{sp3}–C_{sp3} bond is spatially confined along the internuclear axis of the carbon — carbon bond. In contrast to σ bonds, π bonds are regions of delocalized electronic charge distribution above and below the interatomic axis. Because of this delocalization, the electron density of π bonds is much more mobile than the one of the σ bonds. This electron distribution can also be skewed by substituents; the extent of redistribution is measured by the dipole moment, and the ease of redistribution in response to an externally applied electric field by the (hyper)polarizability. If the perturbation to the molecular electronic distribution caused by an intense optical field is asymmetric, a quadratic nonlinearity results (*Chemla et al. 1981*). Virtually all significantly interesting nonlinear optical organic molecules exhibit π bond formation between various nuclei.

The polarizability of a crystal is generally composed of contributions from the lattice components (atoms, molecules) and the interaction between these components. Whereas the second effect is dominant in inorganic materials with their weakly polarizable atoms and complexes, the first contribution is dominant in organic materials because of the weak intermolecular bonding (Van der Waals, dipole-dipole interactions, hydrogen bonds). Nonlinear optical effects of molecular crystals depend mainly on the polarizability of the electrons in the π bonding orbitals in contrast to inorganic materials where lattice vibrations (occurring in a frequency range of typically 1 MHz up to 100 MHz) play a dominant role. Therefore organic materials are well suited for high-speed applications, e.g. nonlinear optics with ultrashort pulses or high data rate electro-optic modulation.

The optical nonlinearity of organic molecules can be increased by either adding conjugated bonds (increasing L, see below) or by substituting donors and acceptors (Figure 2.1). The addition of the appropriate functionality at the ends of the π system can enhance the asymmetric electronic distribution in either or both the ground state and excited state configurations. Functional groups are divided into two categories based on their ability to accept or donate electrons into the π electron system. Common donor groups typified by amines are of predominately p-character, often with an available electron pair on a p orbital. Acceptor groups usually have more s character orbital bonding as in the nitro, NO_2, and nitrile, $-C\equiv N$, functionalities.

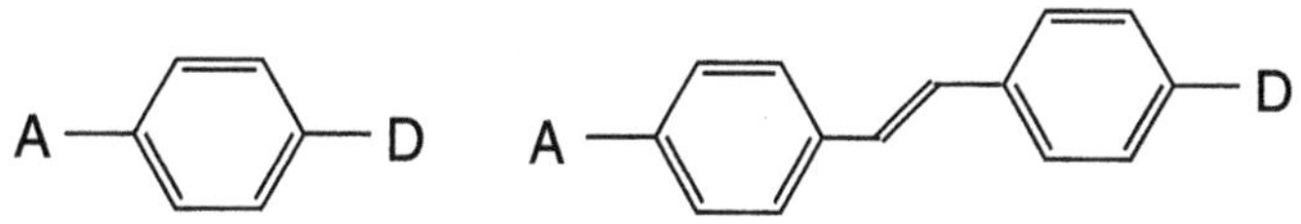

A—⬡—D A—⬡—/—⬡—D

(i) (ii)

Figure 2.1 Typical organic materials for nonlinear optical effects of second-order. These are donor (D)-acceptor (A) substituted molecules with π electron ring systems. (i) one ring systems (benzene analogues) (ii) two ring systems (stilbene analogues).

A simple physical picture for the effectiveness of para substitution in leading to highly polarizable charge transfer molecules can be given in terms of Mulliken resonance structures of the ground and excited states, see Figure 2.2 (*Zyss et al. 1987*). When an oscillating electric field is applied to the π electrons of a donor-acceptor molecule, charge flow is enhanced towards the acceptor whereas the motion in the other direction (towards the donor) is highly unfavorable: the charge transfer molecules have a very asymmetric response to an applied electric field.

The presence of conjugation and donor and acceptor groups generally introduces, however, an undesirable effect, the so-called transparency-efficiency trade-off. The increase of conjugation by linking double bonds and donor and acceptor substituents leads to a shift of the absorption edge towards longer wavelengths. Therefore it can generally be said that the higher the nonlinearity of such materials the more the absorption edge is shifted towards the red.

2.2 STRUCTURE OF MOLECULES FOR ELECTRO-OPTICS AND NONLINEAR OPTICS

Most of the materials studied up to now that allow the fabrication of crystals are nitroaniline and nitropyridine derivatives. Examples for nitroaniline derivatives are:

2-methyl-4-nitroaniline (MNA)	*Levine et al. 1979*
2-methyl-4-nitro-N-methylaniline (MNMA)	*Sutter et al. 1988a*
methyl-(2,4-dinitrophenyl)-aminopropanoate (MAP)	*Oudar et al. 1977b*
N-(4-nitrophenyl)-(L)-prolinol (NPP)	*Zyss et al. 1984*
4-(N,N-dimethylamino)-3-acetamido-nitrobenzene (DAN)	*Twieg et al. 1983, Baumert et al. 1987, Kerkoc et al. 1989*

Examples for nitropyridine derivatives are:

3-methyl-4-nitropyridine-1-oxide (POM)	*Zyss et al. 1981*
2-cyclooctylamino-5-nitropyridine (COANP)	*Günter et al. 1987, Bosshard et al. 1989*
2-(N-prolinol)-5-nitropyridine (PNP)	*Twieg et al. 1983, Sutter et al. 1988b*
(-)2-(α-methylbenzylamino)-5-nitropyridine (MBANP)	*Bailey et al. 1988*

Detailed structures and optical and nonlinear optical properties are given in Chapters 3 and 8.

D^-=⟨⟩=A^+ D–⟨O⟩–A D^+=⟨⟩=A^-

higher state |2> ground state |0> first excited state |1>

Response to ac-field:

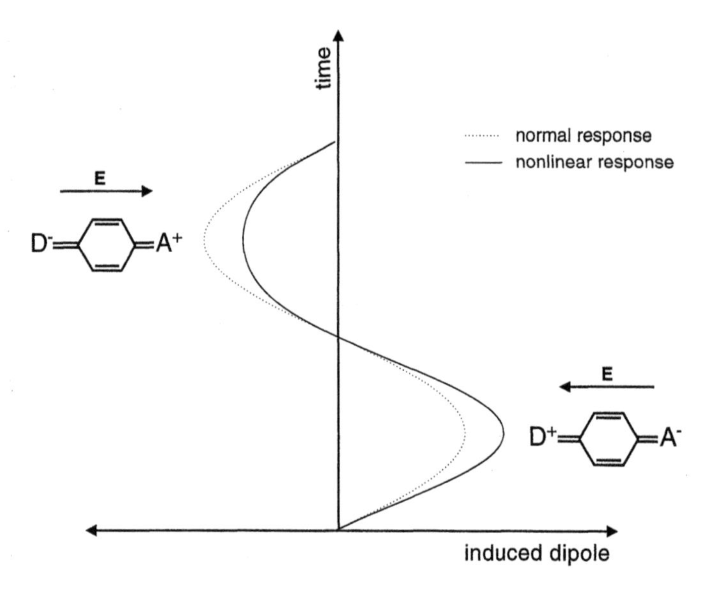

Figure 2.2 Simple picture of the physical mechanisms of the nonlinearity of conjugated molecules. |0> neutral state, |1> situation where one electron is transferred from the donor to the acceptor –> maximum charge transfer state, |2> situation where one electron is transferred from the acceptor to the donor: very unlikely. We get an asymmetric electronic response of the polarization on application of an optical field for molecules with donor and/or acceptor groups in contrast to the symmetric electronic response for centrosymmetric molecules (adapted from *Zyss et al. 1987*).

Longer molecules with two or more rings (i.e. stilbene or azobenzene derivatives) have also been examined, particularly for incorporation into polymeric media. Very recently it was shown that the replacement of benzene rings by thiophene rings can lead to a substantial increase in the second-order molecular polarizabilities (*Rao et al. 1993a,b, Jen et al. 1993*). These systems with two or more rings usually show higher molecular second-order polarizabilities β (or first-order hyperpolarizabilities, see Chapter 3 for the exact definition of β). In such cases the nonlinearity of the assembly (e.g. in crystals) may be reduced because the molecules need more space and therefore the nonlinearity per unit volume becomes smaller. In addition the absorption wavelength is usually shifted towards longer wavelengths.

The relevant properties of the substances mentioned here (second-order polarizability, absorption edge, structure...) will be discussed in Chapters 3 and 8.

As was mentioned above, large π-electron systems that can easily be moved upon application on electric fields and strong donor-acceptor substitutions are necessary for high optical molecular second-order nonlinearities. In the following, important parameters influencing these molecular properties will be discussed.

2.2.1 Effect of substituents of the molecules on the second-order polarizability

Strong donor/acceptor substituents in a para configuration can enhance the optical nonlinearity considerably. As an example Dulcic et al. (*1978a*) have examined several para-disubstituted benzenes and found an increase of the nonlinearity with increasing donor and acceptor strength (Table 2.1, Figure 2.3).

Table 2.1 Molecular second-order polarizabilities $\beta[10^{-40}m^4/V]$ for various para-disubstituted benzenes determined with electric-field induced second-harmonic generation (EFISH, see Chapter 7 for the description of the experiment, *Levine et al. 1975*) in DMSO, λ = 1064 nm.

A—(benzene ring)—D

acceptor	donor			
	CH_3	OCH_3	NH_2	$N(CH_3)_2$
CN	12	20	56	60
NO_2	35	67	197	218

For the most common donors and acceptors the ranking of their strengths can thus be written as follows:

donor	$N(CH_3)_2 > NH_2 > OCH_3 > OH$
acceptor	$NO > NO_2 > CHO > CN$

Other substitution patterns are generally much less effective and multiple substitution of the same functional group typically is of little benefit in maximizing the hyperpolarizability. These results are in qualitatively good agreement with various measures of donor-acceptor inductive and resonant contributions for several physical properties of substituted benzene molecules (*Nicoud et al. 1987*).

The substitution of a hydrogen atom in unsaturated molecules has essentially two effects: the induced and the mesomeric effect. The induced effect arises from the different electronegativity of the substituent. The electric field of the substituent group induces a charge polarization of the σ bonds and induces other polar bondings. This effect is of short range due to charge screening of the σ bonding molecular framework. It is observed in both saturated and unsaturated molecules. The mesomeric effect is observed in conjugated molecules, particularly for aromatic systems. Because of the charge delocalization and mobility of the π electron system, there is a net flow of electron charge density towards or away from the substituted atom. The mesomeric effect therefore characterizes the strength of a π electron charge transfer only. The associated mesomeric dipole moment, $\Delta\mu_m$, is the difference in dipole moment between the conjugated molecule and a corresponding saturated molecule with the same substituents. The sign of $\Delta\mu_m$ provides a means of

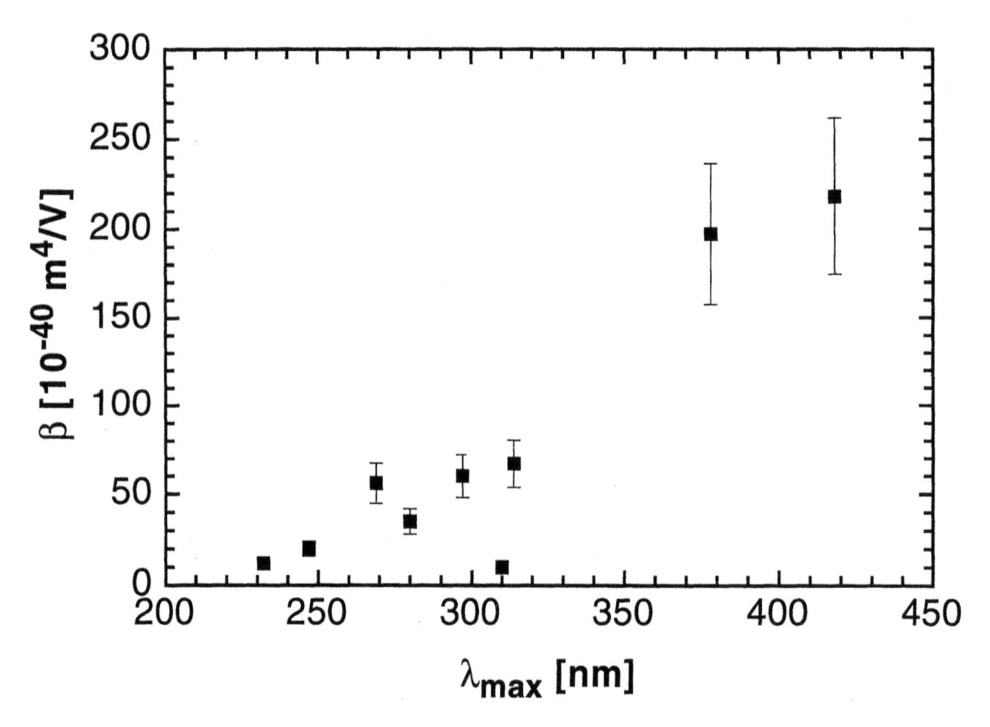

Figure 2.3 Molecular second-order polarizabilities β for various para-disubstituted benzenes versus wavelength of maximum absorption λ_{max}. Measurement: Electric-field induced second-harmonic generation (EFISH, Levine et al. 1975) in DMSO, λ = 1064 nm (data from *Dulcic et al. 1978a*).

classifying substituent groups into donor (D(+)) or acceptor (A(–)) functionalities, depending on whether charge is drawn into or out of the conjugated system. The extension of these ideas to nonlinear optical effects in substituted benzene molecules (EIF model) has been given by Oudar and Chemla, (*Oudar et al. 1975, Chemla et al. 1975*). For longer conjugated systems, charge displacement may extend quite far through the delocalized π electron system, and so it is found that $\Delta\mu_m$ scales roughly with the size of the conjugated molecule (*Zyss et al.1987*).

Other acceptor groups that have been investigated include the polycyanovinyl groups (*Singer et al. 1989*), and the sulfonyl group (*Ulman et al. 1990*). The particularly strong effectiveness of the polycyanovinyl groups as an acceptor functionality arises in part due to extension of the π electron system. In addition to the increase in nonlinearity, the associated red-shift of the spectrum was also observed in these experiments. These substitution effects will also be discussed in Chapter 8.

2.2.2 Effect of conjugation length of the molecules on the second-order polarizability

As previously discussed, the length of the conjugated molecular system has a dramatic effect on the magnitude of the second-order polarizability of organic molecules. This effect has been investigated by a number of authors and will be illustrated with two examples.

Huijts et al. (*1989*) looked at the length dependence of the second-order polarizability in conjugated organic molecules (EFISH measurements in 1,4-dioxane). They used p-methoxynitrobenzene (MNB), 4-methoxy-β-nitrostyrene (MNS) and α-p-methoxyphenyl-ω-p-nitrophenyl polyene (MPNPn with n = 1, 2, 3, 4, 5) (see Figure 2.4). Note that MPNP1 (4-methoxy-4′-nitrostilbene) is also known as MONS (see e.g. *Möhlmann et al. 1990*).

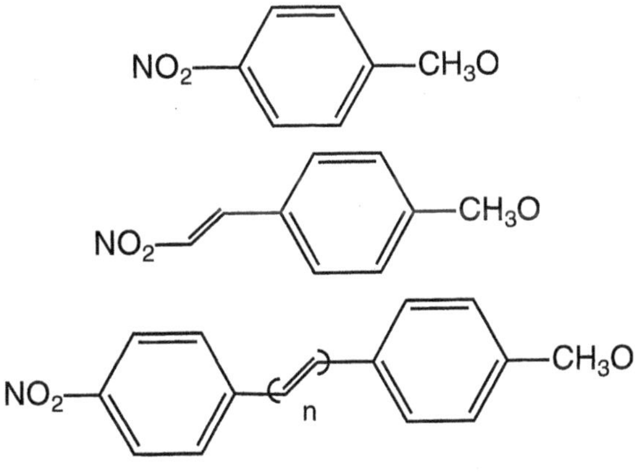

p-methoxynitrobenzene (MNB)

4-methoxy-β-nitrostyrene (MNS)

α-p-methoxyphenyl-ω-p-nitrophenyl polyene (MPNPn with n = 1, 2, 3, 4, 5)

Figure 2.4 Chemical structure of the para-substituted conjugated organic molecules used by *Huijts et al. 1989*.

They found a cubic dependence on the length of the conjugated π-electron system after correction for resonance enhancement (Figure 2.5), i.e. $\beta_o \propto n_\pi^3$, where n_π varies from two to nine (n_π: number of π bonds between the methoxy (CH_3O) and nitro (NO_2) groups, and where a phenyl ring was counted as two bonds). See Chapter 3 for the definition of the first-order hyperpolarizability at zero frequency, β_o. As already mentioned, the increase of the hyperpolarizability with increasing conjugation length inevitably leads to an increase of the molecular volume and loss of transparency.

A second example of the effect of the conjugation length on the first-order hyperpolarizability was illustrated for polyphenyls by Ledoux et al., see Figure 2.6 (*Ledoux et al. 1991*). They synthesized polyphenyl molecules with the number of rings, n, varying from $n = 1 - 4$ and measured the molecular first-order hyperpolarizability β by EFISH (in chloroform solution, at $\lambda = 1064$ nm).

They observed an increase in β_o with increasing n (Figure 2.7). When β_o is scaled with the molecular length, L(L: length of molecule along donor-acceptor axis), it can be seen that a maximum value of $\beta_o L$ is reached for $n = 3$ and decreases for larger n (Figure 2.7). This means that molecules with $n > 3$ are not suitable for macroscopic samples since the loss in the nonlinear optical susceptibilities (see Chapter 3) which are proportional to the number of molecules/unit cell, would be too large.

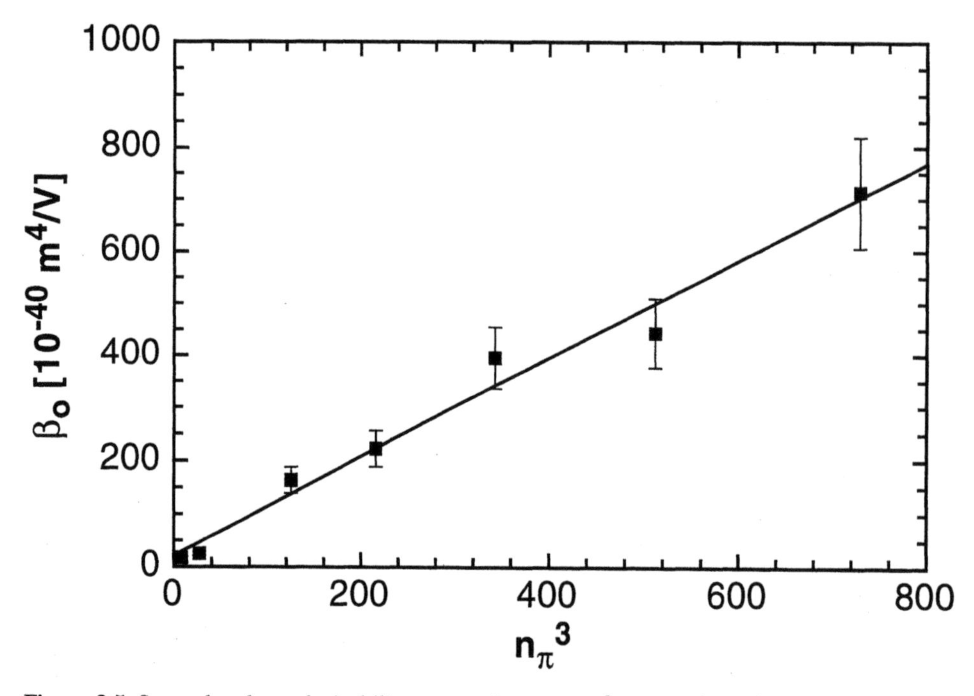

Figure 2.5 Second-order polarizability at zero frequency β_o versus the cube of n_π. The line is a least-square fit to the experimental data (after *Huijts et al. 1989*) using a linear relationship between n_π^3 and β_o.

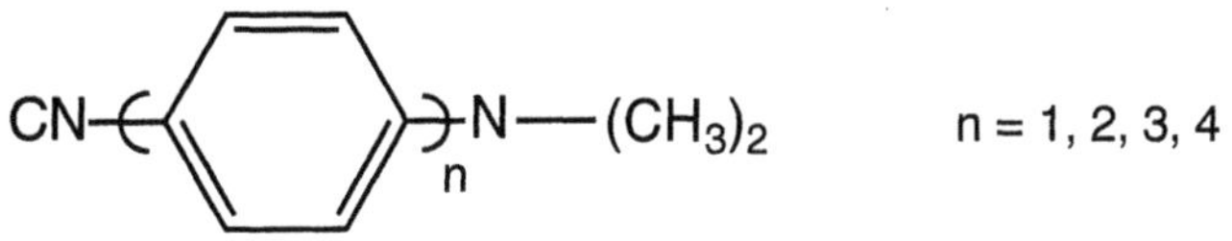

CN-(C₆H₄)$_n$-N—$(CH_3)_2$ n = 1, 2, 3, 4

Figure 2.6 Polyphenyls used to show the effect of conjugation length on the molecular second-order polarizability β_o.

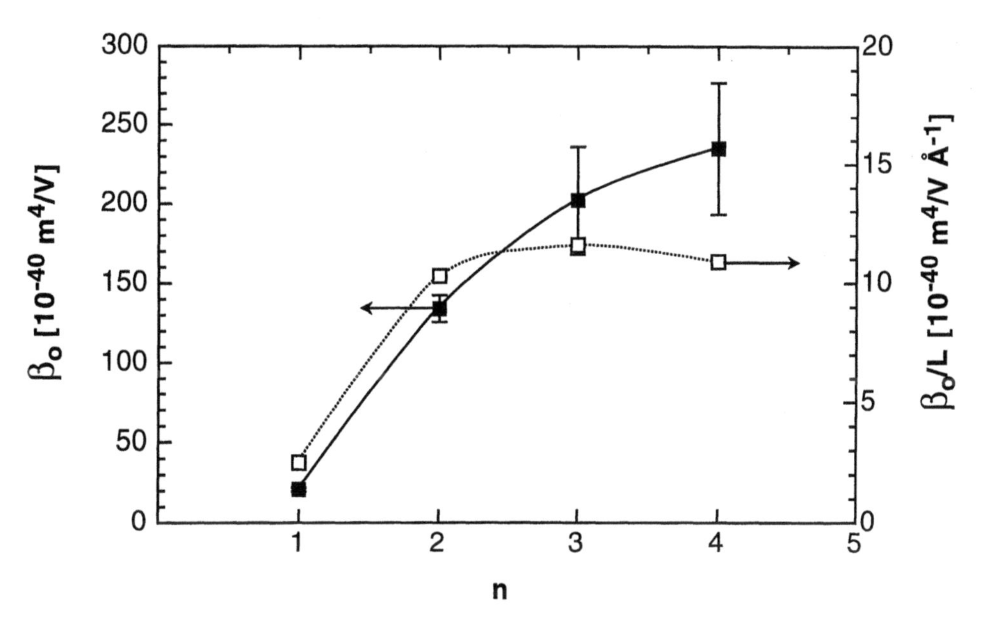

Figure 2.7 β_o and β_o/L as a function of n. The solid and dashed lines are a guide to the eye. Data points are from *Ledoux et al. 1991*.

2.2.3 Effect of planarity of the molecules on the second-order polarizability

Besides optimized donor-acceptor groups and conjugation length the planarity of the molecules is another important aspect to be considered, which is especially important for two- or multi-ring systems. The extent of the planarity also influences the size of the π-electron system and the mobility of the electrons. The effect of twist angles between rings (e.g. biphenyl ring systems) can therefore considerably reduce the charge transfer contribution. As an example of the influence of planarity on the second-order optical nonlinearity, Leslie et al. (*1987*) have examined values of β for the compounds shown in Table 2.2 using the solvatochromic method. The incorporation of the OH group forces the molecule into planarity and therefore leads to an increase of β. Unfortunately the authors give neither the wavelength of maximum absorption λ_{max} nor the twist angle.

Table 2.2 Influence of the planarity of the molecules on the second-order polarizability β.

molecules	β [10^{-40} m^4/V]
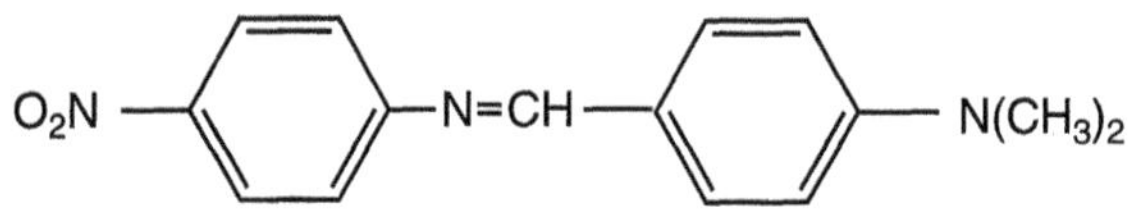 $O_2N-C_6H_4-N=CH-C_6H_4-N(CH_3)_2$	98
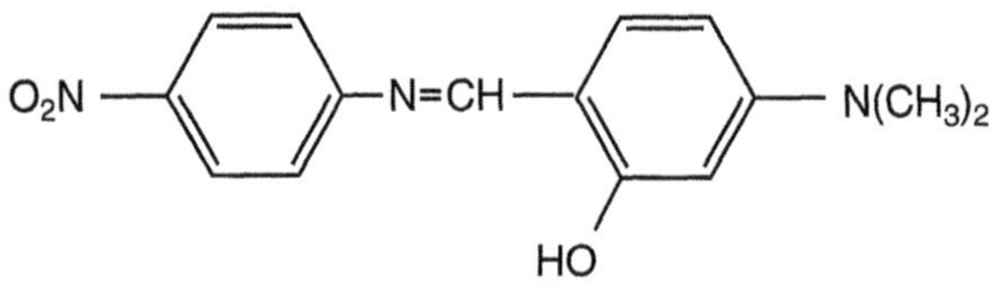 $O_2N-C_6H_4-N=CH-C_6H_3(OH)-N(CH_3)_2$	258

Twisted conformations of molecules have also been investigated by other authors (see e.g. *Akaba et al. 1980, 1985, Ezumi et al. 1974*). They also discuss the connection between twist angle and intramolecular charge transfer effects.

As was illustrated above, the influence of substitution, conjugation length, and planarity of the molecules on the second-order polarizability are all very important aspects to be considered when searching for optimized structures for highly efficient electro-optic and nonlinear optical effects.

More examples of interesting molecules useful for second-order nonlinear optical effects will be given in Chapters 3 and 8.

2.3 BASIC MOLECULAR UNITS FOR THIRD-ORDER NONLINEAR OPTICAL EFFECTS

In contrast to second-order nonlinear optical molecular systems, there are few rational strategies for optimizing the third-order nonlinear optical response of molecular materials. Unlike second-order materials, there exist no molecular symmetry restrictions for the observance of a third-order nonlinear optical response. Furthermore, the two-level state model that has proved so useful for analyzing second-order materials is not sufficient for describing the third-order nonlinear optical response (*Wu et al. 1989, Pierce 1991*). A minimum of three states (ground state and two excited states) are necessary to characterize the third-order response adequately. Due to the inherent difficulties of characterizing excited states of molecular systems, very few molecules have been fully analyzed both experimentally and theoretically.

Most of the experimental basis for optimizing the third-order optical response of molecular materials can be understood and interpreted in terms of an early simple model due to Rustagi et al. (*1974*). For extended linear chains of conjugated molecules, they showed that the π electrons are reasonably well modelled in a first approximation by a free electron gas. In this model, the third-order molecular polarizability, γ, scales as the fifth power of the conjugation length L, i.e. $\gamma \propto L^5$. Concurrent experimental work of the characterization of some alkanes and conjugated molecules (polyenes and cyanines) by third harmonic generation showed the enhanced nonlinearity of the delocalized π electron systems, see Figure 2.8 (*Hermann et al. 1974*).

As seen in Figure 2.8, the optical properties of saturated molecules (e.g. alkanes) are usually very well explained by bond additivity arguments. In this

Alkanes

C_nH_{2n+2}

Polyenes

$$(CH_3)_2\text{-}\overset{+}{N}=CH\text{-}(CH=CH)_{n-1}\text{-}\overset{-}{N}\text{-}(CH_3)_2$$

S, C — CH (= CH - CH)$_{n-2}$ = C, S, + N, - N, C_2H_5, C_2H_5

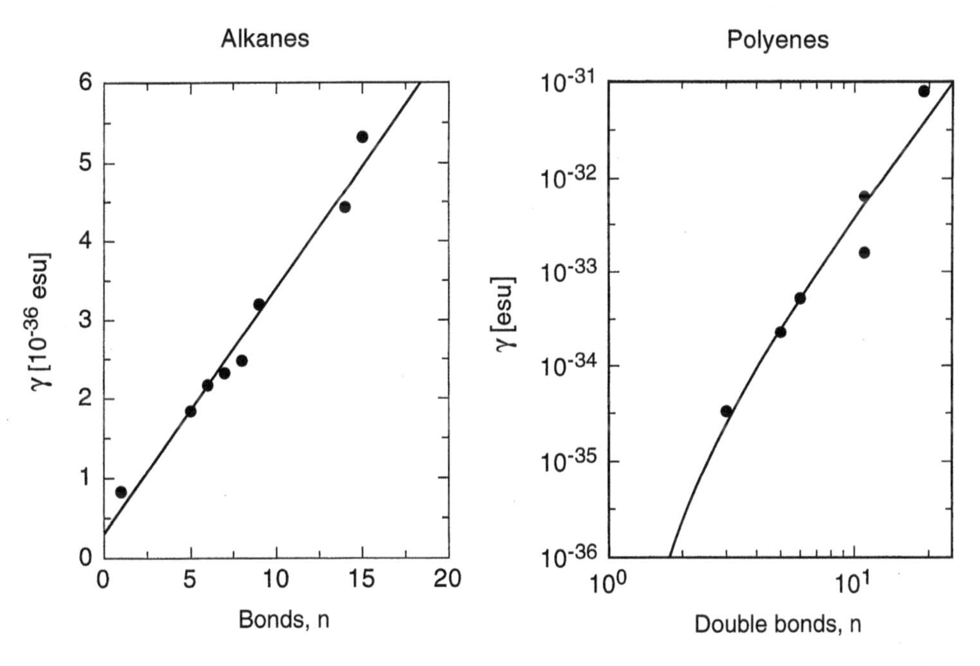

Figure 2.8 Third-order susceptibilities of alkanes and polyenes vs bond number (after *Hermann et al. 1974*).

model it is assumed that the molecular hyperpolarizability can be expressed as the sum of the individual hyperpolarizability of each unit. For a homologous molecular series, a linear dependence of the molecular hyperpolarizability on the number of carbon atoms in the molecule is observed. This has been verified for several homologous series of saturated compounds by Stevenson et al. (*1987*).

For unsaturated molecules one observes the highly nonlinear response shown in Figure 2.8 for polyenes due to delocalized π electrons present in these compounds. More extensive calculations have been performed for the polyene and cyanine molecules and this has given insight into excited state contributions to the nonlinear optical polarizability of these materials (*Heflin et al. 1988, Pierce 1991*). The length dependence of the molecular hyperpolarizabilty as predicted by these calculations is nearly the same as that predicted by the free electron gas model, however.

The current theoretical understanding of structure-property relations for third order-optical nonlinearities is rather poor. Clearly, there is a need for more reliable and cost-effective methods for semiempirical calculations, especially for inclusion of heavy atoms that have weakly bound, polarizable electrons. More experimental work is also needed on systematic studies of functional groups that will enhance the hyperpolarizability of molecules.

2.3.1 Effect of structural order on the third-order polarizability

Most organic molecules can be considered as quasi one-dimensional in consideration of their optical properties. This axis usually has extensive conjugation which considerably enhances the nonlinear optical response due to the electron delocalization of the π bonds, as has been previously discussed. Although there are no particular bulk symmetry requirements for a non-zero third-order susceptibility, ordered materials are expected to have larger tensor components along the axis of conjugated bonding in the molecule (a factor of 5 increase in the totally ordered state compared to the randomly oriented material). These structural ordering effects on the third-order nonlinear susceptibility will be illustrated with two examples.

Wong et al. (*1986*) investigated the third-order nonlinear optical properties of a liquid crystalline material, N-(p-methoxybenzylidene)-p-butylaniline, MBBA, (see Figure 2.9). This molecule has an isotropic to nematic phase transition at 61°C. In the nematic phase the molecules have long-range orientational order while possessing short-range positional order. The bulk phase is uniaxial and has infinite rotational symmetry about this axis. The orientational order of both phases was probed by third-harmonic generation and the results are displayed in Figure 2.10.

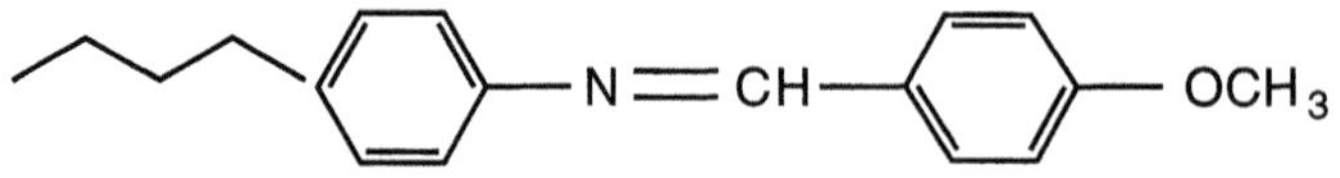

Figure 2.9 Liquid crystalline N-(p-methoxybenzylidene) p-butylaniline, MBBA.

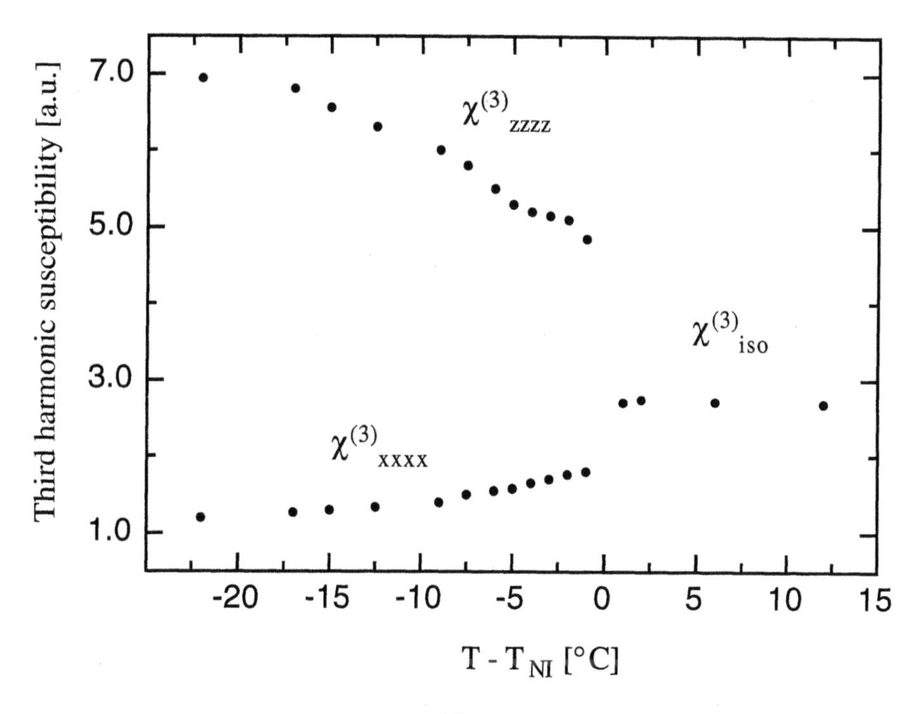

Figure 2.10 Third-harmonic susceptibility $\chi^{(3)}$ of MBBA as a function of temperature (after *Wong et al. 1986*).

In the nematic phase, the lowest order nonvanishing order parameter, $\langle P_2 \rangle$, the Legendre polynomial of order 2, was found to be ≈ 0.6 ($T - T_{NI}$ = 22°C), where T_{NI} is the temperature of the nematic to isotropic phase transition.

As can be seen from the figure, there is a substantial increase in the third-order susceptibility along the director axis (z direction) in the nematic phase. This is due to the long-range molecular order of the nematic phase as compared to the random molecular distribution of the isotropic phase, and the one-dimensional nature of these molecules.

A second example of the effect of structural order on third-order susceptibility has been given by Mittler-Neher et al. (*1992*). They prepared Langmuir-Blodgett multilayer films of poly [bis (*m*-butoxyphenyl)silane] which has a rod-like conformation. Because of the rod-like nature of the molecules, the deposition process induced a high degree of orientational order to the film. A third-harmonic study of the angle, θ, between the polarization of the input beam as a function of the dipping direction, (see Figure 2.11), showed a high degree of anisotropy in $\chi^{(3)}$, in agreement with the linear optical properties and x-ray analysis (*Embs et al. 1991a*).

A major difficulty in making a full characterization of potentially interesting third-order organic materials is their tendency to be highly intractable and semicrystalline. Linear conjugated polymeric materials are often insoluble in mild

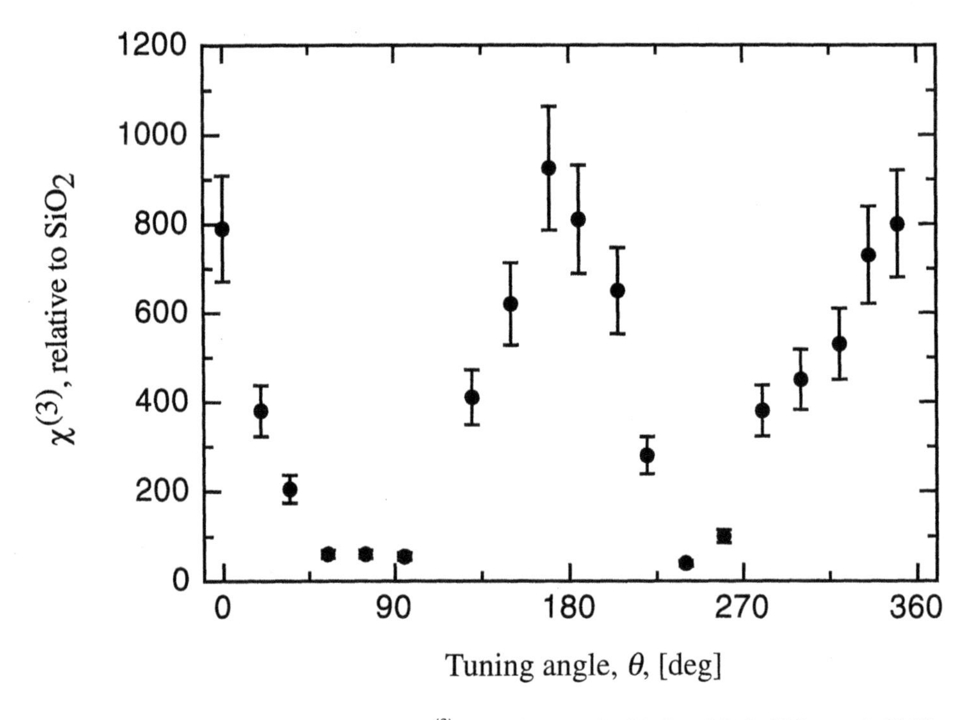

Figure 2.11 Third-order susceptibility $\chi^{(3)}$ vs tuning angle θ (after *Mittler-Neher et al. 1992*).

organic solvents and therefore are not easily processed into potential device structures. The highly conjugated and often aromatic nature of these molecules leads to crystallite formation which results in poor optical properties.

Two possible approaches to improving the processing of these materials are: (*i*) a soluble polymer precursor, and (*ii*) chemical derivatisation for improved processability. An example of the first approach is the Feast synthesis of polyacetylene (*Feast 1990a, Feast et al. 1990b*). In this synthesis, a soluble precursor of polyacetylene was prepared and subsequently converted by high temperature annealing to the final product. An example of the second method is illustrated by the work of Jenekhe et al. (*1989*). In this approach, fused-ring heterocyclic rigid rod polymers were solubilized by complexation with Lewis acids. This has allowed the formation of optical quality thin films of these materials (*Vanherzeele et al. 1991*).

Further examples of organic materials of current interest for their third-order nonlinear optical effects are discussed in Chapters 3 and 8.

3. NONLINEAR OPTICS AND ELECTRO-OPTICS

The real starting point for nonlinear optics dates back to the early sixties when the first lasers became available to all scientists. This is not quite a coincidence but rather a full consequence of the fact that all nonlinear optical phenomena are of at least a second order effect and therefore require high off-resonance electromagnetic field intensities to produce an appreciable result. The intensities achieved by focusing a laser beam are not the only striking feature but also demonstrate the high degree of coherence which renders it possible for small contributions from widely separated parts of an extended medium to sum up to a collective effect.

Nonlinear optical effects can be used for the generation of new optical frequencies, not available with existing lasers, in particular compact blue coherent light sources. The optical nonlinearity results from an anharmonic response of the bound electrons driven by optical frequency fields of the laser radiation. One of the most important effects is optical frequency doubling, a special case of sum frequency generation. The conversion from two single photons of frequency ω to one of 2ω is generally not very efficient, but in the special case where the second harmonic and the fundamental wave travel in phase through the nonlinear optical material (phase-matching), conversion efficiencies of tens of percents are possible. It is therefore clear that all new substances are judged according to whether they are phase matchable or not.

Electro-optic materials in which the optical properties can be changed by electric fields are of basic importance for electro-optic devices. In contrast to pure electronics or photonics where the unit information carriers are electrons and photons respectively, in optoelectronics one has both electrons and photons as information vehicles and they can interact with each other within the electro-optic material. Optoelectronic devices can replace many of the well-known purely electronic functions in communication and measurement systems: propagation, deflection, modulation, amplification, analog-digital conversion, sampling and storage.

In this chapter we will define all quantities related to nonlinear optical and electro-optic effects. Simple physical models describing the wavelength dependence of the second-order nonlinearities and the relation between microscopic and macroscopic quantities will be summarized. A formula describing the dispersion of the electro-optic figure of merit will be discussed. In addition we will give a simple model to describe upper limits of the electro-optic and nonlinear optical response in molecular crystals. Resonance effects either due to electronic dispersion or due to acoustic and optic phonons will be discussed.

In this work we denote the macroscopic coefficients with upper case indices, the microscopic ones with lower case indices.

3.1 MOLECULAR SECOND-ORDER POLARIZABILITIES

The origin of nonlinear optical effects lies in the nonlinear response of a material to an electric field $\vec{E}$ (see Figure 2.2). The polarization $\vec{p}$ of a molecule induced by an external field $\vec{E}$ can be described by (assuming summation over common indices (Einstein summation convention))

$$p_i = \mu_{g,i} + \varepsilon_o\left(\alpha_{ij}^{(1)} E_j + \beta_{ijk}^{(2)} E_j E_k + \gamma_{ijkl}^{(3)} E_j E_k E_l + \ldots\right) \quad i, j, k, l = x, y, z \tag{3.1}$$

(x,y,z) = molecular coordinate system

$\mu_{g,i}$ is the ground state dipole moment, α_{ij} is the polarizability tensor, β_{ijk} is the second-order polarizability or first-order hyperpolarizability tensor, and γ_{ijkl} is the second-order hyperpolarizability tensor. E is the electric field strength at the location of the molecule. For molecules in a solid or solution this electric field has to be calculated from the external field by taking into account appropriate local field corrections. For this local field correction the Lorentz approximation is most often used.

3.1.1 Wavelength dependence of the second-order polarizabilities

Assuming a two-level model (see e.g. *Oudar et al. 1977c*) , where only terms involving either the ground or the first excited state of the molecule are considered, the microscopic second-order polarizability β can be calculated quite easily. This model holds quite well for many organic compounds since the energy difference between the ground state and the higher excited states is usually much higher than the photon energy (whereas the energy required to reach the first excited state lies close to the photon energy).

For many molecules with strong nonlinearities along a single charge transfer axis (donor-acceptor groups at the end of π-electron systems) it is often assumed that one component β_{zzz} of the hyperpolarizability tensor along this axis is sufficient to describe the nonlinearity in first approximation. Using the simple two-level model for the molecular hyperpolarizability with one ground state g and one excited state e, β_{zzz} can be written as (for the case of sum-frequency generation, see e.g. *Teng et al. 1983*)

$$\beta_{zzz}(-\omega_3, \omega_1, \omega_2) = \frac{1}{2\varepsilon_o h^2} \cdot \frac{\omega_{eg}^2\left(3\omega_{eg}^2 + \omega_1\omega_2 - \omega_3^2\right)}{\left(\omega_{eg}^2 - \omega_1^2\right)\cdot\left(\omega_{eg}^2 - \omega_2^2\right)\cdot\left(\omega_{eg}^2 - \omega_3^2\right)} \cdot \Delta\mu \cdot \mu_{eg}^2 \tag{3.2}$$

where ω_{eg} denotes the resonance frequency of the transition and where ω_1, ω_2, and ω_3 are the frequencies of the three interacting waves. Note that ω_{eg} of the nonlinear optical molecules is red shifted in a dielectric medium with $\varepsilon > 1$ due to local field effects. $\Delta\mu = \mu_e - \mu_g$ is the difference between excited and ground state electric dipole moments, and μ_{eg} is the transition dipole moment between excited and ground state. One can derive the appropriate hyperpolarizability describing the electro-optic effect by setting $\omega_3 = \omega$, $\omega_1 = \omega$, and $\omega_2 = 0$.

Introducing the dispersion free hyperpolarizability β_o extrapolated to infinite optical wavelengths away from the electronic resonances

$$\beta_o = \frac{3}{2\varepsilon_o \hbar^2} \frac{\Delta\mu \cdot \mu_{eg}^2}{\omega_{eg}^2} \tag{3.3}$$

one obtains for optical frequency-doubling:

$$\beta(-2\omega, \omega, \omega) = \frac{\omega_{eg}^4}{\left(\omega_{eg}^2 - 4\omega^2\right)\left(\omega_{eg}^2 - \omega^2\right)} \beta_o \tag{3.4}$$

and for the linear electro-optic effect:

$$\beta(-\omega, \omega, 0) = \frac{\omega_{eg}^2\left(3\omega_{eg}^2 - \omega^2\right)}{3\left(\omega_{eg}^2 - \omega^2\right)^2} \beta_o \tag{3.5}$$

3.2 MACROSCOPIC SECOND-ORDER NONLINEAR OPTICAL EFFECTS

The macroscopic polarization $\vec{P}$ of a nonlinear medium under the influence of an external field $\vec{E}$ can be described by the following equation (using again the Einstein summation convention)

$$P_I = P_{0I} + \varepsilon_o\left(\chi_{IJ}^{(1)} E_J + \chi_{IJK}^{(2)} E_J E_K + \chi_{IJK}^{(3)} E_J E_K E_L + \ldots\right) \quad I, J, K, L = 1, 2, 3 \tag{3.6}$$

P_{0I} is the spontaneous polarization and $\chi^{(n)}$ describes the nonlinear optical susceptibility of nth order. If we look at two fields with frequencies ω_1 and ω_2 interacting in a nonlinear medium, the component of the polarization at the sum frequency $\omega_3 = \omega_1 + \omega_2$ is given by

$$P_I^{\omega_3}(\vec{r}) = \varepsilon_o \frac{1}{2}\left[\chi_{IJK}(-\omega_3, \omega_1, \omega_2) E_J^{\omega_1}(\vec{r}) E_K^{\omega_2}(\vec{r}) + \chi_{IJK}(-\omega_3, \omega_2, \omega_1) E_J^{\omega_2}(\vec{r}) E_K^{\omega_1}(\vec{r})\right] \tag{3.7}$$

where

$$E_I(\vec{r}, t) = Re\left[E_I^{\omega}(\vec{r}) e^{-i\omega t}\right], \quad P_I(\vec{r}, t) = Re\left[P_I^{\omega}(\vec{r}) e^{-i\omega t}\right] \tag{3.8}$$

The negative sign of ω_3 indicates an outcoming photon. In the case of frequency-doubling we have (*Boyd et al. 1968*)[1]

---

[1] In the case of frequency-doubling there appear the additional terms $\chi_{IJK}(-2\omega_1, \omega_1, \omega_1)$ $E_J^{\omega_1} E_K^{\omega_1}$ and $\chi_{IJK}(-2\omega_2, \omega_2, \omega_2) E_j^{w_2} E_K^{w_2}$. If ω_1 and ω_2 both approach ω the electric fields become $E^{\omega_i} \to \frac{1}{2} E^{\omega}$. Together with $\lim_{\omega_i \to \omega} \chi_{IJK}(-\omega_3, \omega_1, \omega_2) = \lim_{\omega_i \to \omega} \chi_{IJK}(-\omega_3, \omega_2, \omega_1)$ this leads to $\frac{1}{2}\chi_{IJK}(-2\omega, \omega, \omega) = d_{IJK}(-2\omega, \omega, \omega)$

$$P_I^{2\omega} = \varepsilon_o \frac{1}{2} \chi_{IJK}(-2\omega, \omega, \omega) E_J^{\omega} E_K^{\omega} \tag{3.9}$$

and

$$P_I^{2\omega} = \varepsilon_o d_{IJK}(-2\omega, \omega, \omega) E_J^{\omega} E_K^{\omega} \tag{3.10}$$

The nonlinear optical coefficient d_{IJK} is symmetric in J and K and can be written in a contracted notation

$$d_{IJK} = d_{IKJ} = d_{IM} \text{ where } M = \begin{cases} J, \text{ for } J = K \\ 9-(J+K), \text{ otherwise} \end{cases} \tag{3.11}$$

In the case of non-absorbing, dispersionless materials the Kleinman relation (*Kleinman, 1962*) yields

$$d_{IJK} = d_{IKJ} = d_{JKI} = d_{KIJ} = \ldots \tag{3.12}$$

3.2.1 Phase-matching

For the ideal case of monochromatic Gaussian beams slightly focused on a thin plate, the second-harmonic intensity, neglecting pump depletion of the input wave, is given by (*Armstrong et al. 1962*)

$$I^{2\omega} = \frac{2\omega^2}{\varepsilon_o c^3 (n^{\omega})^2 n^{2\omega}} d_{eff}^2 L^2 (I^{\omega})^2 sinc^2\left(\frac{\Delta k L}{2}\right) \tag{3.13}$$

where $\text{sinc}(x) = (\sin x)/x$, L is the sample thickness, d_{eff} is the effective nonlinear optical susceptibility, c is the vacuum velocity of light and $\Delta k = |\vec{k}_3^{2\omega} - \vec{k}_1^{\omega} - \vec{k}_2^{\omega}|$ is the phase mismatch between the fundamental and second-harmonic waves with wave vectors $\vec{k}_1^{\omega}$, $\vec{k}_2^{\omega}$, and $\vec{k}_3^{2\omega}$ respectively. We can see from Eq. (3.13) that the conversion efficiency $\eta = I^{2\omega}/I^{\omega}$ is proportional to I^{ω}, $(d_{eff})^2$, L^2 and the phase synchronization factor $sinc^2(\Delta kL/2)$. This last factor is largest for $\Delta k \cdot L = \pi L/l_c = 0$. For e.g. type I second-harmonic generation (see below) and for a laser beam incident perpendicularly to one of the dielectric axis, the condition $\Delta k \cdot L = 0$ corresponds to

$$l_c(\theta) = \frac{\lambda}{4\left|n^{2\omega}(\theta'^{2\omega}) cos\theta'^{2\omega} - n^{\omega}(\theta'^{\omega}) cos\theta'^{\omega}\right|} = \infty \quad \text{and} \tag{3.14}$$

$$\rightarrow \quad \boxed{n^{2\omega}(\theta'^{2\omega}) = n^{\omega}(\theta'^{\omega})} \quad \text{where} \tag{3.15}$$

$$n(\theta') = \sqrt{\frac{1}{\frac{\cos^2(\theta')}{n_I^2} + \frac{\sin^2(\theta')}{n_J^2}}} \tag{3.16}$$

The condition in Eq. (3.15), $\Delta k \cdot L = 0$, is called phase-matching. The θ' are the angles inside the crystal. In the case of phase-matching the fundamental and second-harmonic waves travel at the same velocity in the crystal, and constructive interference of the frequency-doubled light generated at different positions in the crystal is possible. The second-harmonic generation efficiency varies as the thickness squared. Therefore non-phase-matched $\Delta k \cdot L \neq 0$ interactions are between 10^{-6} to 10^{-4} times as efficient as phase-matched interactions over distances of e.g. 1 cm.

Since most materials usually have a positive dispersion of the refractive indices ($n^{2\omega} > n^{\omega}$), phase-matching is most often achieved using birefringence. Organic materials are well suited in this respect, due to their large anisotropy and strong dispersion in the refractive indices, e.g. $n_b - n_c = 0.162$ at $\lambda = 550$ nm for 2-cyclooctylamino-5-nitropyridine (COANP, *Bosshard et al. 1989*). Phase-matching can either be achieved by angle-tuning or temperature-tuning.

There exist two different types of phase-matching: type I phase-matching (generation of 2ω with two photons of the same polarization) and type II phase-matching (generation of 2ω with two photons polarized orthogonally to each other).

In addition we can distinguish between critical and noncritical phase-matching. In the case of angle-tuning the distinction is easily demonstrated: depending on the direction of the incoming beam with respect to the optical indicatrix, the phase-matching condition is very sensitive to this direction and the phase-matching condition is only fulfilled for beams incident within the acceptance angle $\Delta\theta'_{PM}$. If the incoming beam is propagating along a dielectric axis of the crystal, this acceptance angle is much larger than in the other case

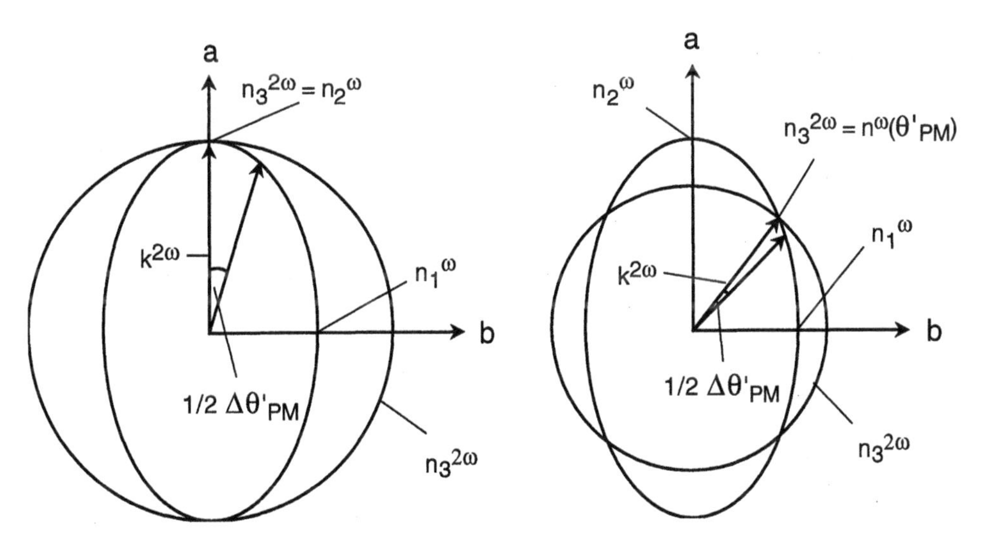

Figure 3.1 (a) Non-critical phase-matching, (b) critical phase-matching illustrated with the normal surfaces.

when the beam is propagating in any direction. This is illustrated in Figure 3.1 (the normal surfaces (that is, the surfaces giving the refractive indices as a function of the wave-normal direction) are drawn exaggeratedly, the fundamental wave being polarized in the ab-plane, the second-harmonic wave is polarized along c). The first case is called noncritical, the second case critical phase-matching.

The interaction length for angle-tuned phase-matched frequency-doubling is limited due to the walk-off of the second-harmonic wave from the fundamental beam. Figure 3.2 shows an example for type I phase-matching in a crystal of orthorhombic point group symmetry, using the nonlinear optical coefficients d_{31} and d_{32}. For this configuration the beam walk-off angle ρ is given by (*Boyd et al. 1965a*)

$$\tan\rho = \frac{1}{2}(n_3^{2\omega})^2\left(\frac{1}{(n_2^{\omega})^2} - \frac{1}{(n_1^{\omega})^2}\right)\sin(2\theta'_{PM}) \tag{3.17}$$

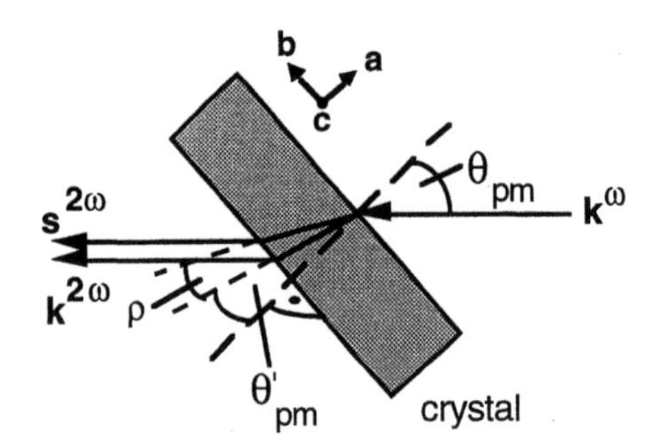

Figure 3.2 Walk-off angle ρ for a p-polarized fundamental and a s-polarized second-harmonic wave for type I phase-matching.

For a Gaussian beam with focal radius w_o this limits the effective interaction length to

$$l_{int} = \frac{\sqrt{\pi}\, w_o}{\rho} \tag{3.18}$$

Obviously noncritical phase-matching is desirable because the beam walk-off is zero and therefore the effective interaction length is infinite.

3.3 LINEAR ELECTRO-OPTIC EFFECTS

Electro-optic effects are most commonly described by considering field-induced changes (deformations, rotations) of the optical indicatrix. In noncentrosymmetric substances (see e.g. *Yariv 1975*) linear electro-optic effects dominate that are described by the tensor r_{IJK} (Pockels electro-optic coefficient):

$$\Delta\left(\frac{1}{n^2}\right)_{IJ} = \Delta\left(\frac{1}{\varepsilon}\right)_{IJ} = r_{IJK}E_K \tag{3.19}$$

In Eq.(3.19) summation over common indices is assumed. Using the field-induced polarization $P_K = \varepsilon_o(\varepsilon_{KL} - \delta_{KL})E_L$, the electro-optic effect is frequently expressed as

$$\Delta\left(\frac{1}{n^2}\right)_{IJ} = f_{IJK}P_K \tag{3.20}$$

where $f_{IJK}[\mathrm{m}^2/\mathrm{C}]$ are the linear polarization optical (PO) coefficients. The connection between these coefficients is given by

$$r_{IJL} = \varepsilon_o f_{IJK}(\varepsilon_{KL} - \delta_{KL}) \tag{3.21}$$

$$\delta_{KL} = \begin{cases} 1, & \mathit{for}\ K = L \\ 0, & \text{otherwise} \end{cases}$$

and where ε_{KL} is the dielectric tensor at the frequency of the modulating field. In the linear approximation the change in the refractive index Δn_I is approximately given by (see e.g. *Yariv 1975*):

$$\Delta n_I = -\frac{n_I^3 r_{IIK}E}{2}, \qquad I = J \tag{3.22}$$

An important parameter for the characterization of electro-optic crystals for modulator applications is the half-wave voltage, V_π, which is the voltage required to obtain a π phase shift in the transmitted beam in an appropriate modulator configuration (see e.g. *Yariv 1975*):

$$V_\pi = \frac{\lambda}{(n^3 r)_{eff}}\frac{d}{L} = v_\pi \frac{d}{L} \tag{3.23}$$

where $(n^3 r)_{eff}$ is the effective electro-optic coefficient. Low modulation voltages require a high value of $n^3 r$ as well as large geometry factors L/d achieved using the transverse electro-optic effect where the light propagation direction is perpendicular to the electric field. The reduced half-wave voltage is often used for characterizing an electro-optic material as a sample dependent quantity.

A figure of merit (*FM*) for linear electro-optic effects to be used for amplitude modulators can be defined as (*Günter 1976*)

$$FM = n^3 r_{IJK} \quad \text{with } n = n_I \text{ for } I = J \quad \text{or } n = \sqrt{2}\, n_I n_J/\sqrt{n_I^2 + n_J^2} \text{ for } I \neq J \tag{3.24}$$

This *FM* will be discussed below.

3.3.1 Phase-matching

In the case of electro-optic modulation, the situation is different as compared to frequency-conversion experiments. As long as the frequency of the modulating electric field is much smaller than that of the light field, the phase-matching condition is always fulfilled. Phase differences between the modulating and light fields may become important for modulation frequencies above 10^9 Hz.

3.4 RELATION BETWEEN MICROSCOPIC AND MACROSCOPIC NONLINEAR OPTICAL COEFFICIENTS

Bergman and Crane (*Bergman et al. 1974*) proposed a model to relate microscopic with macroscopic properties of inorganic nonlinear optical crystals based on bond hyperpolarizabilities. Their theory was then applied to organic crystals by Zyss and Oudar (*Zyss et al. 1982*). In their model, except for local field corrections, all contributions to the optical nonlinearity due to intermolecular interactions are neglected and only intramolecular contributions are taken into account. This leads to a simple dependence of the nonlinear optical susceptibilities for frequency-doubling on structural parameters and molecular hyperpolarizabilities:

$$d_{IJK} = N\frac{1}{n(g)} f_I^{2\omega} f_J^{\omega} f_K^{\omega} \sum_{s}^{n(g)} \sum_{ijk}^{3} \cos(\theta_{Ii}^s)\cos(\theta_{Jj}^s)\cos(\theta_{Kk}^s)\beta_{ijk}(-2\omega,\omega,\omega) \qquad (3.25)$$

For the electro-optic coefficients one obtains in a similar way

$$r_{IJK} = -N\frac{1}{n(g)}\frac{4}{n_I^2(\omega)\,n_J^2(\omega)} f_I^{\omega} f_J^{\omega} f_K^{o} \sum_{s}^{n(g)} \sum_{ijk}^{3} \cos(\theta_{Ii}^s)\cos(\theta_{Jj}^s)\cos(\theta_{Kk}^s)\beta_{ijk}(-\omega,\omega,0) \qquad (3.26)$$

θ_{Ii}^s are the angles between dielectric and molecular axes I and i in the unit cell, N is the number of molecules per unit volume, $n(g)$ is the number of symmetry equivalent positions in the unit cell (e.g. $n(g) = 4$ for COANP), s denotes a site in the unit cell, $f_I^{\omega,2\omega}$ are local field corrections and β_{ijk} is the molecular second-order polarizability. As mentioned above, it is appropriate for molecules with strong nonlinearities along a single charge transfer axis to assume a one-dimensional hyperpolarizability β_{zzz} along such an axis.

The nonlinear optical and electro-optic properties of organic materials have been interpreted in terms of the oriented gas model. Measured values of β_{zzz} could be reasonably well compared with the measured d_{IJK} and r_{IJK}. If the intermolecular interactions do not contribute significantly to the optical nonlinearity, one generally only has to take into account a red shift of the resonance frequency when describing macroscopic electro-optic or nonlinear optical coefficients with microscopic hyperpolarizabilities. The local field correction factors $f^{\omega,2\omega,0}$ are most often calculated within the Lorentz model (*Boettcher 1952*):

$$f^{\omega,2\omega,0} = \frac{(n^{\omega,2\omega,0})^2 + 2}{3} \tag{3.27}$$

which can only be regarded as the simplest approximation. A numerical example may illustrate the importance of local field effects: if we vary the refractive index from 1.7 to 2.4, the local field correction (Lorentz) changes from 1.6 to 2.6. Since f appears in the third power, this yields possible correction from 4.1 to 18! Other models for the local field, where a molecule is divided into submolecules (*Hurst et al. 1986*) and the contribution from each submolecule is taken into account and multiple lattice sums are calculated (*Cummins et al. 1976*), are quite elaborate and, surprisingly, the simple Lorentz correction still yields the best results.

The electronic contributions to the electro-optic effect r^e_{IJK} can be directly calculated from the nonlinear optical coefficients, d^{EO}_{IJK} using (*Boyd et al. 1968*)

$$r^e_{IJK} = \frac{-4}{\left(n_I(\omega)\cdot n_J(\omega)\right)^2} d^{EO}_{IJK} \tag{3.28}$$

With the one-dimensional two-level model for the charge transfer transition d^{EO}_{IJK} is given by (combining Eqs.(3.2) and (3.25))

$$d^{EO}_{IJK} = d^{-\omega;\omega,0}_{IJK} = \frac{f^{\omega}_I f^{\omega}_J f^{0}_K}{f^{2\omega'}_K f^{\omega'}_I f^{\omega'}_J} \cdot \frac{(3\omega^2_{eg} - \omega^2)\cdot(\omega^2_{eg} - \omega'^2)\cdot(\omega^2_{eg} - 4\omega'^2)}{3(\omega^2_{eg} - \omega^2)^2\cdot\omega^2_{eg}} \cdot d^{-2\omega';\omega',\omega'}_{KIJ} \tag{3.29}$$

Equations (3.28) and (3.29) allow the comparison of nonlinear optical coefficients for second-harmonic generation measured at fundamental frequency ω' with the electro-optic coefficients measured at frequency (of light) ω.

3.5 DISPERSION OF THE ELECTRO-OPTIC COEFFICIENTS AND ITS FIGURE OF MERIT

If intermolecular contributions to the macroscopic optical nonlinearities are neglected, the electro-optic coefficients can directly be related to the microscopic hyperpolarizabilities of the molecules that represent basic units of the molecular crystal. This allows the description of the wavelength dependence of the electro-optic coefficients in terms of the wavelength dependence of the corresponding molecular quantities. Using the results from sections 3.1 and 3.4 we obtain for the electro-optic coefficients r_{IJK}

$$r_{IJK}(\omega) = K_1 \cdot \frac{1}{n^2_I(\omega)n^2_J(\omega)} \cdot \frac{n^2_I(\omega)+2}{3} \cdot \frac{n^2_J(\omega)+2}{3} \cdot \frac{3\omega^2_{eg} - \omega^2}{(\omega^2_{eg} - \omega^2)^2} \tag{3.30}$$

where K_1 is a parameter depending on the dielectric constant ε, the number of molecules per unit volume N, the difference between excited and ground state dipole moments $\Delta\mu$, and the square of the transition dipole moment between excited and ground state, μ_{eg}^2, of the material.

A simple expression describing the wavelength dependence of the electro-optic figure of merit for organic and inorganic materials can be derived (*Bosshard et al. 1993*). Equation (3.30), which describes the dispersion of r_{IJK}, can be modified to

$$r_{IJK}^{\omega} = K_2 \frac{-4}{n_I^2(\omega)n_J^2(\omega)} f_I^{\omega} f_J^{\omega} f_K^{0} \frac{\omega_{eg}^2(3\omega_{eg}^2-\omega^2)}{(\omega_{eg}^2-\omega^2)^2}$$

$$= K_3 \frac{-4}{n_I^2(\omega)n_J^2(\omega)} \frac{n_I^2(\omega)+2}{3} \frac{n_J^2(\omega)+2}{3} \frac{\omega_{eg}^2(3\omega_{eg}^2-\omega^2)}{(\omega_{eg}^2-\omega^2)^2} \tag{3.31}$$

where K_2 and K_3 are parameters depending on ε, N, $\Delta\mu$, and μ_{eg}^2 of the material. For $I = J$ we get

$$n_I^3 r_{IIK}^{\omega} = K_3 g(\omega) \frac{\omega_{eg}^2(3\omega_{eg}^2-\omega^2)}{(\omega_{eg}^2-\omega^2)^2} \quad \text{with} \quad g(\omega) = \frac{-4}{n_I(\omega)}\left(\frac{n_I^2(\omega)+2}{3}\right)^2 \tag{3.32}$$

The function $g(\omega)$ contains the local field correction factors at frequency ω (Lorentz approximation) and the refractive indices given from Eqs. (3.24) and (3.31). Its frequency dependence is generally much weaker than the one of the last term in Eq. (3.31), which originates from the dispersion of the molecular second-order polarizability. Typical values for the refractive index n_I for an organic material optimized for electro-optics (i.e. the charge transfer axis of the molecules is nearly parallel to the axis I) are in the order of $n_I = 1.8$ in the infrared and $n_I = 2.4$ near the absorption edge, which gives $g = -6.8$ in the infrared and $g = -11.2$ near the absorption edge. At the same time the molecular second-order polarizability increases by several 100 percent. This shows that for a rough approximation of the dispersion of the electro-optic figure of merit one can assume g to be constant. One then obtains the approximation for the electro-optic figure of merit

$$n_I^3 r_{IIK}^{\omega} \cong F \frac{\omega_{eg}^2(3\omega_{eg}^2-\omega^2)}{(\omega_{eg}^2-\omega^2)^2} \tag{3.33}$$

where $F = K_3 \times g$. Equation (3.33) contains only two parameters, F and ω_{eg}. ω_{eg} is reasonably well known for most of the nonlinear optical organic compounds from absorption measurements or from the coefficients describing the dispersion of the refractive indices (Sellmeier parameters). F can be found once we know ω_{eg} and $n^3 r$ at one wavelength. Thus Eq. (3.33) allows one to plot an approximation of the electro-optic figure of merit for many organic electro-optic materials if we assume that the electro-optic response is dominated by electronic contributions (see also Chapters 2.1 and 3.1).

Figure 3.3 shows the electro-optic figure of merit as a function of wavelength for a number of organic and inorganic compounds. Table 3.1 gives the parameters used for the calculation of the dispersion of different organic compounds. The optimized benzene and stilbene compounds in Table 3.1 are described in detail in Chapter 3.7.

The dispersion of the inorganic ferroelectric materials in Figure 3.3 was estimated using a dispersion formula based on (*Wemple et al. 1972*)

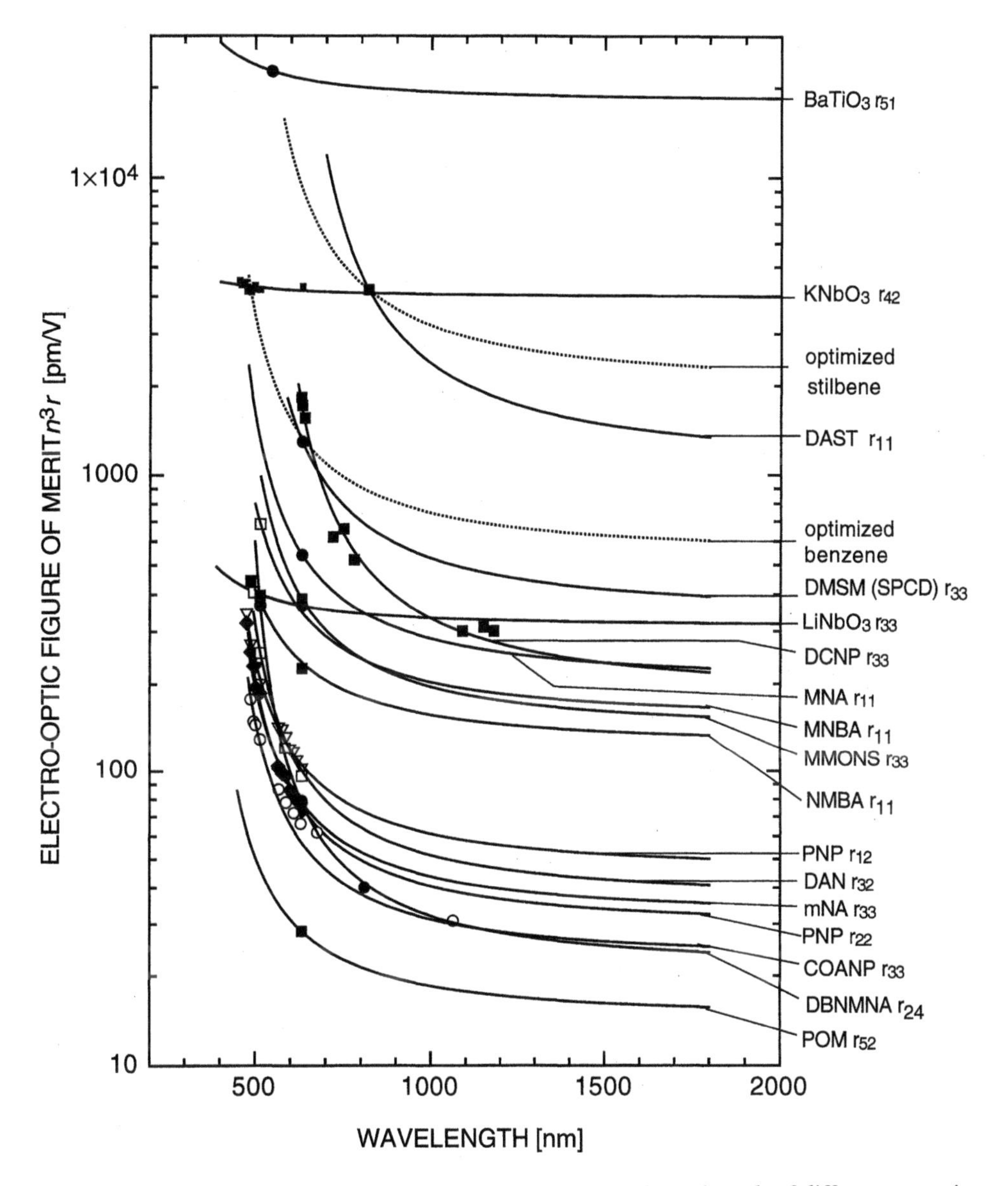

Figure 3.3 The electro-optic figure of merit as a function of wavelength of different organic and inorganic compounds. The curves are drawn over the transmission range of the respective materials.

Table 3.1 Examples of organic compounds that have been investigated for their electro-optic response. Parameters used for the calculations in Figure 3.3 are indicated.

material	r[pm/V], n	$\lambda_{eg} = 2\pi c/\omega_{eg}$ of bulk [nm]	Reference
DAST 4′-dimethylamino-N-methyl-4-stilbazolium tosylate 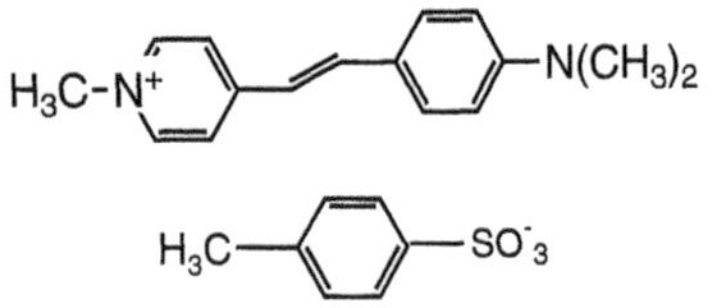	r_{11}(820 nm) = 400 n_1(820 nm) = 2.19	600 assumed: 100 nm below cut-off wavelength at 700 nm	*Perry et al. 1991*
DMSM (SPCD) 4′-dimethylamino-N-methyl-4-stilbazolium methylsulfate 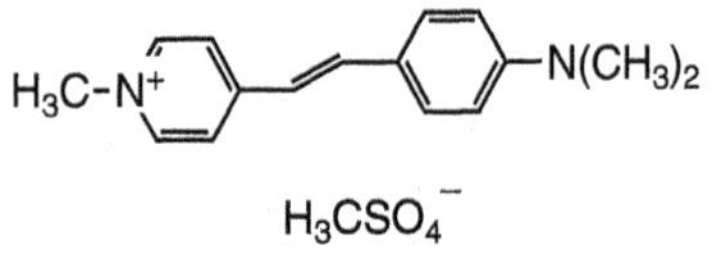	r_{33}(632.8 nm) = 430 $n_3^3 r_{33}$(632.8 nm) =1300 n_3(632.8 nm) = 1.55	460 estimated from absorption peak at 2.7 eV.	*Yoshimura 1987* *Yoshimura et al. 1989*
DCNP 3-(1,1-dicyanoethenyl)-1-phenyl-4,5-dihydro-1H-pyrazole 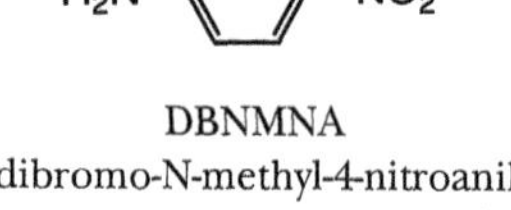	r_{33}(632.8 nm) = 87 n_3(632.8 nm) = 2.7	531 from the fit of the measured values of n^3r using Eq.(3.33)	*Allen et al. 1988*
MNA 2-methyl-4-nitroaniline 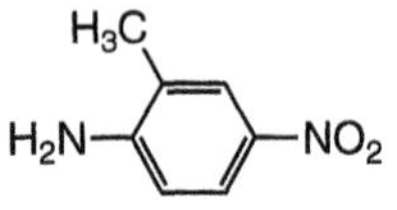	r_{11}(632.8 nm) = 67 n_1(632.8 nm) = 2.00	414 from the Sellmeier coefficients of n_1 (*Morita et al. 1987*)	*Lipscomb et al. 1981* *Morita et al. 1987*
DBNMNA 2,6-dibromo-N-methyl-4-nitroaniline 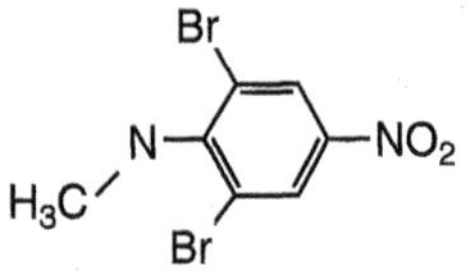	n^3r(514.5 nm) = 367 n^3r(632.8 nm) = 79 n^3r(810 nm) = 40	458 from the fit of the measured values of n^3r using Eq.(3.33)	*Nahata et al. 1990*
MMONS 3-methyl-4-methoxy-4′-nitrostilbene 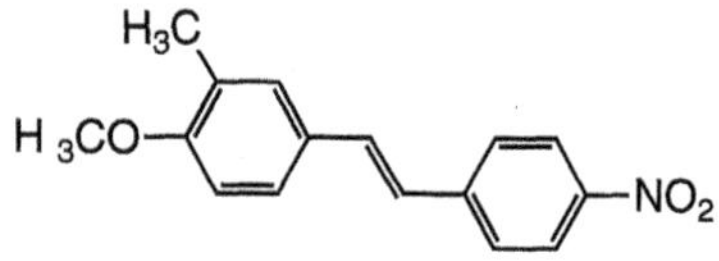	r_{33}(632.8 nm) = 39.9 n_3(632.8 nm) = 2.129	421 from the Sellmeier coefficients of n_3	*Bierlein et al. 1990*

Table 3.1 Continued.

material	r[pm/V], n	$\lambda_{eg} = 2\pi c/\omega_{eg}$ of bulk [nm]	Reference
MNBA 4′-nitrobenzylidene-3-acetamino-4-methoxyaniline 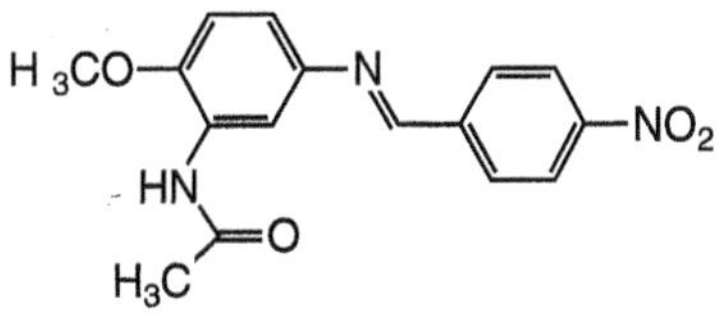	$n_1^3 r_{11}$ = 685 pm/V at 514.5 nm	390 nm from the Sellmeier coefficients of n_1	*Knöpfle et al. 1994*
NMBA 4-nitro-4′-methylbenzylidene aniline 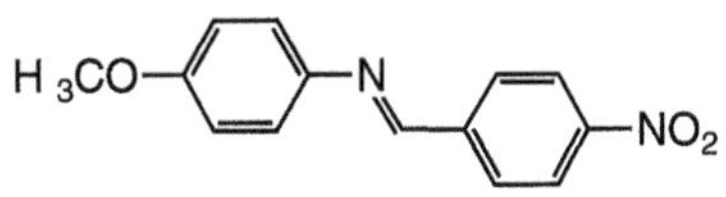	r_{11}(488 nm) = 37.2 r_{11}(514.5 nm) = 36.4 r_{11}(632.8 nm) = 25.2 n_1(488 nm) = 2.283 n_1(514.5 nm) = 2.216 n_1(632.8 nm) = 2.078	353 nm from the fit of the measured values of $n^3 r$ using Eq.(3.33)	*Bailey et al. 1992*
PNP 2-(N-prolinol)-5-nitropyridine 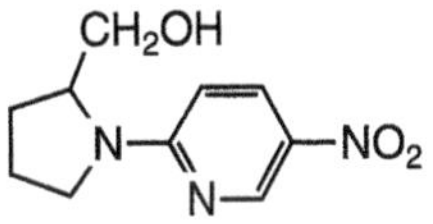	r_{22}(514.5 nm) = 28.3 r_{22}(632.8 nm) = 12.8 r_{12}(514.5 nm) = 20.2 r_{12}(632.8 nm) = 13.1 n_1(514.5 nm) = 2.164 n_1(632.8 nm) = 1.994 n_2(514.5 nm) = 1.873 n_2(632.8 nm) = 1.788	404 from the Sellmeier coefficients of the refractive indices	*Sutter et al. 1988b* *Bosshard et al. 1993*
DAN 4-(N,N-dimethylamino)-3-acetamidonitrobenzene 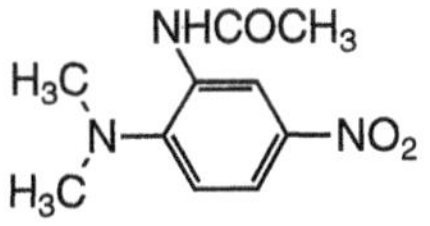	r_{32}(632.8 nm) = 13 n_3(632.8 nm) = 1.949	419 from the Sellmeier coefficients of n_3	*Kerkoc et al. 1989* *Kerkoc 1991*
mNA meta-nitroaniline 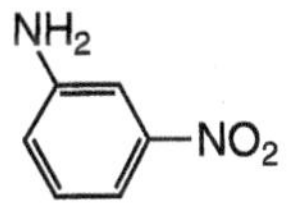	r_{33}(632.8 nm) = 16.7 n_3(632.8 nm) = 1.675	400 estimated	*Stevenson 1973*
COANP 2-cyclooctylamino-5-nitropyridine 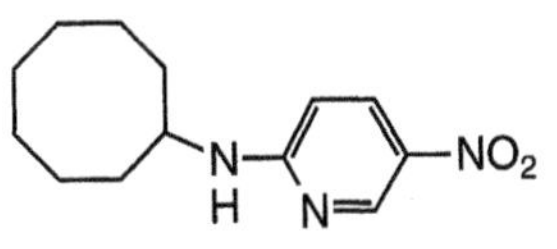	r_{33}(514.5 nm) = 28 r_{33}(632.8 nm) = 15 n_3(514.5 nm) = 1.715 n_3(632.8 nm) = 1.647	410 from the Sellmeier coefficients of n_3	*Bosshard et al. 1989* *Bosshard et al. 1993*

Table 3.1 Continued.

material	r[pm/V], n	$\lambda_{eg} = 2\pi c/\omega_{eg}$ of bulk [nm]	Reference
POM 3-methyl-4-nitropyridine-1-oxide (structure: CH_3; $^-O{-}N^+$; NO_2)	$r_{52}(632.8\text{ nm}) = 5.2$ $n_2(632.8\text{ nm}) = 1.92$ $n_3(632.8\text{ nm}) = 1.64$	358 from the Sellmeier coefficients of the refractive indices	*Sigelle et al. 1981*
optimized benzene (λ_{eg} (solid state) = 401 nm) (structure: D, A)	$r(501\text{ nm}) = 210$ $n(501\text{ nm}) = 2.2$	401	*Bosshard et al. 1993*
optimized stilbene (λ_{eg} (solid state) = 479 nm) (structure: D, A)	$r(579\text{ nm}) = 1150$ $n(579\text{ nm}) = 2.4$	479	*Bosshard et al. 1993*

$$r = K'(1-n^{-2})^2\left[(1-K) + (1+K)\left(\frac{\lambda_{eg}}{\lambda}\right)^2\right], \text{ with } K' = \frac{\phi}{E_d}(\varepsilon - 1) \tag{3.34}$$

where K is the dispersion constant, ϕ is the linear polarization potential, and E_d is the dispersion energy. The parameters used for the inorganic materials are given in Table 3.2. A comparison with the organic crystals is given in Chapter 8.

The electro-optic coefficients of the materials listed in Figure 3.3 range over several decades. n^3r values between some 10 pm/V up to several 1,000 pm/V are

Table 3.2 Parameters used for the calculations in Figure 3.3 for the inorganic materials.

material	r(632.8 nm) [pm/V]	K, K', λ_{eg} of bulk	n(632.8 nm)	reference
$KNbO_3$	$r_{42} = 380$	from fit through the measured points with $K_{42} = -1$ and $\lambda_{eg} = 255.2$ nm $K' = 580$ pm/V	$n_b = 2.329$ $n_c = 2.169$	*Günter 1976* *Zysset et al. 1992*
$LiNbO_3$	$r_{33} = 32.2$	$\lambda_{eg} = 208$ nm (from Sellmeier parameters) $K' = 46$ pm/V (calculated from point) $K = -1$ (arbitrary)	$n_e = 2.20$	*Landolt-Börnstein 1979a*
$BaTiO_3$	$r_{51} = 1640$ (at 546 nm)	$\lambda_{eg} = 215$ nm (from Sellmeier parameters) $K = -1$ (arbitrary) $K' = 815$ pm/V	$n_e = 2.37$ $n_o = 2.44$ (at 546 nm)	*Landolt-Börnstein 1979a*

found. It must also be noted that the electro-optic figure of merit is increased by one decade for wavelengths close to the absorption edge where resonance effects become dominant.

3.6 CONTRIBUTION OF ACOUSTIC AND OPTIC PHONONS TO THE LINEAR ELECTRO-OPTIC EFFECT

The electro-optic coefficients used in Figure 3.3 are all unclamped coefficients, i.e. the frequency ω' of the applied electric field is low ($\omega' < 10$ kHz). For organic compounds with dominating electronic contribution, the dependence of r on ω' is weak and dominated by the electronic charge transfer between donor and acceptor substituents, which corresponds to frequencies between 10^{15} and 10^{16} Hz. The strongest electro-optic organic compounds will especially show an electronic contribution to r that is much larger than the elastic contribution. Lattice vibrations contribute only weakly to the electro-optic effect in organics as comparisons between electro-optic and nonlinear optical coefficients show. For inorganic compounds r shows much stronger resonance peaks at frequencies below 10^{13} Hz since the electro-optic coefficient has large contributions from lattice vibrations (Figure 3.4).

For electro-optic broad band modulators where an applied field is used to change the phase or amplitude of a carrier light wave, the strong dispersion of the electro-optic coefficients around 10^6 Hz gives rise to distortions of the modulator properties. A flat frequency response up to the maximum modulator frequency (around several GHz) is preferred, such as is found in organic materials.

The electro-optic coefficients of inorganic compounds can be strongly increased when the crystal is close to a phase transition. This fact is used in many of the good electro-optic ferroelectrics, which exhibit phase transitions close to the room temperature. A consequence of this fact is the strong temperature dependence of the electro-optic effect for these compounds. For example, in $BaTiO_3$ the coefficient r_{51} decreases by nearly 50% when the temperature is increased from 20°C to 40°C. This strong dependence on the ambient temperature may be inconvenient for many applications. The purely electronic electro-optic effects of organic compounds can be expected to be much less sensitive to the operating temperature.

This comparison shows that organic crystals are well suited for electro-optic applications. They can show electro-optic coefficients up to 200 pm/V. This is somewhat lower than the coefficients of many inorganic compounds. However, the electro-optic effect in organics shows only weak dispersion with increasing frequency of the modulating field over the whole technically useful range up to several GHz and only little dependence on ambient temperature.

Lattice vibrational contributions due to acoustic and optic phonons will now be discussed in more detail. The free electro-optic coefficient r^T contains three contributions: the electronic part (r^e) and the effects from optic (r^o) and acoustic (r^a) phonons (*Wemple et al. 1972*):

$$r^T = r^e + r^o + r^a \tag{3.35}$$

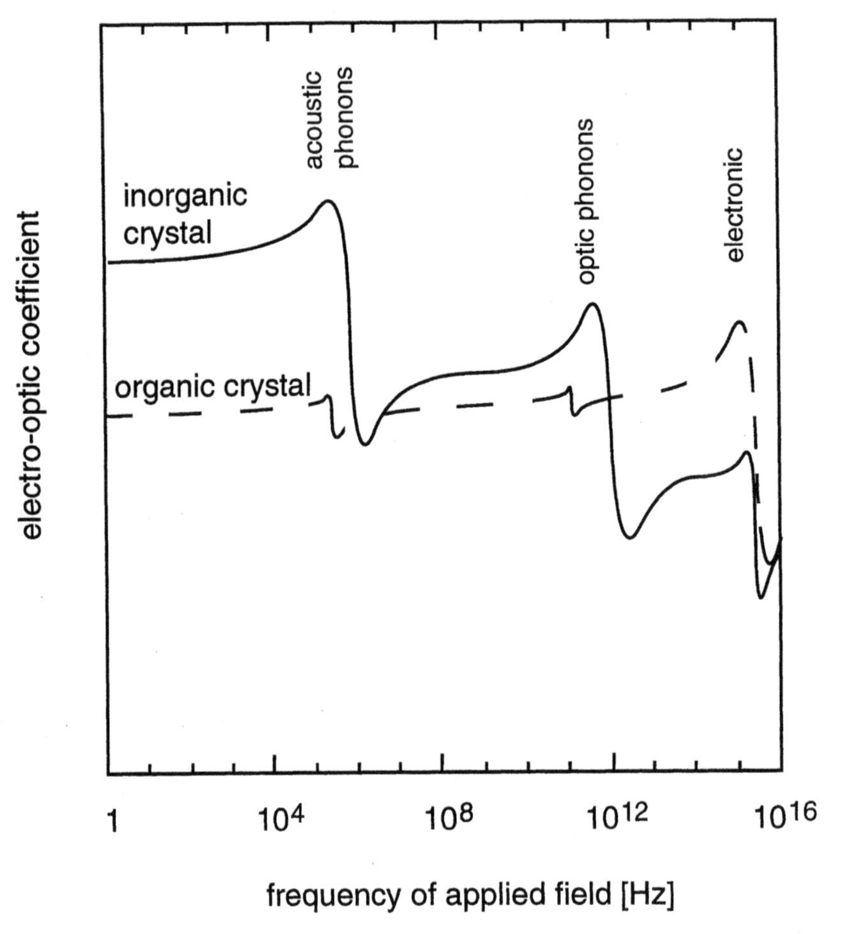

Figure 3.4 Simplified dispersion of the electro-optic coefficients with the applied electric field frequency for organic and inorganic compounds.

To illustrate the various physical processes contributing to the electro-optic effect, we consider field-induced changes of the optical dielectric constant allowing for optic and acoustic vibrational modes (i.e. Raman and Brillouin scattering processes). The optical dielectric constant then depends on long wavelength optic mode displacements Q^n, elastic deformations u_{LM}, and the electric field E_K. In the linear approximation we get for the second-order susceptibility d'_{IJK} (*Günter 1980, Fousek 1978*)

$$4\,d'_{IJK} = \frac{d\varepsilon_{IJ}}{dE_K} = \left.\frac{d\varepsilon_{IJ}}{dE_K}\right|_{Q=u=0} + \sum_n \frac{d\varepsilon_{IJ}}{dQ^n}\frac{dQ^n}{dE_K} + \frac{d\varepsilon_{IJ}}{du_{LM}}\frac{du_{LM}}{dE_K} \tag{3.36}$$

$$4\,d'_{IJK} = 4\,d^{EO}_{IJK} + \sum_n \rho^n_{IJ}\frac{\Delta_n \chi_{LK}}{e^n_L} + p^*_{IJLM}a_{LMK} \tag{3.37}$$

where ρ_{IJ} is the Raman tensor, nonvanishing only if Q^n is a polar mode, e_L^n is an effective charge, $\Delta_n \chi_{LK}$ is the n-th mode contribution to the static susceptibility, p^*_{IJLM} is related to the photoelastic tensor and a_{LMK} is the piezoelectric a tensor. The first term in Eq.(3.37) is purely electronic in origin and was defined in Eqs.(3.28) and (3.29). The second term is due to the optic modes and is nonzero only for vibrational modes being both Raman and infrared active. The third term in Eq.(3.37) arises from the photoelastic contribution, i.e. from lattice strains induced by the inverse piezoelectric effect driven by the applied electric field. It should be mentioned that all three contributions can have either sign.

The acoustic mode contribution can be described by considering elasto-optic effects due to strains $s_{JK} = a_{IJK}E_I$ induced by the inverse piezoelectric effect for electric fields E_I (*Nye 1967*).

In measurements of electro-optic coefficients, the applied electric field first induces a change of the refractive index via the electro-optic coefficient r (clamped electro-optic effect). In addition the applied field can also induce a strain (inverse piezoelectric effect), which leads to a change in the refractive index via the elasto-optic coefficients p_{IJMN} defined by

$$\Delta\left(\frac{1}{n^2}\right)_{IJ} = p_{IJMN}s_{MN} \tag{3.38}$$

This leads to

$$\Delta\left(\frac{1}{n^2}\right)_{IJ} = (r^s_{IJK} + p_{IJMN}a_{KMN})E_K = r^T_{IJK}E_K \tag{3.39}$$

At low frequencies (even below 1 MHz) acoustic and optic phonons can be excited. Above the fundamental acoustic resonances the acoustic modes are partially clamped and above the 'Reststrahlen' region (around 10^{12} Hz), the optic mode contributions are also clamped.

It should be noted that the applied electric field (e.g. along the optic wave vector K) can also cause a piezoelectrically induced change of the light path ΔL_J along the propagation direction (*Ducharme et al. 1987*):

$$\Delta L_J = a_{KJ}E_K L_J \tag{3.40}$$

This can contribute considerably to the electrically-induced phase shift in an interferometric measurement, even away from the acoustic resonances. This has been shown, for example in $BaTiO_3$ (*Ducharme et al. 1987*).

3.7 LIMITS OF THE ELECTRO-OPTIC AND NONLINEAR OPTICAL RESPONSE IN MOLECULAR CRYSTALS

Equations (3.2) and (3.26) describe the electro-optic coefficient r in terms of molecular and crystallographic parameters and the refractive indices. They also describe its dispersion by the frequency dependence of the second-order polarizability β and the local field factors.

In order to find an estimate of the maximum electro-optic coefficient of a material based on charge transfer molecules, the limits of the following parameters must be known (*Bosshard et al. 1993*):

— *The molecular second-order polarizability* β (Eq. (3.2)). This is the parameter that will have the strongest influence on the electro-optic coefficient. It primarily depends on the shape and size of the molecule and the nature of its donor and acceptor substituents. It also depends strongly on the frequency of the optical fields and shows resonance enhancement near the charge transfer transition.

— *The structural parameters* N (number of molecules per unit volume) and θ^s_{Ii} (angle between molecular charge transfer axis and reference system) or θ_p (angle between molecular charge transfer axis and polar crystal axis).

— *The refractive indices* n_I, which are used in Eqs. (3.26) and (3.28) and which give an estimate for the local field factors f_I. They are up to about 2.2 for stilbenes and around 2.0 for benzenes in spectral regions of moderate absorption.

The first two parameters in this list, the hyperpolarizability and the molecular density, are discussed in more detail in the following two sections, 3.7.1 and 3.7.2.

3.7.1 Maximum second-order polarizabilities

As already mentioned, the second-order polarizability β depends on the strength of donor and acceptor substituents as well as on the extension of the π-electron system (conjugation length). An increase in conjugation length leads to an increase in β. However, since crystals of larger molecules have a lower molecular density N, there exists an optimum size for nonlinear optical chromophores. In the case of polyphenyls (Figure 3.5) theoretical calculation shows that the maximum macroscopic nonlinear response is reached for two or three ring systems (*Morley 1989*).

For both types of material the increase of the hyperpolarizability with increasing conjugation length was clearly observed. When looking at the relevant hyperpolarizability per unit volume the situation was different. For the polyphenyls,

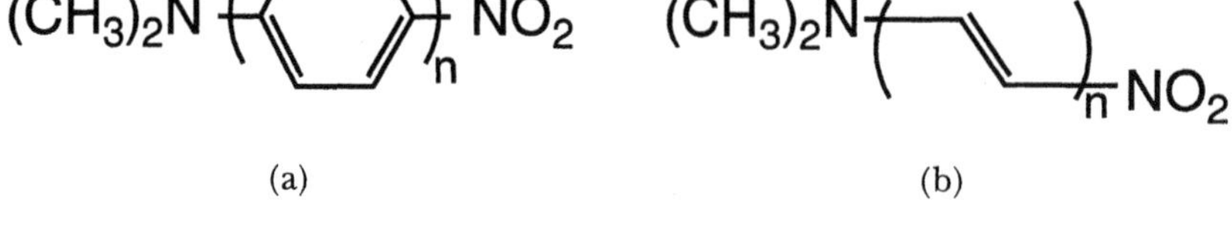

$(CH_3)_2N$ —(phenyl)$_n$— NO_2 (a) $(CH_3)_2N$ —(CH=CH)$_n$— NO_2 (b)

Figure 3.5 Structure of (a) polyphenyl and (b) polyenes.

the highest nonlinearity per unit volume was reached for two and three ring systems ($\beta_o/V \approx 61$ pm/V, β_o is the zero frequency hyperpolarizability) with lower values for systems with fewer or more rings. For polyenes the hyperpolarizability per unit volume continuously increased until a plateau was reached at around n = 20. This shows that, at least for ring systems, a trade-off between nonlinearity and molecular length has to be made (see also Chapter 2.2.2). Two or three rings per molecule seem to be ideal. However, the transparency has not yet been taken into account. It has a much stronger influence on the applicability of a molecule as a nonlinear optical than as an electro-optic material.

The influence of donor and acceptor strength on β is twofold: strong donors and acceptors shift the transition frequency ω_{eg} to the red (which leads to an increase of $\beta(-\omega, \omega, 0)$ and $\beta(-2\omega, \omega, \omega)$ at a given frequency $\omega < \omega_{eg}$) and have an influence on the dipole transition moment μ_{eg} and the change $\Delta\mu$ of dipole moment between ground state and first excited state (Eq. (3.2)).

For a semiempirical estimation of β we used experimental results published by Cheng et al (*1989, 1991*). They took an experimental approach to understand the structural factors that are responsible for the molecular nonlinear optical mechanisms and to optimize structures using this knowledge. The measurements were performed at λ = 1.9 μm to be truly nonresonant. These authors have measured the hyperpolarizabilities of a large number of benzene and stilbene derivatives (Figure 3.6) by electric field induced second harmonic generation (EFISH).

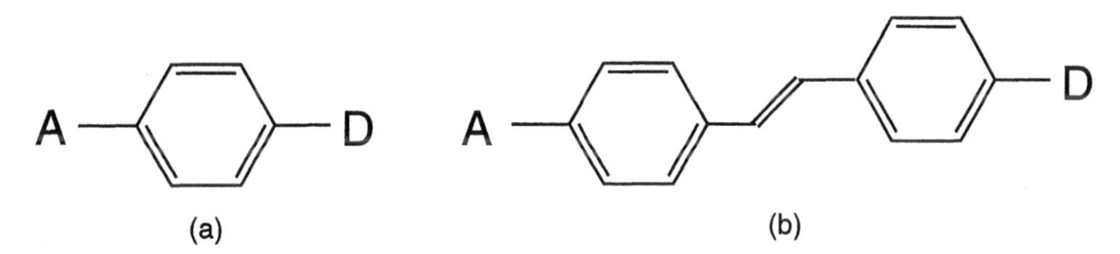

Figure 3.6 Representatives of intramolecular charge transfer molecules (A: acceptor substituent, D: donor substituent). a: benzene, b: stilbene.

They found a remarkable dependence of the second-harmonic hyperpolarizability β on the charge transfer transition frequency ω_{eg}. Figure 3.7 shows a plot of the hyperpolarizability measured at a fundamental wavelength of 1.9 μm as a function of the maximum absorption wavelength λ_{eg}.

These measurements show that, in good approximation, the second-order polarizability for a given type of molecule (benzene or stilbene derivatives) depends mainly on the resonance frequency $\omega_{eg} = 2\pi c/\lambda_{eg}$ (absorption maximum) and not on any further details of the donor or acceptor.

When transferring the data in Figure 3.7 to solids it should be noted that due to higher dipole-dipole interactions a shift $\Delta\lambda$ of $\lambda_{eg}(= 2\pi c/\omega_{eg})$ towards the red is observed: λ_{eg}(solid) = λ_{eg}(solution) + $\Delta\lambda$. For benzenes we have $\Delta\lambda \approx 40$ nm, for stilbenes $\Delta\lambda \approx 60$ nm (for solutions in 1,4-dioxane). Such red shifts can be deduced from a comparison of Sellmeier coefficients and absorption measurements in solutions of 1,4-dioxane of various compounds (*Bosshard et al. 1993*). Using this correction and the relation (3.4), we could calculate the low frequency limit β_0 of the hyperpolarizability as a function of λ_{eg} from the data in Figure 3.7. Once we know these two parameters (λ_{eg} and β_0), we can calculate the hyperpolarizabilities $\beta(-\omega, \omega, 0)$ and $\beta(-2\omega, \omega, \omega)$ at the desired wavelengths using the relations (3.4) and (3.5) (*Bosshard et al. 1993*).

To take full advantage of the resonance enhancement, the operating wavelength should lie close to the absorption edge. As a rule of thumb, the absorption edge for benzene or stilbene derivatives lies 100 nm from its solid state peak absorption λ_{eg}^{solid}, i.e. $\lambda_{eg}^{solid} = \lambda_c - 100$ nm. From this information the maximum value of β for the electro-optic and the nonlinear optical coefficients at the absorption edge λ_c can be calculated using Eq. (3.5). The result of the calculations is shown in Figures 3.8a, b. Results from fits of data points are summarized in Table 3.3.

These maximum values were calculated for benzene and stilbene derivatives. For three ring systems they can be expected to be even higher. However, not enough experimental data is available on such systems to carry out the corresponding calculations. But, even if these compounds may have higher β values, a volume trade-off must be taken into account as described above. Also,

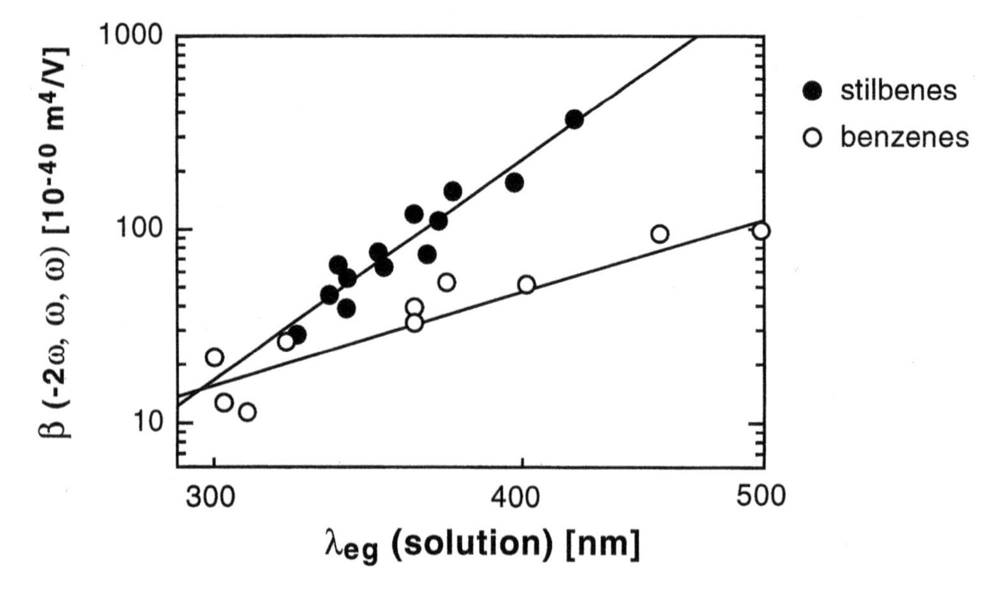

Figure 3.7 Dependence of the second-harmonic hyperpolarizability on the resonance wavelength (in solution) for benzenes and stilbenes. Fundamental wavelength: 1.9 μm. (Doubly logarithmic plot.) The empirical relation between λ_{eg} and β is $\beta = 4.46 \times 10^{-9} \times \lambda_{eg}^{3.85}$ (for the benzenes) and $\beta = 4.52 \times 10^{-22} \times \lambda_{eg}^{9.11}$ (for the stilbenes), where λ_{eg} is given in nm and β in 10^{-40} m^4/V (data from *Cheng et al. 1989*).

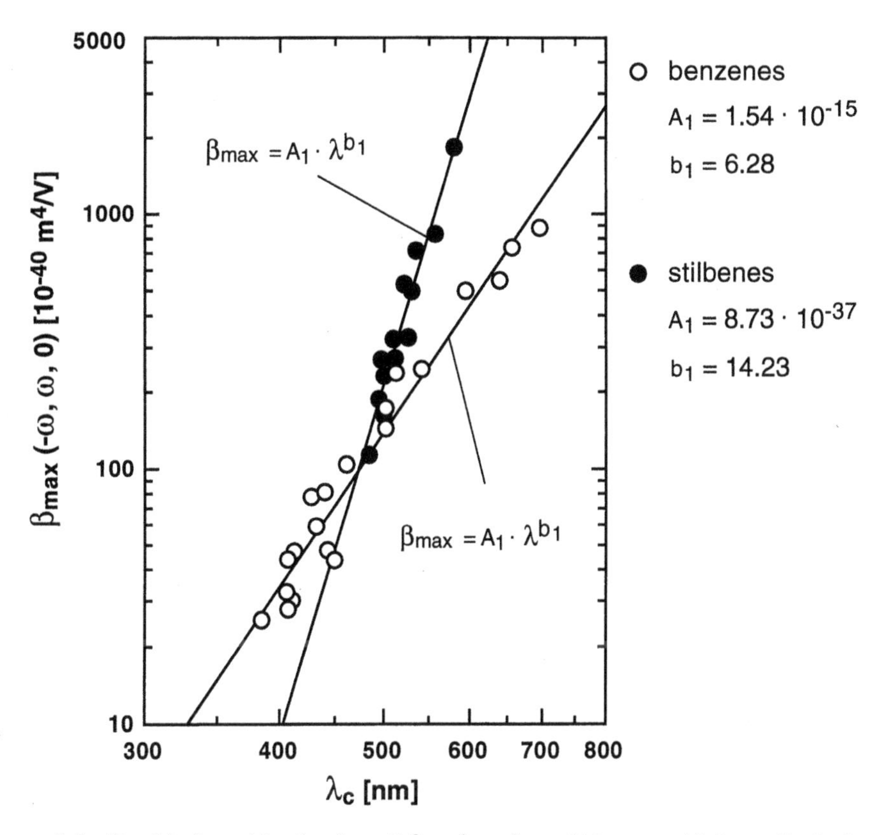

Figure 3.8a Doubly logarithmic plot of β vs $\lambda_c = \lambda_{eg} + 100$ nm + $\Delta\lambda$ for p-disubstituted benzenes ($\Delta\lambda$ = 40 nm) and 4-4′-disubstituted stilbenes ($\Delta\lambda$ = 60 nm) for electro-optics (using data points from *Cheng et al. 1989*, in SI units (β_{max} in 10^{-40} m^4/V, λ in nm)).

many-ring systems tend to be chemically unstable, which makes them unsuitable for many applications.

3.7.2 Limitations due to crystal structure

The relations between microscopic and macroscopic nonlinear optical or electro-optic response (Eqs. (3.25) and (3.26)) depend on the structural parameters N and θ_p. N is the number of molecules per unit volume and θ_p the angle between the molecular charge transfer axis and the polar crystalline axis.

In order to maximize a coefficient of r_{IJK} (or d_{IJK}), it is easy to see from Eq. (3.26) (or Eq.(3.25)) that θ^s_{Ii} (or θ_p) should be close to zero. Such a configuration optimizes the diagonal coefficient r_{III} (or d_{III}). In other words, the charge transfer axes of all molecules should be parallel.

For nonlinear optical applications using bulk crystals the situation is different. For effects such as e.g. frequency-doubling, phase-matching considerations come into play. Materials in which phase-matching is not achievable cannot be used for

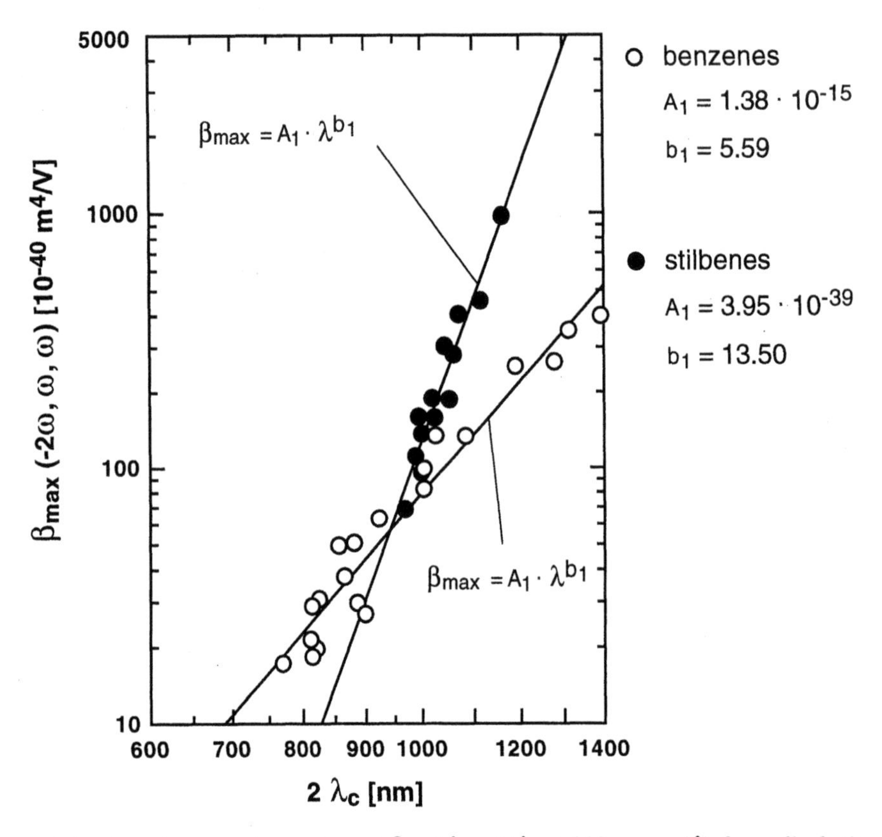

Figure 3.8b Doubly logarithmic plot of β vs $2\lambda_c = 2(\lambda_{eg} + 100\ \text{nm} + \Delta\lambda)$ for p-disubstituted benzenes ($\Delta\lambda$ = 40 nm) and 4-4′-disubstituted stilbenes ($\Delta\lambda$ = 60 nm) for frequency-doubling (using data points from *Cheng et al. 1989*, in SI units (β_{max} in 10^{-40} m^4/V, λ in nm)).

practical applications. The optimum angle θ_p^{PM} = 54.7° (or 125.3°) between the charge-transfer axis (assumed one-dimensional) and the polar axis of the crystal (for the most favorable crystal point groups) for phase-matchable nonlinear optical coefficients was determined by Zyss (*Zyss et al. 1982*). In addition highly efficient nonlinear optical interactions are only possible if noncritical phase-matching can be achieved and if one of the crystal axes coincides with the propagation direction.

If quasi phase-matching in crystals and waveguides, or phase-matching using modal conversion in waveguides is used, highest conversion efficiencies are obtained for the same orientation of the molecules as for electro-optics.

The molecular density N can be estimated from typical densities and molecular weights for benzene and stilbene derivatives. A typical benzene derivative with a nearly optimized structure is 2-methyl-4-nitroaniline (MNA) (*Lipscomb et al. 1981*), a typical stilbene is 3-methyl-4-methoxy-4′-nitrostilbene (MMONS) (*Tam et al. 1989, Bierlein et al. 1990*).

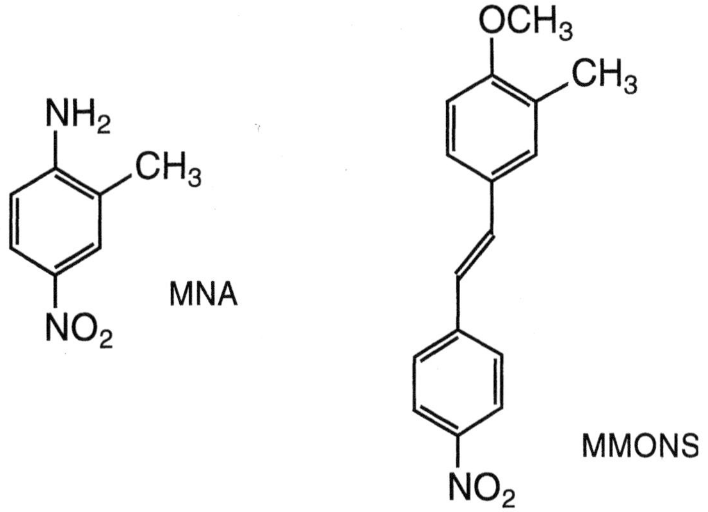

Figure 3.9 2-methyl-4-nitroaniline (MNA) and 3-methyl-4-methoxy-4′-nitrostilbene (MMONS).

The molecular density for MNA is $N \approx 6 \cdot 10^{27}$ m^{-3}, for MMONS it is $N \approx 3 \cdot 10^{27}$ m^{-3}. Since the packing in most molecular crystals is very tight, these numbers do not vary much for other molecules.

3.7.3 Maximum electro-optic and nonlinear optical coefficients

We used the optimized hyperpolarizabilities β_{max} to find the upper limits of the nonlinear optical susceptibilities d_{IJK} and electro-optic coefficients r_{IJK} (*Bosshard et al. 1993*). If all intermolecular contributions (except for local field corrections) to the nonlinearity are neglected, optimized structures for electro-optic and nonlinear optical applications can easily be obtained (see Eqs.(3.25) and (3.26)). For a series of compounds, such as e.g. para-disubstituted benzene derivatives, we chose a small molecule to use the corresponding number of molecules per unit volume and refractive index in order to get an approximation for the upper limit for electro-optic and nonlinear optical effects. These values were then used for all compounds in this series. Since MNA and MMONS are well characterized, we chose the values from MNA (*Lipscomb et al. 1981, Morita et al. 1987*) ($N \approx 6 \cdot 10^{27}$ m^{-3}, $n^{2\omega}$ = 2.2, n^{ω} = 2.2 – 0.3) for the benzene derivatives and MMONS (*Bierlein et al. 1990*) ($N \approx 3 \cdot 10^{27}$ m^{-3}, $n^{2\omega}$ = 2.4, n^{ω} = 2.4 – 0.3) for the stilbene derivatives, as was already mentioned in section 3.7.2 (see Figure 3.9).

Taking these requirements into account and using Eqs.(3.25) and (3.26) one obtains the upper limits for the nonlinear optical and electro-optic coefficients d_{max} (phase-matched frequency-doubling in crystals), $d^{\parallel}_{max}$ (e.g. for quasi phase-matching) and r_{max} as shown in Figures 3 10a, b.

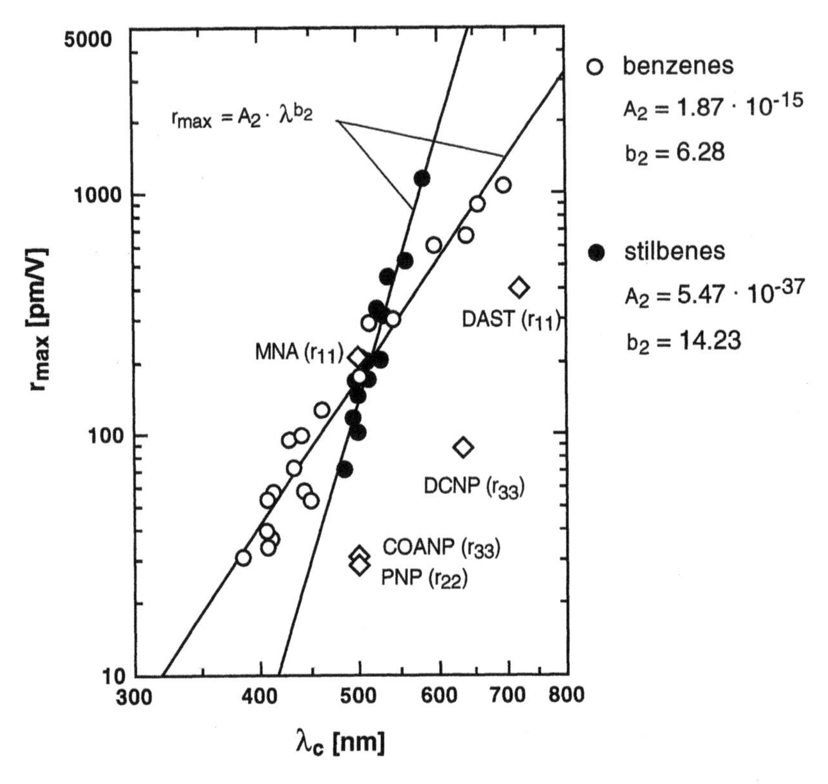

Figure 3.10a Doubly logarithmic plot of r_{max} vs $\lambda_c = \lambda_{eg} + 100$ nm + $\Delta\lambda$ for p-disubstituted benzenes ($\Delta\lambda = 40$ nm) and 4-4′-disubstituted stilbenes ($\Delta\lambda = 60$ nm) for electro-optics (using data points from *Cheng et al. 1989*, in SI units (r_{max} in pm/V, λ in nm)). Examples of electro-optic coefficients of different organic compounds are also marked.

The values are again calculated at λ_c. The results from fits of the data points are given in Table 3.3.

It can be seen that e.g. r is proportional to a high power of the absorption edge λ ($r \sim \lambda^{14}$ for stilbenes and $r \sim \lambda^{6.3}$ for benzenes). Shifting the absorption edge to longer wavelengths by adding stronger donors and acceptors will increase the maximum value of the electro-optic coefficient. However, it should be noted that the absorption edge can only be shifted up to values around 700 to 800 nm for very strong donors and acceptors. Also, the addition of strong donors and acceptors increases the ground state dipole moment of a molecule, which often leads to centric crystal structures.

For the estimate of the maximum electro-optic response at longer wavelengths, away from the absorption edge, the dispersion of the electro-optic coefficient was discussed more carefully in section 3.5.

We see that most materials known up to now are far from being optimized. The major problem is not the molecular but the crystal engineering. There are already many new molecules with exceptionally large nonlinearities. Few of them could,

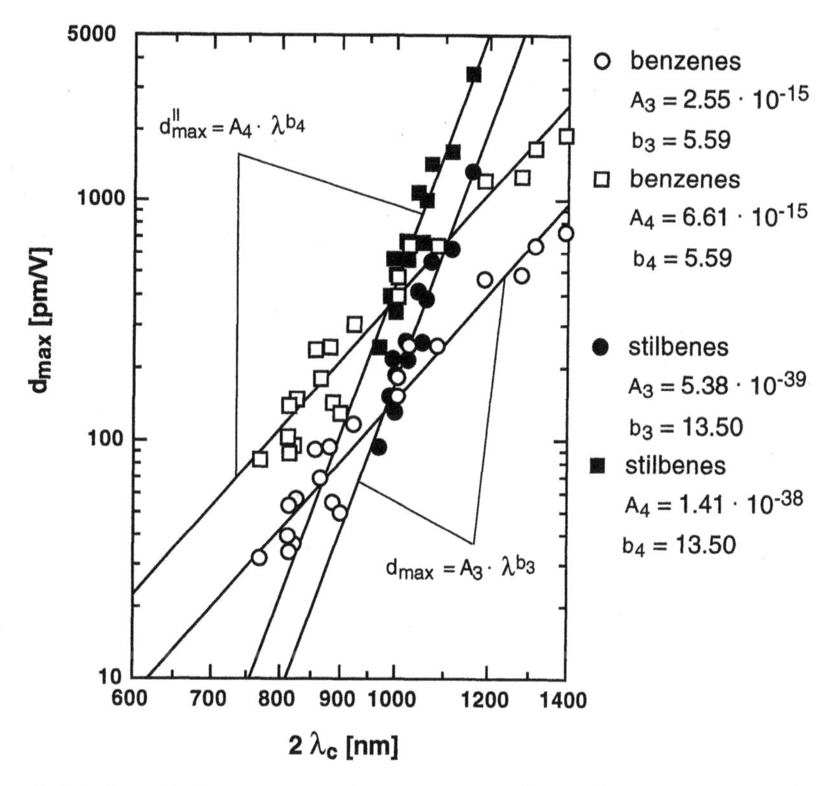

Figure 3.10b Doubly logarithmic plot of d_{max} vs $2\lambda_c = 2(\lambda_{eg} + 100\ \text{nm} + \Delta\lambda)$ for p-disubstituted benzenes ($\Delta\lambda$ = 40 nm) and 4-4′-disubstituted stilbenes ($\Delta\lambda$ = 60 nm) for frequency-doubling (using data points from *Cheng et al. 1989*, in SI units (d_{max} in pm/V, λ in nm)).

however, be used to grow optimized noncentrosymmetric crystals. They may be better applied in LB films or polymers where structural problems can be overcome more easily. On the other hand, one should remember that in these systems the density of molecules is decreased with respect to crystals leading to smaller macroscopic nonlinearities.

3.8 RESONANCE EFFECTS IN CRYSTALS

3.8.1 Electronic dispersion

The optical nonlinearity of the organic materials known up to now arises mainly from the nonlinearity of the molecules. This indicates that resonance effects in crystals also originate from molecular resonances. Such resonances can occur if the wavelength of one of the interacting beams is near the absorption wavelength of the molecule. This behavior can easily be understood from Eq.(3.2). Such resonant enhancements of nonlinear optical and electro-optic effects cannot

Table 3.3 (a) Results from fits of molecular second-order susceptibilities β_{max}, electro-optic and nonlinear optical coefficients r_{max}, d_{max} and $d^{\parallel}_{max}$. (b) Examples for selected wavelengths. The highest value comes either from the theoretical fit or from the measured hyperpolarizability (*) (from *Cheng et al. 1989*) if that value is higher for a specific wavelength.

(a)

electro-optics (λ in nm, β in 10^{-40} m^4/V, r in pm/V)		phase-matched frequency-doubling (λ in nm, β in 10^{-40} m^4/V, d in pm/V)	
benzenes	stilbenes	benzenes	stilbenes
$A_1 = 1.54 \cdot 10^{-15}$	$A_1 = 8.73 \cdot 10^{-37}$	$A_1 = 1.38 \cdot 10^{-15}$	$A_1 = 3.95 \cdot 10^{-39}$
$b_1 = 6.28$	$b_1 = 14.23$	$b_1 = 5.59$	$b_1 = 13.50$
$A_2 = 1.87 \cdot 10^{-15}$	$A_2 = 5.47 \cdot 10^{-37}$	$A_3 = 2.55 \cdot 10^{-15}$	$A_3 = 5.38 \cdot 10^{-39}$
$b_2 = 6.28$	$b_2 = 14.23$	$b_3 = 5.59$	$b_3 = 13.50$
—	—	$A_4 = 6.61 \cdot 10^{-15}$	$A_4 = 1.41 \cdot 10^{-38}$
		$b_4 = 5.59$	$b_4 = 13.50$
$\beta_{max} = A_1 \cdot \lambda^{b_1}$	$r_{max} = A_2 \cdot \lambda^{b_2}$	$d_{max} = A_3 \cdot \lambda^{b_3}$	$d^{\parallel}_{max} = A_4 \cdot \lambda^{b_4}$

(b)

λ_c [nm]	$\beta_{max}(-\omega, \omega, 0)$ [10^{-40} m^4/V]	$2\lambda_c$ [nm]	$\beta_{max}(-2\omega, \omega, \omega)$ [10^{-40} m^4/V]	
500	170*[1]......220[2]	1000	100*[1]......140*[2]	—
550	250[1]......860[2]	1060	110[1]......270[2]	—
630	590[1]......6000[2]	1300	350[1]......4300[2]	—
	r_{max}[pm/V]		d_{max}[pm/V]	$d^{\parallel}_{max}$[pm/V]
500	140[2]......210*[1]	1000	180*[1]......190*[2]	470*[1]......490*[2]
550	300[1]......540[2]	1060	210[1]......370[2]	540[1]......980[2]
630	710[1]......3700[2]	1300	650[1]......5900[2]	1700[1]......15400[2]

[1] benzenes
[2] stilbenes

usually be used for applications because of the large absorption present. In monolayers of LB films, however, extremely large nonlinearities can be generated in this way.

Electronic dispersion effects in molecules have been discussed in detail in section 3.5.

3.8.2 Other dispersion effects

As was discussed in section 3.6, acoustic as well as optic phonons can contribute to the linear electro-optic effect, especially in inorganic materials. However, at least acoustic phonon contributions can also appear in organic crystals as is shown in

Figure 7.20. When varying the modulation frequency of the applied voltage resonance enhancements of up to a factor of 33 for $|n_2^3 r_{22} - n_1^3 r_{12}|$ of PNP (modulation frequency ν = 210.4 kHz) have been observed (*Bosshard et al. 1993*).

For electro-optic broad band modulators where an applied field is used to change the phase or amplitude of a carrier light wave, the strong dispersion of the electro-optic coefficients around 10^6 Hz (see Figure 3.4) gives rise to distortions of the modulator properties. A flat frequency response up to the maximum modulator frequency (around several GHz) is preferred. Resonance effects due to optic phonons have not yet been observed in organic crystals.

4. CRYSTAL GROWTH OF ORGANIC MATERIALS

4.1 INTRODUCTION

Single crystals of organic materials are critically important to experimental work on the nonlinear optical, electro-optic and photorefractive phenomena to be reviewed here. To produce either bulk, thin platelet and fiber crystals or epitaxial layers, different and, in some cases, rather sophisticated growth techniques are required. Depending on the properties to be investigated or to be optimized, knowledge on crystal defects is additionally of importance. Most of the optical applications need highly pure materials, showing a low density of all kinds of structural defects. In the case of organic photorefractive crystals a homogeneous doping by a sensitizer molecule must be achieved. Altogether, this gives rather strong constraints to be fulfilled by growth procedures to be applied to new and generally less characterized organic compounds, exhibiting promising optical effects which have been observed for first-obtained crystals. As in other rapidly developing fields of solid state physics, the *materials science* part of the endeavor preferentially concentrates on two different approaches:

I So-called *exploratory crystal growth techniques* to obtain first and reasonably good crystals during the search and testing of new materials.

II *Advanced crystal growth techniques* to produce quality crystals showing a size in the range of 0.5 to 1 cm^3.

As any professional crystal grower knows, the first approach may be fast and does not need complicated equipment. The second, however, can be slow and painstaking. Last, but not least, the performance of optical experiments depends on oriented, polished and coated cuts. Not much has been published on this issue in the field of molecular crystals.

For the convenience of the non-specialized reader, let us first indicate some introductory literature on the fundamentals and the general methods of crystal growth, as well as current reviews on the growth and basic characterization of molecular crystals, including numerous nonlinear optical and electro-optic materials.

4.1.1 Reviews on the fundamentals and the general methods of crystal growth

J. J. Gilman: 'The Art and Science of Growing Crystals', Pergamon Press, Oxford (1963)
B. Pamplin (ed.): 'Crystal Growth', Pergamon Press, Oxford (1975)
J. C. Brice: 'Crystal Growth Processes', Wiley, London (1986)
F. Rosenberger: 'Fundamentals of Crystal Growth I', Springer Series in Solid-State Sciences vol. 5 (1979)
W. A. Tiller: 'The Science of Crystallization: Macroscopic Phenomena and Defect Generation (vol. I), Microscopic Interfacial Phenomena (vol. II)' Cambridge University Press, Cambridge, 1991

P. M. Dryburgh, B. Cockayne and *K. G. Barraclough* (eds.): 'Advanced Crystal Growth', Prentice Hall, London (1987)

K. Th. Wilke and *J. Bohm*: 'Kristallzüchtung', H. Deutsch, Frankfurt (1988)

H. Arend and *J. Hulliger* (eds): 'Crystal Growth in Science and Technology', NATO ASI Series B: Physics vol. 210, Plenum Press, New York (1989)

A. A. Chernov: 'Results and Problems in Crystal Growth Studies', *Z. Phys. Chemie (Leipzig)* 941–970 (1988); 'Formation of crystals in solutions', *Contemp. Phys.* 30, 251–276 (1989)

J. A. Dirksen and *T. A. Ring*: 'Fundamentals of Crystallization: Kinetic Effects on Particle Size Distribution and Morphology', *Chem. Eng. Sci.* 46, 2399–2427 (1991)

J. Hulliger: 'Chemie und Kristallzüchtung', *Angew. Chem.* 106, 151–171 (1994), *Int. Ed.* 33,143–162 (1994)

4.1.2 Reviews on the design, the purification, the growth and the characterization of organic crystals

G. J. Sloan and *A. R. McGhie*: 'Techniques of Melt Crystallization', Wiley, New York, (1988)

G. R. Desiraju: 'Crystal Engineering; the Design of Organic Solids', Mat. Science Monogr. 54, Elsevier, Oxford (1989)

G. R. Desiraju (ed.): 'Organic Solid State Chemistry', Elsevier, Amsterdam, (1987)

J. D. Wright: 'Molecular Crystals', Cambridge University Press, Cambridge (1987)

D. S. Chemla and *J. Zyss* (eds): 'Nonlinear Optical Properties of Organic Molecules and Crystals', vol. 1, 2, Academic Press, New York (1987); in particular: *J. Badan, R. Hierle, A Perigaud and P. Vidakovic*: 'Growth and Characterization of Molecular Crystals' (chap. II–4, pp. 297–356)

N. Karl: 'High Purity Organic Molecular Crystals', in Crystals (Growth, Properties and Applications) vol. 4, pp. 1–100, *H.C. Freyhardt* (ed.), Springer, Berlin (1980)

R. S. Tipson: 'Crystallization and Recrystallization', in Techniques of Org. Chem. vol. III (part I: separation and purification) (A. Weissberger, ed.), Wiley, New York, (1966)

P. van der Sluis, A. M. F. Hezemans and J. Kroon: 'Computer knowledge base for crystallization', *J. Cryst. Growth* 108, 719–727 (1991)

B. J. McArdle and *J. N. Sherwood*: 'The Growing of Organic Crystals', in Advanced Crystal Growth, *P. M. Dryburgh et al.* (eds.), pp. 179–217, Prentice-Hall, London,(1987)

M. Samoc, J. O. Williams and *D. F. Williams*: 'Molecular Organic Crystals: A Survey of their Purification, Growth and Characterization', *Prog. Cryst. Growth Charact.* 4, 149–172 (1981)

B. G. Penn, B. H. Cardelino, C. E. Moore, A. W. Shields and *D. O. Frazier*: 'Growth of Bulk Single Crystals of Organic Materials for Nonlinear Optical Devices: an Overview', *Prog. Cryst. Growth Charact.* 22, 19–51 (1991)

B. G. Penn, A. W. Shields and *D. O. Frazier*: 'A Preliminary Review of Organic Materials Single Crystal Growth by the Czochralski Technique', NASA Tech. Memo. NASA TM-100341, pp. 1–8 (1988)

N. Karl: 'Growth of Organic Single Crystals', in Materials for Non-linear and Electro-optics (Inst. of Physics, Conference Series Number 103, IOP, Bristol, 1989), *M. H. Lyons* (ed.) pp. 107–118 (1989)

T. Fukuda and *T. Sano*: 'Melt Growth of Organic Crystals for Nonlinear Optical Applications', *J. Jap. Assoc. Cryst. Growth* 16, 26–33 (1989) (Japanese, special issue on organic crystals: from superconductors to proteins)

J. N. Sherwood: 'The Growth, Perfection and Properties of Crystals of Organic Nonlinear Optical Materials', in Organic Materials for Nonlinear Optics. *R. A. Hahn* and *D. Bloor* (eds), Royal Society of Chemistry, special publication No. 69 (1989)

R. T. Bailey, F. R. Cruickshank, D. Pugh and *J. N. Sherwood*: 'Growth Perfection and Properties of Organic Nonlinear Materials', *Acta Cryst.* A47, 145–155 (1991)
H. Klapper: 'X-Ray Topography of Organic Crystals', in Crystals (Growth, Properties and Applications), vol. 13, pp.109-162. *H. C. Freyhardt* (ed.), Springer, Berlin (1991)

4.1.3 Special topics

R. J. Davey: 'The Role of the Solvent in Crystal Growth from Solution,' *J. Cryst. Growth* 76, 637–644 (1986)
D. J. Morantz and *K. K. Mathur*: 'Inhomogeneous Distribution of Impurities in Organic Single Crystals and the Effect of Molecular Parameters', *J. Cryst. Growth* 16, 147–158 (1972)
B. Kahr and *J. M. McBride*: 'Optisch anomale Kristalle', *Angew. Chem.* 104, 1–28 (1992), *Int. Ed.* 31, 1–26 (1992)
J. Jacques, A. Collet and *S. H. Wilen*: 'Enantiomers, Racemates and Resolutions', Wiley, New York, (1981)
Ch. Reichardt: 'Solvents and solvent effects in organic chemistry', Verlag Chemie, Weinheim (1988)

4.2 GROWTH TECHNIQUES FOR ORGANIC CRYSTALS

Crystallization of organic compounds is based on *solution growth* (SG), *melt growth* (MG) and *vapor growth* (VG), depending on:

(i) The production of either bulk (3-D), thin platelet or epitaxial layer (quasi 2-D) and fiber crystals (quasi 1-D);
(ii) the purity of the starting materials;
(iii) the thermal stability and general chemical stability (long-term) of melts, solutions and solids;
(iv) the melting temperature, the solubility and the vapor pressure;
(v) the formation of solution or solvent inclusions and other defects.

(see e.g. *Karl 1980, 1989, Tipson 1966, Sloan et al. 1988)*

4.2.1 Solution growth (SG)

Since most of the organic compounds show melt degradation effects, solution growth techniques are very attractive and represent, in many cases, the only possibility of obtaining large organic single crystals. According to all references we are aware of, growth techniques from solution have most frequently been applied to molecular crystals (for materials, see Table 4.1). However, growth from solution is limited to rather low growth rates (in many cases ~ 0.1 to 1.0 mm/day) and suffers often from *solvent* or *solution* inclusion problems. A statistical analysis on solvent inclusions showed that many organic crystal lattices tend to include solvent molecules preferentially at lattice sites of low symmetry. Among widely used solvent molecules, water shows the highest occurrence (61.4% of crystal structures containing solvents) followed by methylene dichloride (5.9%), benzene (4.7%) and methanol (4.1%). Solvents such as acetonitrile (1.9%) and dimethylsulfoxide (0.5%) which are frequently used in the crystallization of nonlinear optical compounds, exhibit a low probability of being incorporated (*van der Sluis et al. 1989*). For a review on the properties of solvents see *Reichardt 1988.*

While not providing a complete treatise on solution growth techniques, let us introduce some basic procedures and discuss details on corresponding equipment which proved to be feasible in *exploratory or advanced growth* of various types of organic and other materials. For a more complete treatise see *Wilke et al. (Chapter 4, 1988).*

4.2.1.1 Temperature difference technique (TD): transport of dissolved feed material by a thermally induced convection of the nutrient. In its most simplest version the TD technique uses glass tubes provided by a lid (e.g. SOVIREL) and a square cross section (glass ware from e.g. SOCIETE ATA, Geneva) in its central part (see Figure 4.1). The bottom part of the tube (containing solid excess material) is placed e.g. into a heated Al-block at temperature T_2, whereas the growth area showing the square cross section is within a thermostating air bath at $T_1 < T_2$.

The set-up operates as follows. At a $\Delta T = T_2 - T_1$ of typically 1 to 5°C, a closed loop of a thermally enforced circulation of solution establishes a stationary supersaturation around nucleating and growing crystals which grow preferentially in the upper part of the tube. Under favorable conditions spontaneously nucleated

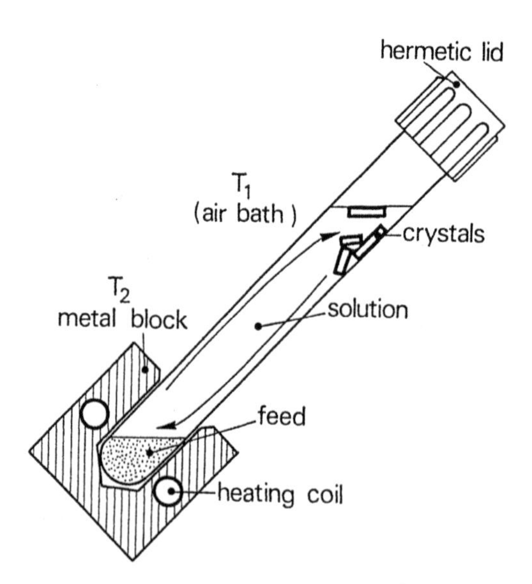

Figure 4.1 Closed test tube (solution growth) apparatus based on a temperature difference procedure. Spontaneous nucleation occurs due to the supercooling of the up-streaming saturated solution. Tubes showing an outer square cross section of 12 × 12, 15 × 15 and 23 × 23 mm^2 have been used (*Arend et al. 1986*).

crystals can reach sizes of up to ~1 cm^3 (Figure 4.2). (For more technical details see *Arend et al.* 1986 and *Wang et al. 1989.*) Taking into account that most exploratory work in crystallizing new organic materials starts out from *solution growth,* the ΔT-method is recommended here as a straightforward technique, particularly when using many tubes assembled into an exploratory matrix (i.e. a 2-dimensional array: y-axis ⇒ different ΔT's; x-axis ⇒ variation of solvent or solid compounds) to produce seed crystals and to explore optimum growth conditions (*Wang et al. 1989*) for subsequent seed growth experiments. In the past, application of the ΔT-method proved to be efficient for a large number of different organic and inorganic materials (with about 90% chance for getting crystals in many cases). In cases where seed growth should be continued to obtain larger crystals than possible from these tubes, an enlarged set-up can be used (*Arend et al. 1986*).

As will be outlined in section 4.4, the growth of solid solutions is always complicated by striations, i.e. an inhomogeneous distribution of dopants. Homogeneous solid solution crystals are generally obtained only for growth systems which operate at a constant supersaturation. Within its stationary regime of operation the ΔT-method provides a constant supersaturation.

Recently ΔT-grown nonlinear and electro-optic crystals are: 2-cyclooctylamino-5-nitropyridine (COANP) (*Wang et al. 1989*), 2-(N-(L or S)-prolinol)-5-nitropyridine (L-PNP or S-PNP) (*Hulliger et al. 1990a*), 4-(N,N-dimethylamino)-3-acetamido-nitrobenzene (DAN) (*Kerkoc et al. 1990*), N-(5-nitro-2-pyridyl)-(S)-phenylalaninol (S-NPPA) (*Sutter et al. 1991a*), thiosemicarbazide cadmium bromide hydrate (TSCC) (*Wang et al. 1988a*) and others.

4.2.1.2 Temperature lowering technique (TL). Various TL techniques using seed

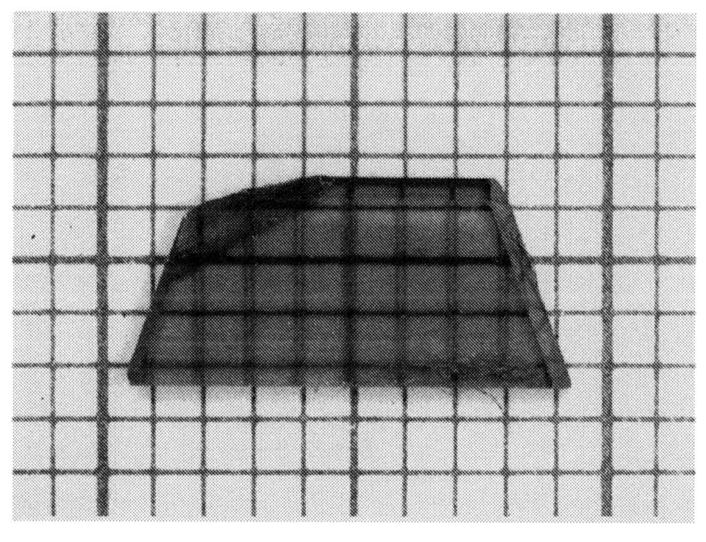

Figure 4.2 A ΔT-grown DAN single crystal (mm scale) using dimethylsulfoxide as the solvent (ΔT ≈ 3°C, 23 × 23 mm^2 tube) (*Kerkoc et al. 1990*).

crystals differ mainly in the hydrodynamics of the nutrient. Simple rotation or a planetary motion of seed crystals as well as the accelerated seed rotation technique have been widely used (for details see *Wilke et al. 1988*). As an example, nearly cm^3-large L-PNP single crystals of optical quality have been grown (Figure 4.3) by a set-up equipped with the accelerated seed rotation technique (*Hulliger et al. 1990a*).

In the course of growing various new nonlinear optical materials we have developed a growth system as shown schematically in Figure 4.4 : a tubular glass set-up is placed into a thermostated water bath allowing for programmed temperature changes (0.01 to 0.1°C/day). A supersaturated solution is driven forward by a slow rotating spiral made out of TEFLON or DELRIN. Properly oriented seeds are placed into an area of nearly laminar flow, which is established by pushing the nutrient first through a diaphragm and afterwards through a cone. Circulation speeds of the order of 0.5 to 1.0 cm/s were used in the case of growing 4′-nitrobenzylidene-3-acetamino-4-methoxyaniline (MNBA) from e.g. N,N-dimethylacetamide (Figure 4.5) and 4′-dimethylamino-N-methylstilbazolium-p-tosylate (DAST) from methanol.

As compared to most of the other commonly used TL techniques, this configuration involves efficient mixing and favorable flow conditions around growing crystals, showing preferentially a pyramidal, a plate-like or a needle-like habitus.

4.2.1.3 Isothermal evaporation (IE) with after-saturation. IE of the solvent or evaporation of its most volatile component is a widely used technique in chemical laboratories. In cases of materials showing a rather low solubility (concentration $c < 5\%$) or a retrograde solubility behavior, as well as for materials sensitive to elevated temperatures, growth by IE provides reasonable conditions. Appropriate equipment

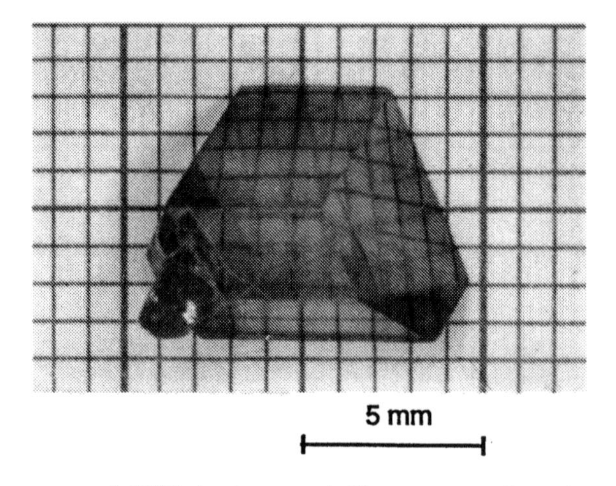

Figure 4.3 An as grown L-PNP single crystal. Temperature lowering procedure and application of the accelerated seed rotation technique. Solvent: water/ethanol (vol. 1:1) with the addition of nearly equimolar amounts of Nd^{3+} to enhance the optical quality of grown L- or S-PNP crystals (*Hulliger et al. 1990a*).

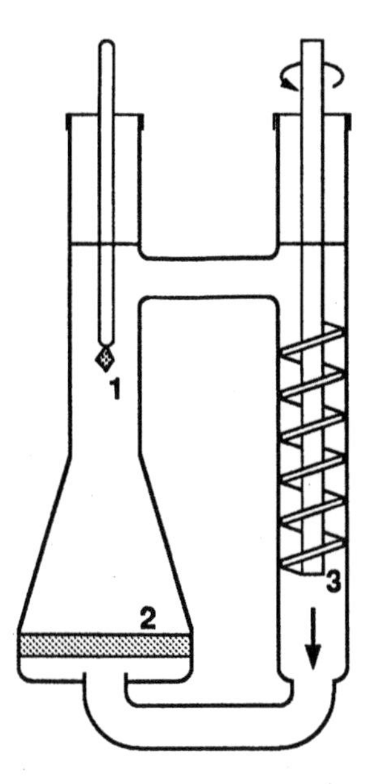

Figure 4.4 Closed glass ware apparatus (placed into a thermostated water bath) for solution growth by lowering the temperature. The nutrient is driven forward by a spiral. After passing through the cone its flow becomes nearly laminar. The set-up provides efficient mixing as well as controlled hydrodynamic conditions around the growing seed crystal (*Hulliger 1994*).

(see Figure 4.6) is placed into a thermostated waterbath kept at constant temperature T_1. Evaporation at T_1 (bath) followed by a condensation at T_2 (cooled finger) leads to a hydrostatically driven circulation of the nutrient which exceeds in the part where the feed material is placed in after-saturation. There is some analogy between the ΔT-method and this modification of the IE procedure: both operate in a stationary regime of a nearly constant supersaturation and both allow for the growth of 0.5–1 cm^3 crystals also for materials exhibiting fairly low solubilities.

4.2.2 Examples of recently solution grown materials

Most of the reported organic nonlinear and electro-optic crystals which are investigated today have been grown either by a temperature lowering technique or by a simple isothermal evaporation procedure (*Badan et al. 1987*). Examples of materials are: urea, methyl-3-nitro-4-pyridine-1-oxide (POM), 2-methyl-4-nitroaniline (MNA), (-)-2-(α-methylbenzylamino)-5-nitropyridine (MBANP), N-(4-nitrophenyl)-(L)-prolinol (L-NPP), 4-N-dimethylaminobenzylidene cyanacetic acid (AMA), 4-chloro-7-nitro-2,1,3-benzoxadiazole (NBD-Cl), 4-amino-benzophenone (4-ABP), N-methoxymethyl-4-nitroaniline (MMNA), 4-Br-4′-methoxychalcone (BMC), dicyanovinyl anisole (DIVA), 3-methyl-4-methoxy-4′-nitrostilbene (MMONS), para-dimethylamino-β-nitrostyrene (MANS), 4-nitro-4′-methylbenzylidene aniline

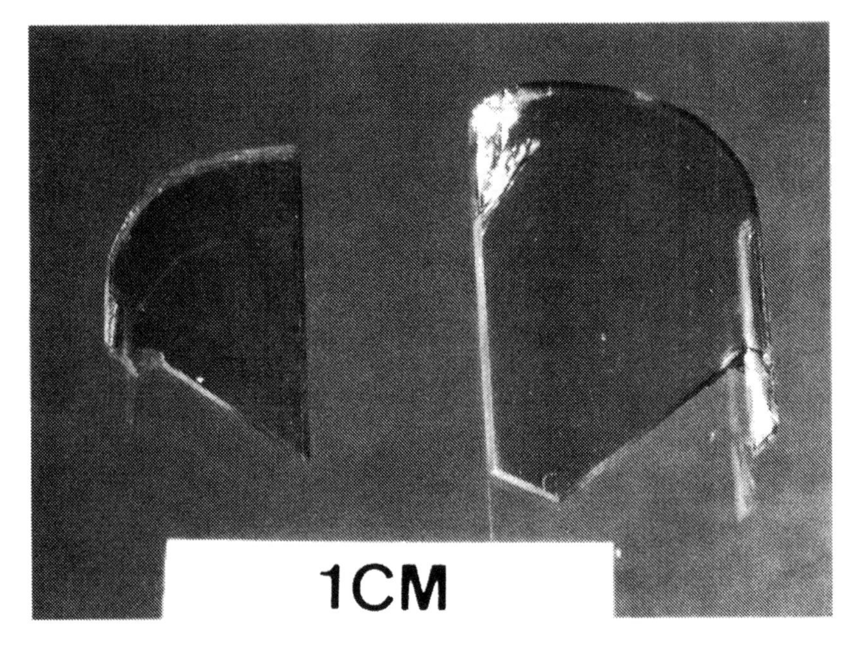

Figure 4.5 As grown MNBA single crystals. Temperature lowering procedure using an apparatus as shown in Fig. 4.4 and N,N-dimethylacetamide as the solvent.

(NMBA), N-methyl-N,N-dimethylamino-stilbazoliummethylsulfate (MSMS), COANP, L- or S-PNP, S-NPPA, DAN, DAST, TSCC, and others (for references see Table 4.1).

4.2.3 Gel growth (GG) of organics

GG is usually performed for inorganic materials which show a very low solubility for a wide range of temperature. In the case of organic materials, crystallization from a gel can be maintained by (i) lowering the temperature of a saturated system (solvent + gel + nonlinear optical material) or (ii) by decreasing the solubility due to the in-diffusion of a precipitation agent into the gel.

What are the advantages to be expected from gel grown organic crystals? It was demonstrated for inorganic materials that some defects related to convection phenomena within the nutrient can be reduced if solvent convections are trapped by a microporous polymer network such as tetramethoxysilane (TMS) gel. In gels-systems growth is only (except at the crystal-gel interface) controlled by a diffusion of material.

Recently, POM and L-NPP have been grown from acetonitrile/water mixtures and TMS gel (~5–7 vol. %), by lowering the temperature and awaiting spontaneous nucleation (*Andreazza et al. 1990a,b*). As compared to earlier solution growth results of POM and L-NPP, these gel-grown crystals displayed improved nonlinear optical properties. Similar experiments have led to single crystals of

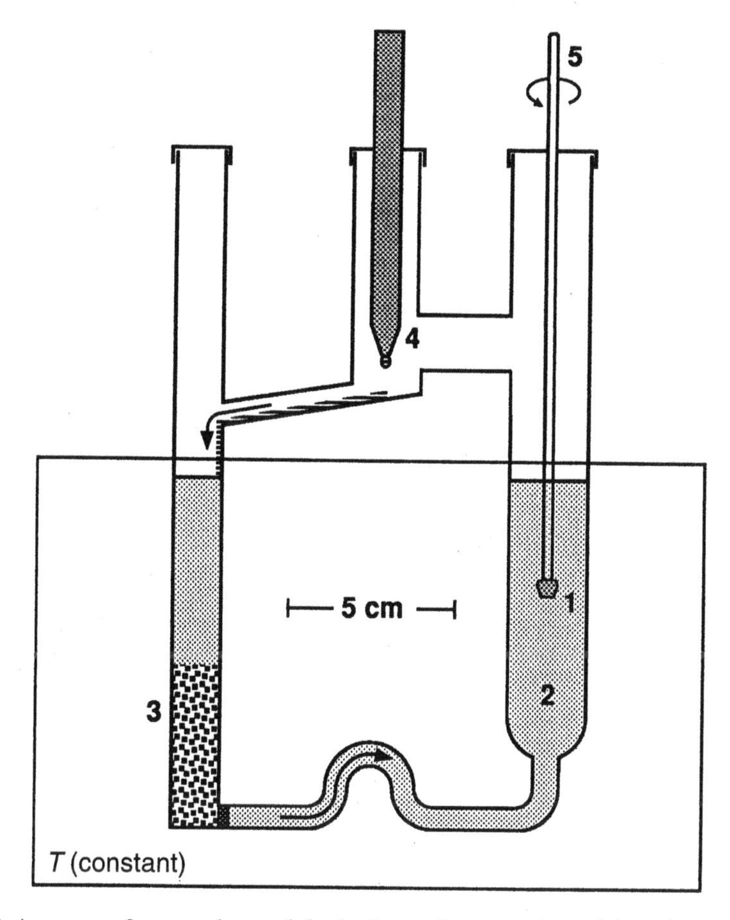

Figure 4.6 Apparatus for crystal growth by isothermal evaporation of the solvent or some of its components: 1: seed crystal, 2: solution, 3: excess material, 4: cooled finger, 5: crystal rotation (if necessary). The set-up is provided by a mechanism of after-saturation. This represents (together with the ΔT-method (section 4.2.1.1)) a tool to grow relatively large crystals from e.g. a solvent which is showing a fairly low solubility (*Hulliger 1994*; for a similar approach see also *Fuith et al. 1984*).

COANP and L- and S-PNP (*Březina et al. 1991*). At present, the control of the spontaneous nucleation and relatively long growth times for sufficiently large crystals (3 to 18 months for PNP to a size of ~0.75 cm^3!) are considered to be the main difficulties of growing reasonably large organic crystals from a gel media.

4.2.4 Melt growth (MG)

Despite the fact that solution growth appears to be quite universal for organic compounds, this technique generally provides low growth rates and is often complicated by solution or solvent inclusion defects. For those organic materials

Table 4.1 Data on bulk crystal growth of some recent NLO and EO active organics

Compounds (abbreviations see text)	growth techniques (temperature range) (abbreviations see below)	growth media	growth times [days]	crystal habitus, maximum sizes [mm^3]	references
COANP	SG-TD	ethanol/water (1:1)	30	~3 × 3 × 9 (plate)	*Wang et al. 1989*
	25 < T < 30°C	acetonitrile or ethylacetate	30	~3 × 10 × 30 (plate)	*Günter et al. 1987*
	GG-TL	ethanol/water + tetramethoxysilane	30–60	~(2–4) × 2 × (10–20) (bars and plates)	*Březina et al. 1991*
	SMG	melt	14–28	~2 × 8 × 10 (plate)	*Hulliger et al. 1990b*
PNP	SG-TL 30 < T < 35°C	ethanol/water (1:1 or 3:7) with addition of $NdCl_3$	30–60	~7 × 8 × 10 (sphenoidal)	*Hulliger et al. 1990a*
	GG-TL T=22°C	ethanol/water +$NdCl_3$ +tetramethoxysilane	30–540	~5 × 5 × 8 (sphenoidal)	*Březina et al. 1991*
DAN	SG-TD 32 < T < 34°C	dimethylsulfoxide	14	~2 × 6 × 15 (plate)	*Kerkoc et al. 1990*
	SG-IE	methanol	90	~3 × 3 × 8 (plates)	*Sherwood 1989*
	BT	melt	—	~(20–30) × 20 (cylindrical)	*Sherwood 1989*
				~(20–30) × 20 (cylindrical)	*Lin et al. 1990a*
NPPA	SG-IE T=20°C	methanol	14	~1 × 2.5 × 3	*Uemiya et al. 1990*
	SG-TD 22 < T < 25°C	ethanol/water (1:1)	30	2 × 3 × 10	*Sutter et al. 1991a*
MBANP	SG-TL	methanol	—	70 × 70 × 100 (prismatic)	*Sherwood 1989*
		toluene	15	5 × 15 × 15	*Kondo et al. 1989b, 1990*
MMNA	SG-IE	tetrahydrofuran	>2	~3 × 5 × 50 (bars)	*Hosomi et al. 1989*
MAP	THM	melt	< 15	~10 × 70 (cylindrical)	*Perigaud et al. 1986*
MMONS	SG-IE T ≈ 25°C	chloroform/ ethanol (1:1)	—	~10 × 10 × 10 (rhombic)	*Bierlein et al. 1990*
BMC	SG-IE	acetone	~30	~4 × 10 × 10 (prismatic)	*Zhang et al. 1990a*
NBD-Cl	SG-TL	diethylether or ethylacetate	30	~(1–2) × 3 × 8 (plates)	*Suzuki et al. 1988*

Table 4.1 Continued.

Compounds (abbreviations see text)	growth techniques (temperature range) (abbreviations see below)	growth media	growth times [days]	crystal habits, maximum sizes [mm^3]	references
CMONS MONS BMONS BONS	SG	various		small crystals	*Wang et al. 1988b*
MNA	VG	vapor		~2 × 2 × 5	*Ono et al. 1989*
	SG	—		thin needles	*Fukuda et al. 1989*
	BT	melt		~4 × 25	*Nakanishi, 1989*
	CZ	melt		(cylindrical)	
	VG	vapor		50 μm × 5 × 10	
	BT	melt		1 × 2 × 2	
DIVA	SG-IE	acetone	2–3	5 × 10 × 10	*Wada et al. 1989*
	VG	vapor	1–2	1 × 2 × 10 (needle)	
4-ABP	SG-TL	dimethylformamide	~20	2 × 4 × 4	*Guha et al. 1988*
POM	GG-TL	acetonitrile/water + tetramethoxysilane	30–80	~5 × 10 × 15 (plate)	*Andreazza et al. 1990a,b*
NPP	GG-TL	e.g. acetonitrile/water + tetramethoxysilane		~3 × 5 × 11 (plate)	*Andreazza et al. 1990a,b*
Urea	SG-TL T < 45°C	methanol	~120 –240	~46 × 50 × 171 (bars)	*Huang et al. 1990*
MNBA	SG-TL	N,N-dimethylacetamide		~2 × 8 × 10	*Gotoh 1990*

SG: solution growth
VG: vapor growth
GG: gel growth
SMG: supercooled melt growth
TD: temperature difference technique
TL: temperature lowering technique
CZ: Czochralski technique
BT: Bridgman technique
IE: isothermal evaporation
THM: travelling heater method

which exhibit a high long term thermal stability, melt growth techniques are more attractive. In these cases much higher growth rates could yield larger crystals within a shorter time than commonly necessary for solution growth.

In the past, organic Bridgman growth and zone-melting have been extensively used (see e.g. *Karl, 1980, Sloan et al. 1988*), whereas Nacken-Kyropoulos and Czochralski growth has been much less applied (*Penn et al. 1988*). Main difficulties in using e.g. the Czochralski technique for organic materials are due to their relatively low thermal conductivity. The removal of the heat of crystallization makes special equipment necessary.

4.2.4.1 Examples of recently grown materials. Recently, Bridgman growth produced 2-adamanthylamino-5-nitropyridine (*Tomaru et al. 1991*), the Kyropoulos technique was used in the case of meta-bromonitrobenzene (m-BNB), mixed crystals of m-chloro- and m-bromonitrobenzene (*Badan et al. 1987*) and a pulling process has been applied in the case of COANP (*Hulliger et al. 1993*).

Further progress on the melt growth of recently developed other nonlinear optical and electro-optic organic compounds is reported for methyl-(2,4-dinitrophenyl)-amino-2-propanoate (MAP), DAN, MNMA, MNA and others (see Table 4.1)

4.2.4.2 Growth from supercooled melt (SCMG). Commonly applied melt growth techniques are in some cases encountered by considerable disadvantages that are related to: (i) the mechanical interaction of the crystal with the container due to a difference in the thermal expansion coefficient (e.g. *Bridgman growth*; stress that is imposed on crystals can be partially avoided by use of a transparent and thin TEFLON container), (ii) thermal gradients due to an anisotropic heat release via the seed (*Kyropoulos* and *Czochralski* growth) and mechanical stress imposed on the growing crystal (softness near the melting point) by gravity and torques (seed rotation).

There exists a further method that seems to avoid most of these problems: it is the growth within a less *supercooled* ($\Delta T \sim 0.1$–$2°C$) melt. In this case the growing crystal is completely surrounded by its melt at a constant supercooling and growth occurs at almost isothermal (the growing crystal faces being the hottest part of the system) conditions. At a considerably low growth rate the heat of crystallization will efficiently be removed throughout the melt. In the past, rather simple equipment produced highly perfect and large crystals (~25 cm^3) of e.g. benzophenone, benzil and salol within only several days (*Scheffen-Lauenroth et al. 1981*).

It has been known for a long time that some organic compounds form extremely stable supercooled liquid states. De Coppet, (see *Wilke et al. 1988*) for example, has shown that a supercooled melt of salol could be kept free of any spontaneous crystallization for more than 35 years! This can be explained by the fact that for such materials the temperature region where the diffusion is high enough for crystallization lies above the region for maximum nucleation rate (in case of excluding kinetic effects due to impurities). Properties of a melt as found for organic compounds allow for seeded growth at the slightly supercooled, i.e. metastable state. Recently we could demonstrate that this method can be applied to COANP (*Hulliger et al. 1990b*). Growth experiments from a supercooled COANP melt have been performed in the apparatus shown in Figure 4.7.

The melt and the growing crystal could be inspected through a set of crossed

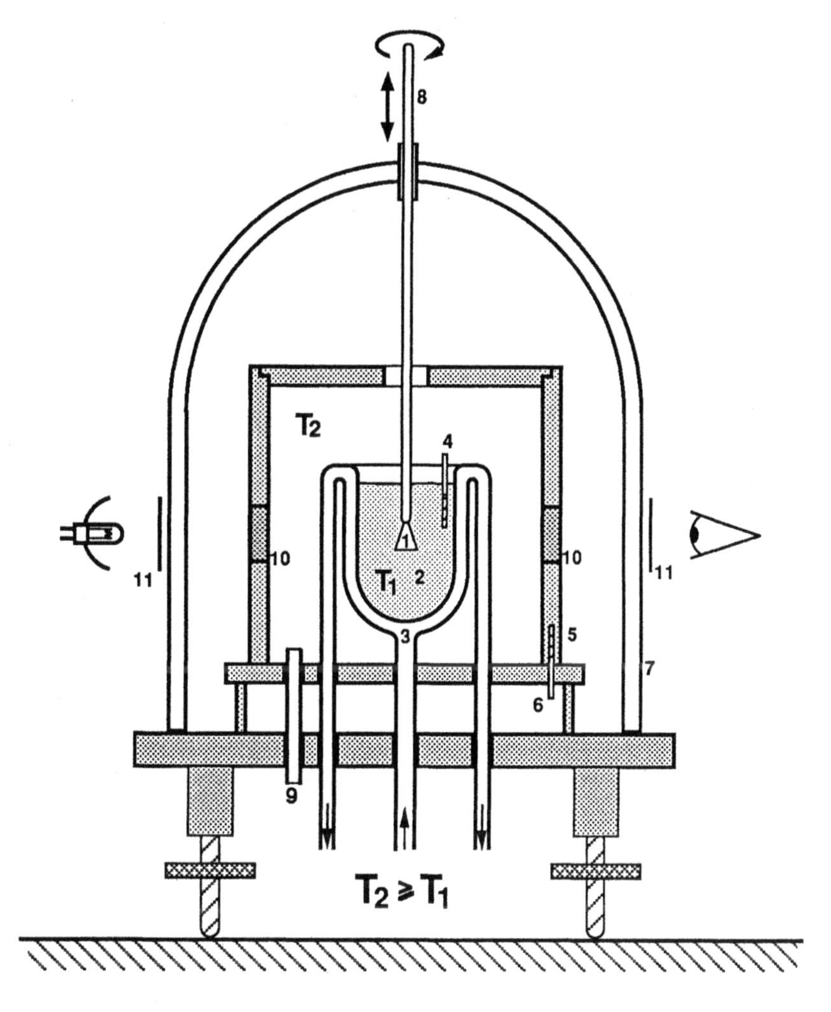

Figure 4.7 Equipment for supercooled melt growth of organic compounds. Crystals can be grown upon a seed for the following conditions: (i) isothermal temperature field, (ii) *Kyropoulos*-type growth ($T_2 < T_{melting}$) or (iii) *Czochralski*-type growth ($T_2 < T_{melting}$). In the case of COANP best results were obtained for $T_1 < T_2$, at supercoolings of $\Delta T = T_{melting} - T_1 \leq 0.3$–$0.5°C$ and at pulling rates of 0.5 to 1 mm/day (*Hulliger et al. 1993*).

polarizers. The possibility of adjusting different temperatures for the melt and the surrounding atmosphere prevented nucleation at the melt surface in the case of a fully immersed seed, or it established growth conditions suitable for crystal pulling. In both cases optical quality crystals have been obtained for supercoolings of $\Delta T = 0.3$ to max 1.0°C and within growth times of several weeks (Figure 4.8) As compared to a plate-like morphology obtained for fully immersed seeds, a surprisingly strong morphological change was observed when pulling conditions were set up. Crystal pulling at a speed of ~1 mm/day has led to a more isometrical habitus suitable for cutting out larger optical samples than possible for previously obtained plates. (Results on the growth of doped organic crystals, e.g. photorefractive materials, are discussed in section 4.4.)

4.2.5 Vapor growth (VG)

Vapor growth is generally known to give crystals of the highest purity and lattice perfection. However, as a consequence of the physical vapor deposition process, low growth rates are normally found. Vapor growth has extensively been used for arenes (see *Karl, 1980*), but up to now rarely in the case of nonlinear optical and electro-optic materials. Most of the organic compounds considered for such

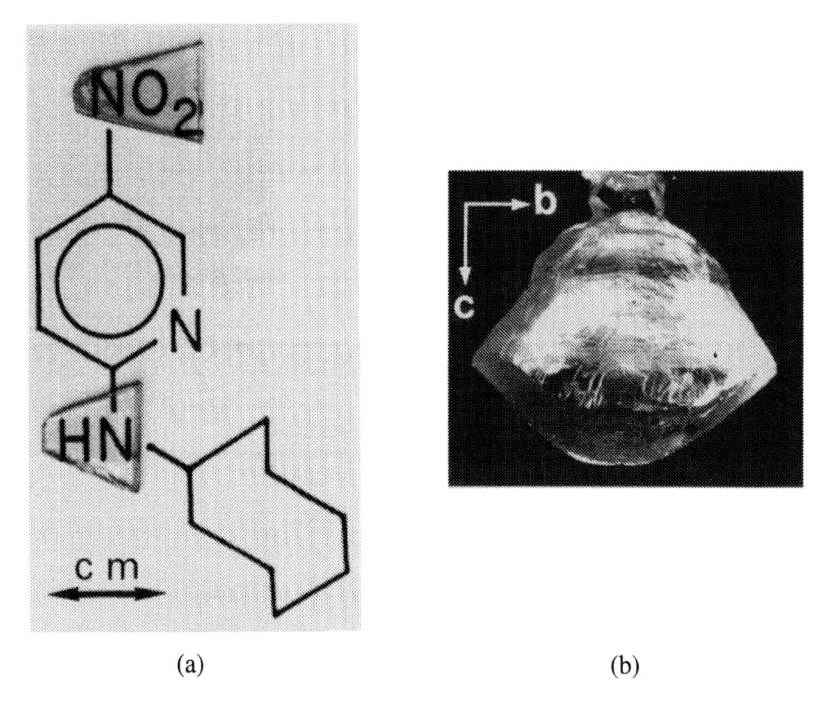

Figure 4.8 Melt grown pure COANP single crystals. (a) A seed has been fully immersed into a supercooled melt at $\Delta T \leq 0.3$°C. Growth time: 14 days. Growth takes place only from one side of the polar 2-fold axis of the mm2 symmetry group (*Hulliger et al. 1990b*). (b) A seed has been contacted to a supercooled melt and pulling was carried out at a speed of 1 mm/day. As a result of the inhomogeneous heat release, pulled COANP crystals showed a pronounced increase of the growth rate for the {100}-faces (*Hulliger et al. 1993*).

applications show a low vapor pressure and VG would most likely not be effective enough to produce crystals of a reasonable size.

4.3 GROWTH OF ORGANIC 2-D AND 1-D OPTICAL WAVEGUIDES

After the discussion of several growth techniques which were mostly applied to produce *bulk crystals* we shall give also a brief review on the methods to obtain quasi *2-D* and *1-D crystals.*

Thin layer and fiber crystals are of interest in planar or channel waveguide configurations which lead to a high confinement of the optical wave in the nonlinear optical or electro-optic active part of a device.

2-D crystals can be grown in the form of thin platelets e.g. between planar glass plates or as epitaxial layers on various possible substrates. 1-D crystals for optical use are mainly grown as crystal cored fibers.

4.3.1 Melt and solution growth of 2-D crystal platelets

By this first approach highly planar ($\sim\lambda/10$) and silane coated plates (kept at a distance of several μm) act as a cladding and impose a quasi 2-D growth morphology. Crystallization has been performed in two ways: (i) from the melt by use of a Bridgman-type configuration or from a supercooled melt and (ii) by isothermal evaporation of the solvent. Both procedures are encountered by characteristic difficulties. Growth from both the melt and the solution is accompanied by a large and negative volume change upon solidification. In the case of a melt, stress and the formation of cracks within the layer arise from a difference in the thermal expansion coefficient for the glass and the organic layer. The use of transparent TEFLON is recommended for such an approach. In contrast, isothermal evaporation of solvent will avoid difficulties related to temperature, but this technique is complicated in regard to a defined lateral loss of the solvent in order to maintain a uniform interface of crystallization to yield large (cm^2) single crystalline areas.

In both cases one of the post-growth tasks will be the removal of one or both glass or TEFLON plates. Several procedures have been developed and yielded finally μm thick platelet crystals. Instead of removing one of the cladding plates, light can be coupled into the layer by use of a grating which is etched into the basic plate prior to the growth.

Figure 4.9 shows a schematic view of a growth chamber suitable for the *in situ* observation of thin layer growth either from the supercooled melt or by isothermal evaporation of the solvent. During growth the formation of the μm thick layer is inspected via crossed polarizers. Application of supercooled melt conditions to a pure melt of e.g. COANP which was fed in between 8.7 μm spaced and silane coated quartz plates produced single crystalline ($\sim$0.5 to 1 cm^2) layers showing multimode waveguiding at the He-Ne wavelength (*Bosshard et al. 1992a*).

Most of the published work on 2-D growth concentrated on materials such as

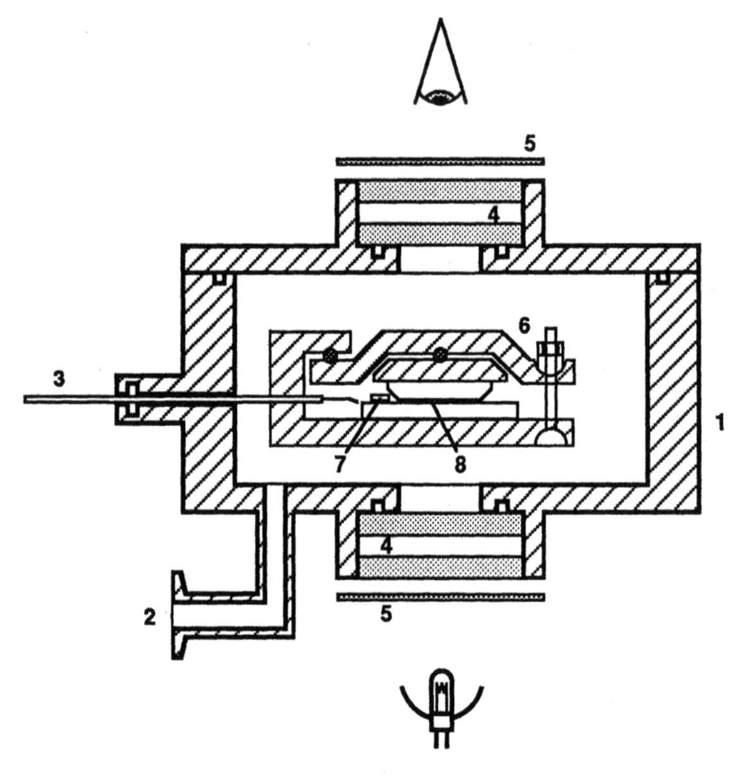

Figure 4.9 Isothermally heated chamber for the growth of thin platelets (or fibers) using either a supercooled melt or the solvent evaporation technique. 1: water heated double-wall chamber, 2: to pump, 3: manipulator to contact a seed to the nutrient, 4: glass windows, 5: polarizers, 6: press to maintain the gap, spaced by the use of μm thin wires, 7: seed, 8: melt or solution.

DSMS (*Yoshimura et al. 1988*), p-chlorphenylurea (PCPU) (*Badan et al. 1987*), MNA (*Kubota et al. 1988, Hiroshi et al. 1987*) and DAN (*Gotoh et al. 1988*). An elegant approach seems to be to set up a multilayered configuration using planar glass plates provided by a grating structure for beam in- and out-coupling. This obviously rules out the problem of removing one of the cladding plates — an operation that often leads to a mechanical damage of the thin crystalline layer. For such a configuration second harmonic generation was demonstrated for MNA when using a wedge-type layer (*Sugihara et al. 1990, 1991a*).

4.3.2 Thin layer growth by molecular beam epitaxy

Work on organic epitaxy up to 1972 is summarized in Landolt-Börnstein (*1972*). However, most of this earlier work does not describe the formation of large area or strip-like films to be used for planar waveguide optics. Recent developments make use of the molecular beam epitaxy (MBE). The MBE process allows for highest control of the growth, especially for multicomponent and multilayer deposits. Present work concentrates on the growth of phthalocyanine or C_{60}-type compounds deposited on alkaline halides and other substrates including epitaxial growth of e.g. hexabromobenzene or 7,7,8,8,-tetracyanoquinodimethane (TCNQ) (*Kobayashi 1991*) and MNBA (*Gotoh et al. 1992*). In the case of MNBA, heteroepitaxy have been studied in using MNBA-Et (Et: ethylcarbonylamino, instead of acetamino for MNBA) as a substrate crystal and a conventional MBE apparatus equipped with Knudsen cells. Epitaxial layers showing waveguiding properties have been grown on cm^2-large MNBA-Et wavers at 80°C and at a rate of 10 Å/s, leading to a final thickness of 4.8 µm (lattice misfit: 1 to 4%) (*Gotoh et al. 1992*).

Organic MBE can be considered as one of the open fields and promising techniques to produce e.g. hetero-layer structures containing semiconductors as well as nonlinear optical organic materials integrated on one chip. However, one of the striking difficulties will be related to the layer substrate interactions as found for any other hetero-epitaxial process (lattice-mismatch). Ideas for tailoring substrates emerged recently from work on Langmuir-Blodgett films (*Tomita et al. 1988, Kondo et al. 1989b*).

4.3.3 Growth of crystal cored organic fibers

At present there is mainly one technique for growing single crystal fibers for organics: Kerkoc (*1991*) reviewed single crystal cored melt growth of e.g. m-NA, 2-bromo-4-nitroaniline (BNA), m-DNB, formyl-nitrophenyl-hydrazine (FNPH), benzil, MNA, NPP, N-(4-nitrophenyl)-N-methylamino-acetonitrile (NPAN), 3,5-dimethyl-1-(4-nitrophenyl)-pyrazole (DMNP), 2-methyl-4-nitro-N-methylaniline (MNMA), COANP and DAN, by *Bridgman-Stockbarger* techniques and by the use of glass capillaries. The basic idea relies on the growth in capillaries (filled by the liquid compound) which are placed into a two-zone furnace. As a modification, CO_2 laser scanning zone melting was also applied (*Badan et al. 1987*).

Under the conditions mentioned above, most of the potentially interesting nonlinear optical compounds crystallize in a way where all second order susceptibility coefficients are small or vanish for optical propagation along the fiber axis. As one of the few known exceptions, DAN fibers involve d_{22}, d_{33} , and d_{34} elements. Inverted Bridgman-Stockbarger growth in sodium glass and quartz glass capillaries resulted in optical quality DAN fibers of several cm in length and crystal diameters between 3 to 10 μm (*Kerkoc et al. 1991*). X-ray diffraction showed that the a-[100] axis is the fiber axis. For polarized light 20 to 30 mm long sections exhibited sharp extinctions showing no cracks or other defects. A 100 μm long section of such a 10 μm thick fiber is shown in Figure 4.10. End faces of DAN fibers have been examined by scanning electron microscopy: little free space between core and glass cladding was left due to the difference in the thermal expansion coefficient.

Summing up: The main difficulty in growing nonlinear optical fibers is to obtain the *proper crystallographic growth direction* with respect to the use of large *d*-tensor elements. Various experimental approaches showed that it is nearly impossible to influence the natural growth direction along the fiber axis. Many interesting

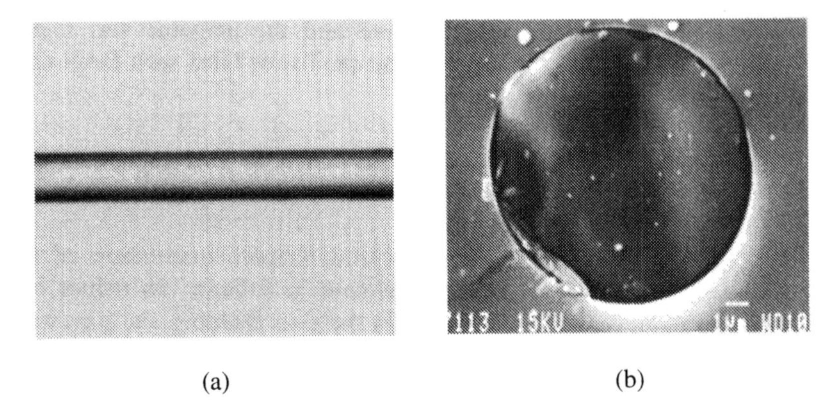

(a) (b)

Figure 4.10 As grown crystal cored DAN fibers (*Kerkoc et al. 1991*). a) Section of a DAN fiber of 10 μm in diameter (crossed polarizers). b) SEM micrograph of an end face (after cleaving the sample) of a DAN crystal cored fiber.

materials (e.g. COANP, PNP and others) are hence ruled out for nonlinear fiber optics grown by a crystal cored technique. (An alternative would be to grow first a planar waveguide, followed by a structuring process yielding finally a stripe-like optical element.) Furthermore, only materials exhibiting sufficiently high thermal stability can be grown by a crystal cored melt technique. It should be mentioned here that growth from an isothermally supercooled melt (see section 4.2.4.2) represents an alternative to the inverted Bridgman-Stockbarger technique discussed above. In this case seed growth could probably lead to a change of the natural growth direction along the capillary axis.

4.4 GROWTH OF PHOTOREFRACTIVE (DOPED) ORGANIC CRYSTALS

Design and engineering of organic photorefractive crystals should start out from (i) donor acceptor substituted π-electron systems showing a maximum nonlinear optical response, (ii) a crystal structure providing a large electro-optic effect (i.e. an intentionally parallel alignment of molecular charge transfer axes), and (iii) a partial overlap of molecular π-orbitals localized at different lattice sites (e.g. stacked configuration of planar π-electron systems allowing also for electronic interaction along the polar axis of the crystal structure). In the case of a low photoconductivity of pure crystals, photoconductivity may be sensitized by an appropriate doping with molecular dyes or other charge-generating moieties which show a strong absorption close to the absorption edge of the molecular host lattice. The design and engineering of organic photorefractive materials thus opens an interesting new area of growing homogeneously doped organic crystals.

Viewed in terms of a binary phase diagram, doping implies growth of solid solutions which is known to be complicated by the formation of growth striations and the presence of constitutional supercooling. Special growth equipment will hence be needed (e.g. ΔT-method, see section 4.2.1.1). Furthermore, the shape of the dopant, its volume as well as intermolecular bonding is determining the acceptability. Optimum size-match between the dopant and the host lattice moieties is often considered as a sufficient condition for the solubility in the solid state. Following criteria given by Kitaigorodski (*1984*) the degree of molecular isomorphism can be defined by a "coefficient of geometrical similarity". Experimental work (*Scott et al. 1974*) on a series of binary organic systems (including substituted arenes and some other compounds and cyclododecane as the host lattice) revealed a quite universal relation between a simple volume overlap factor $V = V^{no}/V^{o}$ (no: non-, o: overlapping volumes) and measured distribution coefficients k_{eff} ($k_{eff} = C_{cryst}/C_{(melt\ or\ solution)}$). A linear relationship in k_{eff} vs V was found for $0.5 \geq k_{eff} \geq 3.\ 10^{-3}$ and $0.24 \leq V \leq 0.92$ ($V = 1$: equally sized molecules). In cases where no specific strongly lattice stabilizing or destabilizing interactions are broken or newly formed, the concept of "geometrical similarity" should allow for a first estimate to what extent doping will take place.

However, most of the strongly absorbing dyes which were used for many other purposes are too big as compared to the molecules we are presently dealing with. Any particular molecular and electro-optic host lattice would, in principle, require the development of a well fitting sensitizer molecule.

4.4.1 Growth techniques for homogeneous solid solutions

The effective distribution coefficient k_{eff} for a given system depends on the growth conditions such as the growth temperature, the growth speed and the hydrodynamic motion of the nutrient, as well as on the nature of the growing crystal faces (for a comprehensive theoretical treatise see *Wilke et al. 1988*). To account for both kinetic and thermodynamic effects, growth has to be performed at a constant supercooling (i.e. constant growth speed). Thereby the use of (i) relatively large volumes of a melt or solution as compared to the growing crystal or (ii) a system involving after-saturation of the dopant can help in getting homogeneously doped crystals.

As we discussed in section 4.2.1.1 the temperature difference technique of solution growth showed that homogeneous solid solutions could be obtained in the case of various inorganic compounds (*Arend et al. 1986*).

Finally we should emphasize here that, in cases where k_{eff} is small (k_{eff} 10^{-2} to 10^{-4}) and at a relatively high dopant concentration in the liquid phase ($C_{liquid} \leq 1$ mol%), the continuous change of the dopant concentration in the liquid and solid phases is small (for a low concentration of the dopant, k_{eff} can be assumed to be constant). Rather, homogeneously doped crystals should be obtained also by techniques such as growth from supercooled melt or from solution (TL or IE). In any case sufficient mixing must be provided whilst growing.

4.4.2 A first example of an organic photorefractive crystal

Recently we have reported on a first example of an organic crystal showing space charge photorefractive effects (*Sutter et al. 1990a*). Supercooled melt growth produced COANP single crystals which were doped by negatively-charged derivatives of the electron acceptor 7,7,8,8-tetracyanoquinodimethane (TCNQ)

(*Hulliger et al. 1991, 1993*). TCNQ is known to form (i) charge-transfer complexes [TCNQ • D] with donor-type molecules D (absorption bands: 700 to 800 nm, D: e.g. aminobenzenes), (ii) $[TCNQ]^{-\bullet}$ in less polar solvents (absorption bands: 650 to 900 nm), and (iii) $[TCNQ]^{-2}_2$ in polar solvents (absorption band: 600 to 700 nm), as well as (iv) 1,6-addition reactions with e.g. primary and secondary amines. Following geometrical arguments as outlined above (section 4.4.1), none of these molecules really fits into the crystal lattice of COANP. Particularly an incorporation of the anions would be unlikely. However, optical spectroscopy demonstrated that melt grown, greenish COANP crystals show an $[TCNQ]^{-2}_2$ concentration of the order of 50 ppm. The observed strong absorption band in the range of 600 to 700 nm was indicative for the $[TCNQ]^{-2}_2$ dimer. Close to the absorption edge of the pure COANP lattice (450 to 650 nm) there appeared another band which so far could not be attributed to any known TCNQ derivative. The same, but much more intense, bands were found for frozen (glassy) melts, revealing that after a few days at T = 75 to 80°C all initially added TCNQ was converted mainly into the dimer $[TCNQ]^{-2}_2$, as well as into some other minor and presently not identified melt components. Surprisingly, only traces of the radical (S = 1/2) $[TCNQ]^{-\bullet}$ were present in such a melt, since neither optical spectroscopy nor paramagnetic resonance could trace significant amounts of this radical which is known to be stable in less polar solvents. During dissolution of 1,000 to 2,000 ppm TCNQ and a melt conditioning at 75 to 80°C (leading to a saturation by the dopants), a dimerization reaction seems to be supported by liquid COANP.

Thin layer chromatography assisted by electrophoresis could separate three additional components that were present in the melt, two of them being anions. Nearly cm^3-large doped crystals have been grown (Figure 4.11) using the apparatus shown in Figure 4.7. As compared to the growth of pure COANP, in this case a higher supercooling was necessary ($\Delta T \geq 1.5$°C) to induce growth on c-

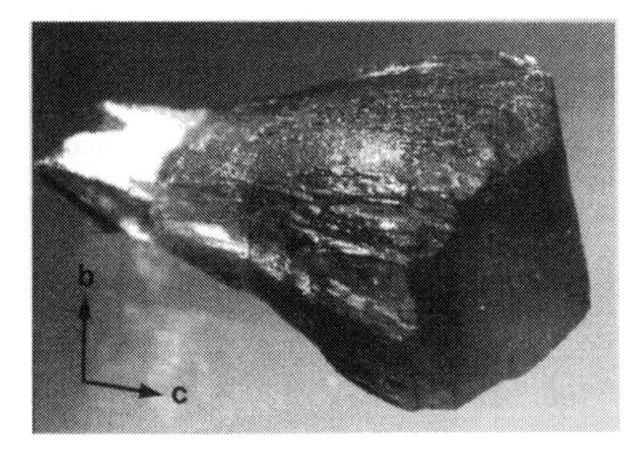

Figure 4.11 As grown $[TCNQ]^{-2}_2$ doped COANP single crystal (pulled from a supercooled melt with nominally ~1,000 ppm TCNQ):~ 8 × 8 × 12 (height) mm^3 (*Hulliger et al. 1993*).

oriented, fully immersed or pulled seeds. In the case of seeds which have been immersed, the as grown doped crystals showed a modified habitus (increased size of the plate thickness). If sufficient stirring was provided, quite homogeneously colored crystals could be produced. To explore the field of possible doping within the described system we have also performed solution and vapor growth in order to provide different concentrations of the involved species which could contribute to the mechanism of the photorefractive effect (for details see *Hulliger et al. 1993*).

4.5 CONCLUDING REMARKS ON THE CRYSTAL GROWTH OF ORGANIC MATERIALS

Recent progress in enhancing nonlinear optical, electro-optic and photorefractive effects of organic materials showed clearly that single crystals of all these compounds play a key role not only in the study of their basic solid state properties, but also for possible applications.

However, it seems that there is no unified, fast and safe approach in how to obtain quality crystals (~cm^3) of any generally less characterized *organic* nonlinear optical compound. In the course of most experimental work in this field, solution growth emerged to be the most widely used technique.

Especially techniques operating on the basis of *lowering the temperature or the evaporation of a solvent component* have been recently applied to materials such as DAN, PNP, COANP, NPPA, MBANP, DANS, MNBA, DSMS, DAST and others (see Table 4.1).

For exploratory work the ΔT-method (section 4.2.1.1) proved to be effective for a variety of organic and inorganic compounds, showing a solubility larger than a few wt% and a positive temperature coefficient of the solubility. In this case reasonably good crystals were obtained within several weeks.

Whereas most of the interesting materials can be grown from solution, there are only a few which were thermally stable enough (long-term stability) to apply melt growth techniques such as Bridgman, zone-melting or supercooled melt growth. As demonstrated recently, quality crystals of e.g. COANP can only be obtained by supercooled melt growth including crystal pulling.

Besides the need of ~cm^3 bulk crystals, thin layer and fiber crystals are similarly of interest. First layers have been obtained for e.g. MNBA or MNA, optical fibers have been produced for DAN.

To support progress on organic photorefractive materials, future organic crystal growth should concentrate on the growth of solid solutions and doped systems. Therefore, the development of techniques operating at a constant supersaturation making use of a continuous release of source material will become more and more important.

5. LANGMUIR-BLODGETT FILMS

5.1 INTRODUCTION

Another method for the preparation of oriented thin films of organic materials is the Langmuir-Blodgett (LB) technique. This technique is based on the fabrication of organic monofilms which are first oriented on a water surface and then subsequently transferred layer by layer onto a solid substrate. The procedure is sketched in Figure 5.1. A more detailed description of the film preparation follows in the next chapters.

The possibility of obtaining monofilms by spreading liquids was originally realized by Rayleigh in 1890(Rayleigh,1890). Langmuir was the first who transferred monolayers to a solid support (*Langmuir 1917*). The first formal and comprehensive report on the preparation of multilayers appeared in 1935 by Langmuir's co-worker Katharine Blodgett (*Blodgett 1935*). The monolayers swimming on a water surface are called Langmuir films. LB films are the films transferred onto a solid support.

Highly anisotropic molecules form centrosymmetric bulk crystals in most cases. Such crystals are inappropriate for second-order electro-optic and nonlinear optical applications. Unfortunately there is no method to predict whether a newly created molecule will crystallize acentrically or not. There is no structure/activity relationship (*Hutchings et al. 1989*), which would allow predictions at least within a series of homologous molecules — a practice used routinely and successfully in many other fields of organic chemistry, e.g. in the development of dyes, medicaments, agrochemicals or liquid crystals.

The LB technique, however, allows the preparation of highly anisotropic thin films due to the initial organization of the molecules on an interface. A further distinct advantage of the LB preparation technique is the easy possibility of changing the environmental parameters of a Langmuir monolayer in a wide range (*Möhwald 1986, 1988a*). This provides many opportunities to alter considerably the ordered states of the films in a well defined way.

Moreover the exploration of series of new molecules for the formation of functional LB films can be done relatively quickly compared to crystal investigations. A lot of basic information of molecular parameters can be obtained simply from Langmuir monolayers on a water surface or from a small number of layers transferred onto a solid support. Additionally only a few milligrams of material per substance are required.

A further advantage of thin films is that they are naturally compatible with waveguide and integrated optical devices. Beam confinement leads to a homogeneous distribution of light intensity all over the length of a waveguide, thus yielding strong internal fields, favorable for large electro-optic and nonlinear optical effects.

The most outstanding advantage of LB thin films for optical guided-wave applications is the homogeneous thickness which can be controlled to better

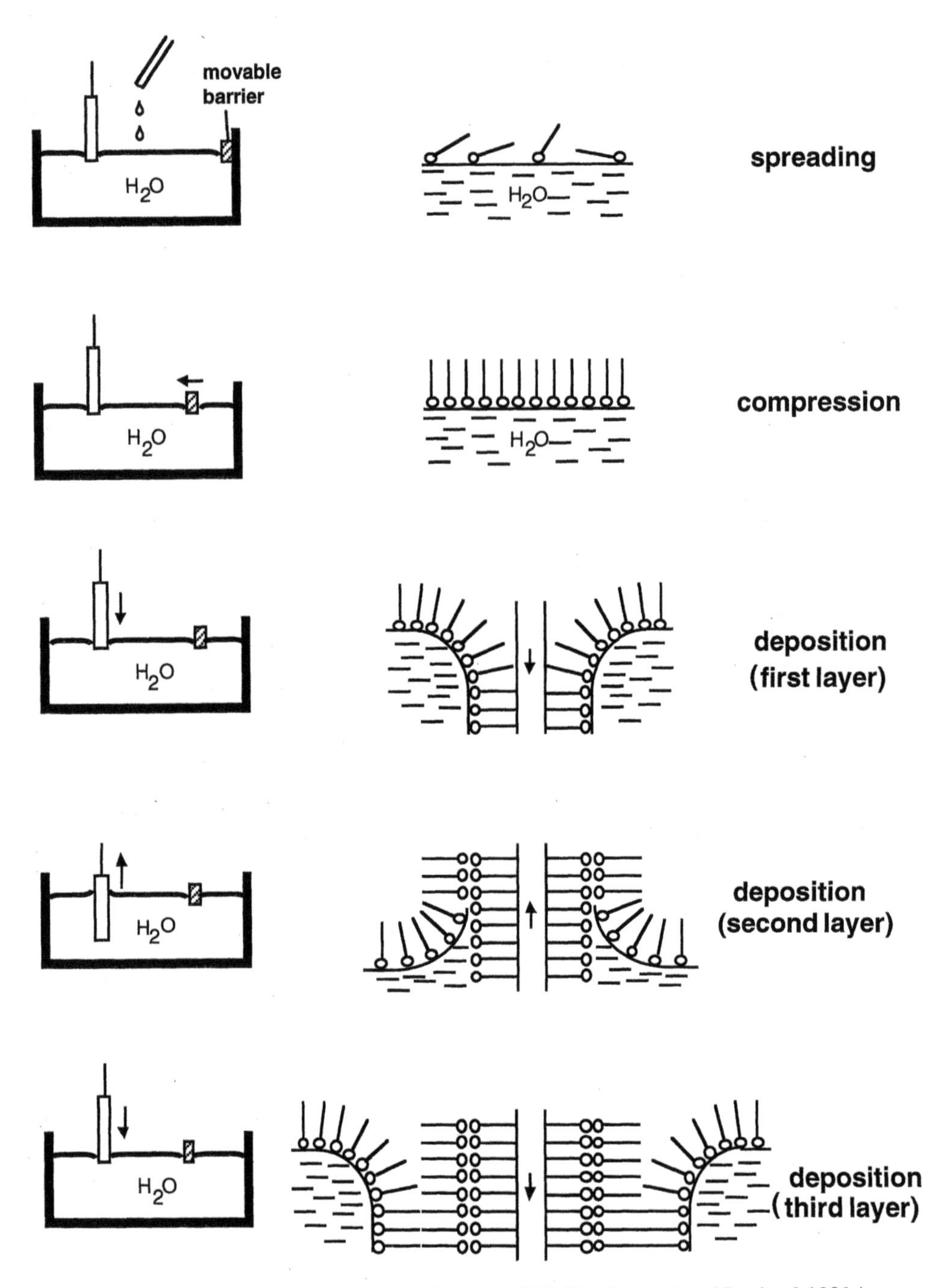

Figure 5.1 Successive steps in monolayer and LB film formation (*Bosshard 1991a*).

than ~4 nm by the number of dipping cycles. Since the effective refractive index of a waveguide can be tuned sensitively by its thickness, phase-matching can be easily achieved with LB films (*Bosshard et al. 1991a,b, Flörsheimer et al. 1992a,b*). Even films prepared by other methods such as spin coating, ion implantation, diffusion techniques or thin film crystal growth could be tuned by selected LB overlayers.

The main problems with LB films for optical applications are difficulties with the preparation of stable, highly anisotropic and homogeneously ordered films thick enough for optical waveguiding. There exist numerous examples of molecules where a highly anisotropic order of the LB films was maintained over the first few monolayers. The order, however, usually decreased or changed with the increasing number of deposited layers. Moreover, most of the presently available LB films which are thick enough for waveguiding show excessively high scattering losses. Consequently, the application of LB films in electro-optic or nonlinear optical devices could not be demonstrated until now. However, there is considerable research in progress which will be reviewed in Chapters 5.2, 5.3 and 8.3. A main topic for discussion will also be the possibility for future improvement.

Additional information on characterization methods, properties and possible applications of LB films can be obtained from the books edited by Roberts (*1990*) and written by Ulman (*1991*). The proceedings of the biannual International Conferences on Organized Molecular Films (LB and self-assembled films) are published in *Thin Solid Films* (*Barraud et al. 1992, Leblanc et al. 1994*). The physical chemistry of surfaces is described in Adamson's well known textbook (*1990*).

5.2 PREPARATION OF DIFFERENT TYPES OF MULTILAYERS

Many amphiphilic molecules are able to organize a monolayer spontaneously when they come into contact with a water surface. Usually these molecules are composed of one or two hydrophobic hydrocarbon tails and of a hydrophilic head group ('lollipop' molecules, *Peterson 1990*). The reason for their self-organization is a strong anisotropic interaction with the subphase via hydrogen bonds, head group charges or dipoles. The tails are responsible for the insolubility of the molecule as a whole. Due to the inter-chain van der Waals forces, the tails play an important role in the order of condensed monolayer states, as well as in holding the film together during and after the LB transfer.

The molecules are usually applied to the water surface in the form of a solution with a volatile solvent of low surface tension. A droplet of the solution placed on the surface of pure water immediately spreads due to the difference of surface tension of both liquids. Thus the amphiphilic molecules are distributed over the surface. After evaporation of the solvent the organic monolayer remains. It can now be manipulated in a defined way, normally using a film balance. A film balance is an apparatus where the lateral pressure π and the area A of the film are controlled by a moveable barrier. The temperature T, the index of pH, the ionic strength and concentration of dyes or reagents (which could be enzymes, initiators for polymerizations, oxidizing agents, antibodies etc.) in the aqueous subphase can be varied.

By changing and adjusting the environmental parameters, ordered states of the monolayer which are best suited for the LB transfer process, as well as for applications, can be achieved. A denser packing of the molecules as the result of a film compression is sketched in Figure 5.3. The hydrocarbon tails and the head groups are tilted in this example where the tilt angles decrease with increasing pressure. The head groups are polar. Their normal dipole components increase with pressure.

Since the properties and the order of such interface layers are very sensitive to impurities, it is important to make sure that the subphase concentration of

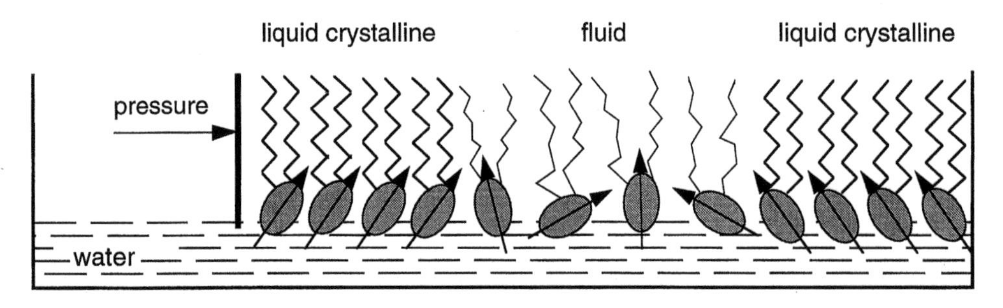

Figure 5.2 Sketch of a film balance with a Langmuir layer. Two phases are coexisting. Head group dipole moments are symbolized by arrows.

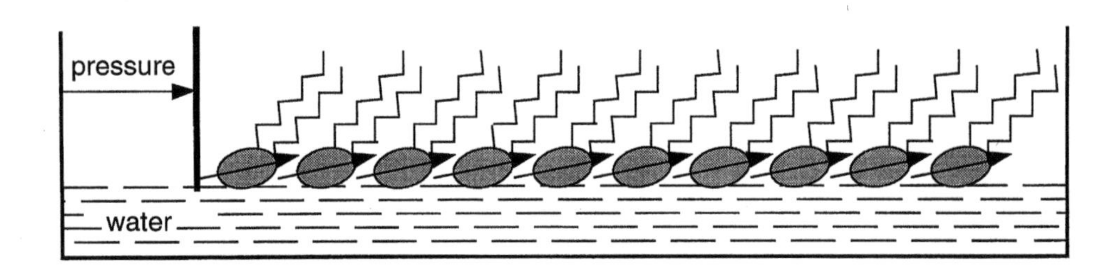

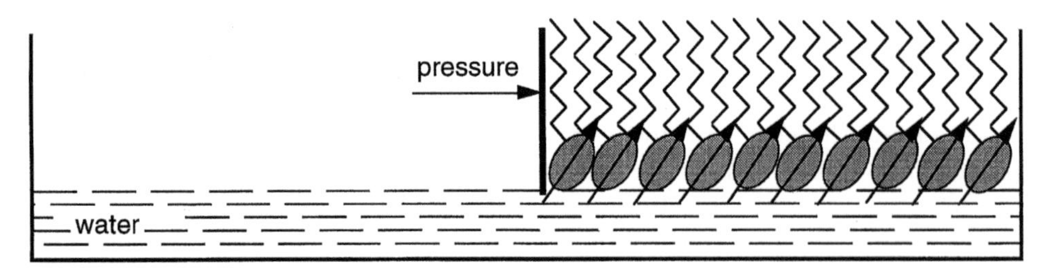

Figure 5.3 Manipulation of order by a lateral pressure change.

unwanted contamination is low enough. Organic impurities in the ppm (parts per million) range which are collected at the interface can totally change the film properties. In particular, charged films are affected by ions. Further experimental details can be obtained from the excellent review written by Kuhn, Möbius, and Bücher (*1972*). The authors also describe the build-up and investigation of multilayers with functional properties not present in the single layers(Kuhn films).

Liquid crystalline or solid monolayer states are preferably suited for LB multilayer deposition. In principle any material with a fairly well defined or prepared surface (either hydrophobic or hydrophilic) can be used as a substrate. In most cases the head to head / tail to tail film type (Y-type) shown in Figure 5.1 is obtained. Using a hydrophobic substrate (e.g. silicon, silanized glass, silver, gold) a monofilm is deposited at each down and upstroke of the substrate, yielding an even number of layers. Using a hydrophilic support (e.g. glass, metals and semiconductors with a native oxide surface) no film is deposited at the first downstroke, yielding an odd number of final layers.

Multifilms can often be better deposited onto hydrophobic substrates. Then an odd number of layers of a functional molecule (which may be desired for phase-matching) can be obtained by transferring another suitable layer during the last withdrawal circle. For this, either the film floating on the water surface must be exchanged before starting the last upstroke procedure, or one employs a special double film balance where two Langmuir layers can be processed independently and both materials can be transferred in optional sequence (*Puterman et al. 1974, Barraud et al. 1983, Girling et al. 1984, Daniel et al. 1985a, Holcroft et al.1985*).

Such film balances are commercially available. A schematic diagram is given in Figure 5.4. Such film balances are of particular importance for the preparation of noncentrosymmetric alternating Y-type multilayers. Due to the initial organization of the amphiphiles on an interface all monolayer phases show vertical dipole

Figure 5.4 Alternating layer trough (schematic). Non centrosymmetric Y-type multilayers can be prepared starting from two Langmuir films of different molecules, one film with normal dipole moment components and the other with dipole moment components of opposite sign.

moment components (see e. g. Figure 5.2). The normal dipoles cancel by pairs in ordinary Y-type multilayers. This is demonstrated in Figure 5.5 and compared with other types of multilayers. In the case of alternating Y-type multifilms two kinds of molecules are chosen, one with normal dipole moment components and the other with dipole moment components of opposite sign, yielding an overall polarization of the multilayer system.

Besides the molecules forming Y-type multifilms, there is a smaller number of molecules which can be deposited as X- or Z-type LB films under certain conditions. In the first case, layers are only deposited during the downstroke of the substrate. In the latter case films are only transferred during the upstroke. X-type films on a hydrophobic substrate and Z-type films on a hydrophilic substrate are sketched in Figure 5.5. These structures are interesting because of their normal overall polarization.

For various LB films of fatty acid salts it has been reported that the deposition behavior can be changed from Y-type to X- or Z-type by altering the subphase pH (*Blodgett 1935, Aveyard et al. 1992*). The reason is the pH dependent degree of dissociation of the acid so that the binding of the acid anion with the metal cation can be controlled by the pH. There were several attempts to favor X- and Z-type deposition of materials which usually form Y-type films. Often the substrate was moved alternately through a Langmuir film of this material and a free water surface. This can be done elegantly using alternating layer troughs. However the strong forces acting on the molecules near the free water surface may disturb the order of the films.

Even in the case of materials which naturally form X- or Z-type films, such as the above mentioned fatty acid salts, there is an extensive controversial discussion on the stability and order of such films. (*Roberts et al. 1990*, Chapter 2, *Ulman 1991*, Chapter 2.1.E, *Peterson 1990*, and references therein). Reorganization during and after film transfer often plays a main role. In recent atomic force microscopic studies (*Schwartz et al. 1992*) reorganization was even observed for Y-type LB films of fatty acid salts submerged under aqueous subphase. Such findings contradict the 'carpet model' of LB transfer (*Peterson 1990*), a model which has been widely accepted so far and describes the deposition to proceed without major changes of the molecular organization.

Another important exception to the 'carpet model' are Y-type films with an in-plane overall polarization where the polar molecules are aligned during the dipping process. The hyperpolarizable 2-docosylamino-5-nitropyridine (DCANP, *Decher et al. 1988, 1989a,b, Klinkhammer 1989*) is the main example. Using this material phase-matched second-harmonic generation could be demonstrated for the first time with LB film waveguides. The material is described in more detail in Chapter 8.3 (Tables 8.3 and 8.4). The frequency conversion experiments are reported in Chapter 9.1.

The mechanism of the ordering process is not yet completely understood. There are, however, examples of other films where an ordering process also occurs during the dipping procedure. These films contain objects with elongated shapes. Such objects are e.g. J-aggregates in films of merocyanine/fatty acid

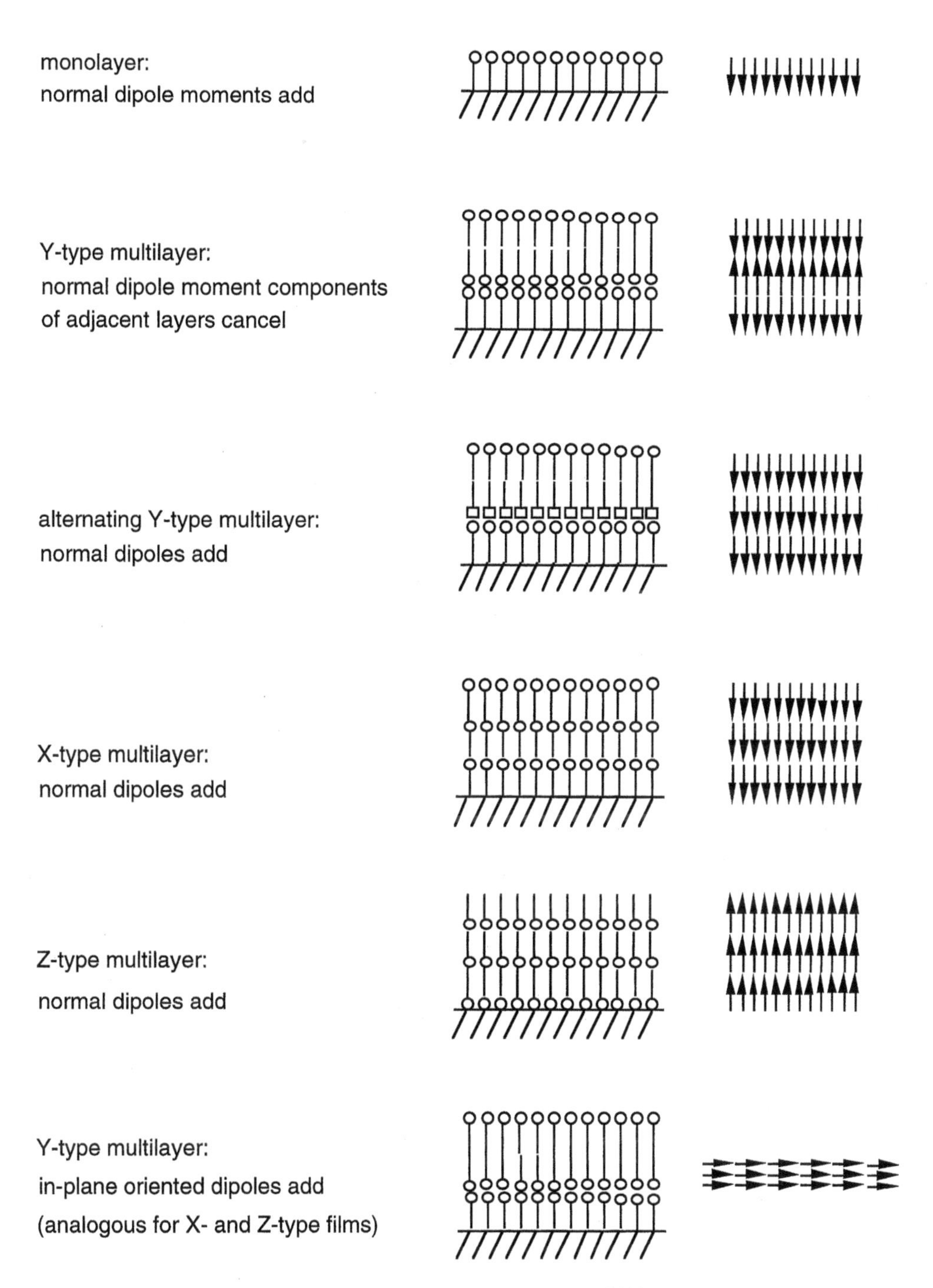

Figure 5.5 Different types of LB multifilms.

mixtures, rigid-rod polymers and oligomers, or needle-like crystals. The ordering of these objects is attributed to the flow of the Langmuir film in the direction of the solid support during the transfer. The velocity gradient of the surface flow produces a torque on the elongated objects and favors their orientation parallel to the flow. The Brownian motion has to be considered as the counteracting influence favoring an isotropic organization. The orientation of the above mentioned J-aggregates could be described quantitatively on the basis of this model for various trough and substrate geometries. The Langmuir films were regarded as a Newtonian liquid (*Minari et al. 1988*) or as a Bingham fluid (*Sugi et al. 1989, Minari et al. 1989, Tabe et al. 1990, Matsumoto et al. 1989*) allowing additionally for plastic behavior. The anisotropy of the J-aggregate assemblies was detected spectroscopically, by Electron Spin Resonance (ESR) measurements and by second-harmonic generation (*Saito et al. 1989, Kuroda et al. 1989, Kajikawa et al. 1991*). These assemblies of aggregates are interesting as photoconductors (*Sugi et al. 1983, 1985a,b*).

An example for rigid-rod polymers are polyimides (*Sorita et al. 1991*). LB films of such polymers are interesting as alignment layers for liquid crystals. Most work on rigid-rod polymers and oligomers has been done on 'hairyrod' molecules. These molecules are one of the most important exceptions from the 'lollipop'-type molecules usually used. They consist of a stiff backbone which is substituted by conformationally mobile side chains wrapping each backbone with a fluid-like skin (*Embs et al. 1991a*). One purpose of this liquid-like skin is to prevent aggregation and hence to reduce waveguide losses. As backbones α-helical polymers (polyglutamates, *Duda et al. 1988a, b, Mathy et al. 1992*), phthalocyaninato-polysiloxane oligomers and polymers (*Orthmann et al. 1986, Caseri et al. 1988, Sauer et al. 1990*) or poly(diphenylsilanes) (*Embs et al. 1991b*) have been used.

The chromophores used in the cited experiments have not been chosen for second-order nonlinear optical or electro-optic processes. It was shown that the 'hairyrod' concept is well suited for the preparation of LB films with a considerable in-plane anisotropy. However, it is not clear so far if polar molecules can be prevented from a random antiparallel alignment. With some of the 'hairy rod' molecules hundreds of anisotropic monolayers could be transferred onto a substrate. Waveguide properties will be discussed in Chapter 8.3.2.

Surface flow introduced by the dipping process was also found to influence the shapes of monolayer domains (*Daniel et al. 1985b*). Nitsch and co-workers (*Kurthen et al. 1991*) recently produced an in-plane anisotropy in mixed Langmuir monolayers of a rigid-rod polymer and β-carotene using a flowing subphase. The lateral pressure in a conventional film balance is produced by a moving barrier. In contrast to this, the Nitsch group developed a flow channel where the streaming subphase compresses a monolayer in front of a stowing arrangement. Such equipment was also used for thermodynamical and kinetic experiments (*Ollenik et al. 1981, Nitsch et al. 1987*). A similar construction (Figure 5.6) could be utilized for future industrial production of LB films since large scale production must work continuously. Recently the Hoechst LB group developed such a machine and prepared polymeric LB films in laboratory scale (*Hickel et al. 1993*).

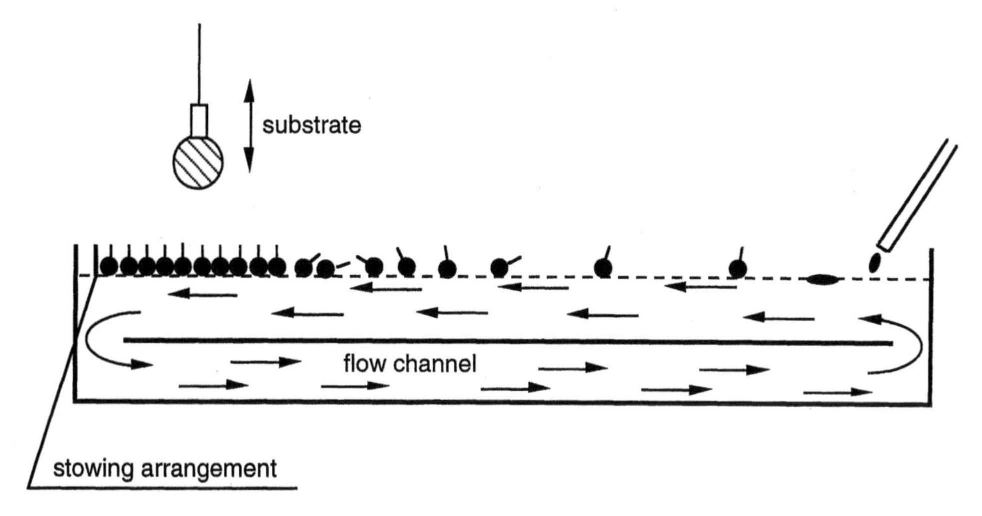

Figure 5.6 Continuously working equipment for multilayer preparation.

With such an apparatus a wafer of e. g. 10 cm in diameter could be coated with 300 monolayers for optical waveguiding within 17 hours assuming a (relatively high) dipping rate of 30 mm min^{-1}. In principle several wafers can be coated in parallel. Using a conventional film balance for research, the compression time for the Langmuir film must also be taken into account. This process needs some minutes or some hours depending on the material. The repetition rate of spreading also depends on the area of the film balance and the required sample area. Furthermore, it should be realized that some kinds of films need a drying time following each upstroke cycle.

The time required for LB preparations is, of course, large as compared to polymer preparation techniques like spin coating, but it is short as compared to crystal growth methods. Furthermore the LB film thickness is naturally defined very precisely and the equipment is cheap as compared to vacuum techniques.

LB multilayers should be prepared under clean-room conditions in order to reduce the incorporation of dust particles during the long time which is needed for film preparation. Moreover one must take care that there are no bacteria present which use the film as a nutrient. Reducing the concentration of dust and bacilli, it was shown that the attenuation of LB film waveguides prepared with a preformed polymer could be decreased from 80 to 11 ± 2 dB cm^{-1} (*Tredgold et al. 1987*).

5.3 MONOLAYERS ON THE WATER SURFACE

Often only minor attention is paid to the monolayer on the water surface. However, as mentioned above it is one great advantage of the LB technology that the order and morphology of Langmuir layers can be systematically varied in a wide range. A key for future improvement of waveguiding properties and molecular order is the film treatment before and during the transfer process.

5.3.1 Basic thermodynamic characterization

Basic characterization of a monolayer is done by recording the compression and expansion isotherms. Of course a thermodynamical experiment does not yield information about the orientation of single molecules and their microscopic organization. However a lot of information about phase states and phase transitions can be obtained.

Most data on the order of monolayers are available on lipid films. The investigation of such amphiphiles has a long tradition in biology, biophysics and physical chemistry. Some lipids, especially some fatty acid salts, are well known for their excellent LB film forming properties (*Blodgett 1935*). Moreover phospholipid monolayers at the air/water interface can be considered as simple model systems for biological membranes (*Phillips 1972, Hoppe et al. 1982*).

The pressure/area diagrams for the isothermal film compression of two phospholides, L-α-dimyristoyl-phosphatidyl-ethanolamine (DMPE) and L-α-dipalmitoyl-phosphatidyl-ethanolamine (DPPE), are depicted in Figure 5.7. The DMPE isotherm looks similar to a van der Waals isotherm. It shows a plateau marked with b in the figure. A plateau is an indication of a first-order phase transition. The textures of such a film can be observed *in situ* on the water surface by fluorescence microscopy staining selectively one of the coexisting phases with a small amount of a fluorescent dye (some per cent or per mill). Monolayer sensitive fluorescence microscopy was developed a decade ago by two independent teams (*Lösche et al. 1983, Weiss et al. 1984*). The first order transition point in Figure 5.7 is marked with (A_c, π_c). The pressure π_c depends on the temperature, the chain length of the molecule and other parameters. The transition pressure decreases in general for decreasing temperature or increasing chain length (van der Waals forces). The longer chains of DPPE with respect to DMPE are the reason why no plateau exists in the isotherm of DPPE in Figure 5.7. The domains seen in the micrographs can show equilibrium features (*McConnell et al. 1988*). Their size and shape may be influenced systematically as will be discussed below.

Information on higher order phase transitions can also be obtained from thermodynamic measurements. If h is an intensive quantity complementary to the order parameter, which describes the phase transition, G denotes the Gibbs' energy and T the temperature, a continuous phase transition of ith order is defined by the vanishing or divergence of at least one of the ith derivatives $(\partial^i G/\partial T^i)_h$ and $(\partial^i G/\partial h^i)_T$, while all the lower derivatives are continuous. In a monolayer system h may be interpreted as the lateral pressure π. If κ is the lateral film compressibility, A_0 a molecular reference area and z the number of spread molecules one can write

$$\kappa = -\frac{1}{A_0}\left(\frac{\partial A}{\partial \pi}\right)_T = -\frac{1}{zA_0}\left(\frac{\partial^2 G}{\partial \pi^2}\right)_T \tag{5.1}$$

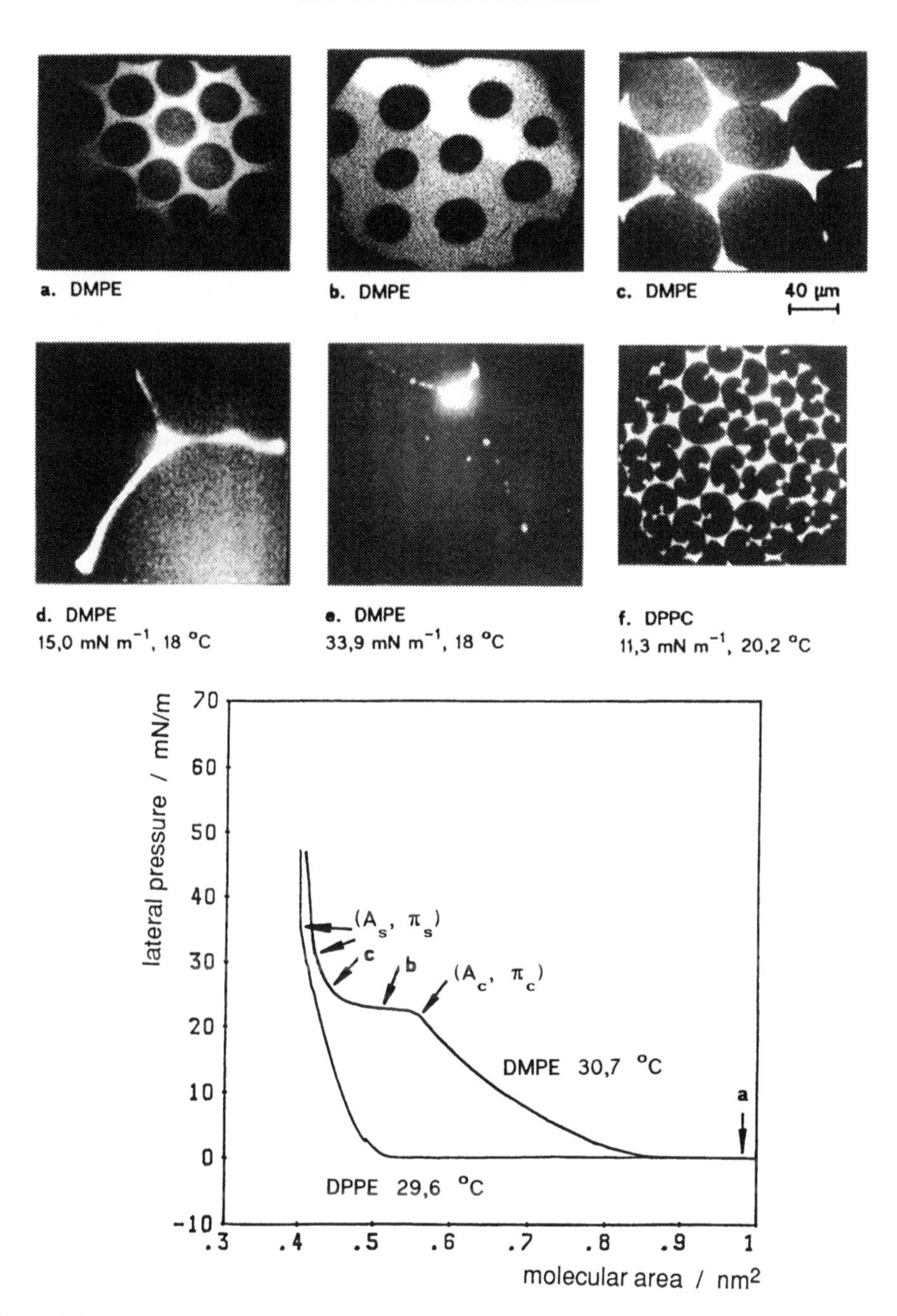

Figure 5.7 Morphology and isotherms of typical phospholipid films. The fluorescence micrographs a–c correspond to the isotherm of L-α-dimyristoyl-phosphatidyl-ethanolamine (DMPE). The images d and e are obtained from a DMPE film compressed at 18°C (15.7 mNm^{-1} below and 5.2 mNm^{-1} above π_s, respectively). On the isotherms the phase transition points (A_c, π_c) and (A_s, π_s) are marked. f gives a comparison to elongated domains (L-α-dipalmitoyl-phosphatidyl-choline, DPPC).

This means e.g. that breaks in an isotherm as the breaks in Figure 5.7 marked with (A_s, π_s) represent phase transitions of second order. Using the Landau theory one deduces that a film compressed to A_s achieves a state of higher symmetry in A_s (*Albrecht et al. 1978*).

5.3.2 Head/chain compatibility

Further phase transitions have been extensively discussed using thermodynamical data (*Albrecht et al. 1978*). Utilizing additional information obtained from recently developed techniques, such as monolayer sensitive electron and X-ray diffraction (*Fischer et al. 1984, Garoff et al. 1986, Helm et al. 1987, Kjaer et al. 1987, Dutta et al. 1987*) and fluorescence microscopy, the lipid phase states have been discussed in terms of hexatic phases (see below) and head/chain compatibility (*Garoff et al. 1986, Bareman et al. 1988, Flörsheimer et al. 1991*).

As already realized by Langmuir (*1933*) the packing of the molecular chains and head groups in condensed monolayer states requires a similar basal surface for the hydrophobic and the hydrophilic part of the molecules. Matching of the areas can be achieved by a different tilt angle of the groups as sketched in Figure 5.3 for a solid or liquid crystalline phase state at two different lateral pressures. However, when the shape of the molecules or the forces in the hydrophobic or in the hydrophilic part of the film do not allow a proper matching of the areas, the lateral range of head/chain compatibility is limited. This might result in sub and superstructures. Langmuir proposed clusters (two-dimensional 'micelles', *Langmuir 1933*) consisting of only a few molecules where the tilt of the hydrocarbon chains increases towards the edge of the cluster due to the crowding apart of the hydrophilic groups.

Compressing DMPE Langmuir films at low temperatures (below 26°C) the growth of liquid crystalline domains was observed by fluorescence microscopy at smaller molecular areas as compared to the onset of the first-order phase transition detected thermodynamically. This discrepancy could be explained by an intermediate phase state with poor head/chain compatibility, allowing only for small clusters and not for large domains which would be visible in the microscope (*Flörsheimer et al. 1988, 1991*). Similar clusters consisting of eight molecules were revealed recently in a scanning tunneling microscopy study of alkyl-cyanobiphenyl derivatives adsorbed on graphite surfaces (*Smith et al. 1990*). A micellar substructure of the domains of Cd stearate monolayers was proposed in order to explain the long range orientational and short range positional order revealed by electron diffraction experiments (*Garoff et al. 1986*).

Simulating lipid domains with arrays of hydrocarbon chains in a molecular dynamics study a lateral periodic modulation of chain order and thickness of the films was obtained (*Bareman et al. 1988*). A periodic modulation of height and density of the (closer packed) chains of Cd arachidate (surface of two or more LB layers) was recently detected by atomic force microscopy ('buckling' superstructure, *Garnaes et al. 1992*).

What can be learnt from order and phase states of lipid films for the development of new materials? It can be expected that the head/chain compatibility is also important for films of synthetic molecules. A proper alignment of the functional groups and a long range orientational order will be favored by a proper control of the head/chain compatibility. Molecules must be designed such that

a) matching of the hydrophobic and hydrophilic parts of a film is possible
b) the orientation of the chromophore is optimized as well as possible for the desired application

Charge-transfer axes lying on the water surface as flat as possible should be preferred for efficient optical frequency-doubling in LB film waveguides utilizing fundamental TE modes. As will be pointed out in Chapter 9.1, frequency-doubling of TE modes is more efficient as compared to TM modes. So far only minor attention has been paid to molecular packing considerations which, however, offer a large potential for future improvement. Some remarkable results obtained by Ashwell and co-workers will be discussed in Chapter 8.3.1.

In many recent publications concerning new promising LB film-forming materials no isotherms were given. There may be several reasons for the omission of such basic information. One reason is that such isotherms often do not refer to any phase transition and therefore are considered to be boring. The lack of phase transitions is an indication that either strong forces between the chains or between the head groups dominate the film. Often there exist two-dimensional aggregates or islands of various size and shape immediately after spreading the molecules. Such films are far away from thermodynamic equilibrium. During a film compression essentially only the randomly oriented islands are pushed together. More possibilities for altering and fine tuning of chromophore orientations and order of the films can be expected for Langmuir films where the forces in the hydrophobic and in the hydrophilic part are similar. Such films should react on changing of environmental conditions as sensitively as the lipids discussed above.

As already mentioned the forces in the hydrophobic region depend on the chain lengths and the temperature. The intermolecular forces in the hydrophilic region of Langmuir films (ionic or hydrogen bonds, etc.) can be influenced by altering the pH or the ionic strength of the subphase. Using an example of Langmuir films and membranes of acidic phospholipids, it was shown that the states of such films can be considerably influenced by altering the ion concentrations of the aqueous phases and the interface charges (*Helm et al. 1986*). Unusual (non-physiological) pH values or ionic strengths should be considered for films of synthetic functional molecules with stronger polar interactions in order to find a window of parameters where the ordered states can be influenced.

Since the film transfer and material properties, such as e. g. the hyperpolarizability and the long-term stability (*Girling et al. 1985*), also strongly depend on the environmental parameters in a first step the conditions should be chosen such that the order of the synthetic film can be influenced. Then one has to try to keep or to freeze the order while changing the environmental parameters for film transfer and further applications.

5.3.3 Short range translational and long range orientational order

A short range translational order combined with a long range correlation in the orientation of the local crystallographic axes was revealed for various lipid films from X-ray (*Helm et al. 1987, 1991 Dutta et al. 1987, Kjaer et al 1989, Jacquemain et al. 1991, Kenn et al. 1991, Tippmann-Krayer et al. 1992*) and electron (*Fischer et al. 1984, Garoff et al. 1986*) diffraction data, from polarization fluorescence microscopy studies (*Moy et al. 1986*) and from controlled fusion experiments with domains under the influence of electric fields (*Flörsheimer et al. 1990*). There is a discussion (*Sackmann et al. 1987, Peterson 1987a, Peterson et al. 1987b, Helm et al. 1987, Schlossman et al. 1991*) whether the positional order in these films is disturbed by high densities of dislocations as described for hexatic mesophases which occur in theories of two-dimensional melting (*Nelson et al. 1979, 1982, Nelson 1983*).

The answer cannot be given by integrating measuring techniques such as diffraction, fusion or light microscopy experiments but only by local probe methods like scanning tunneling or atomic force microscopy (STM, AFM). In fact a long range orientational and short range translational order was recently detected in an AFM study (*Flörsheimer et al. 1994a*) of the hydrophobic surface of Cd arachidate LB multilayers (Y-type films, odd number of layers, transferred onto hydrophilic substrates at a lateral pressure of 30 mN/m from a subphase with pH 6.5 and 10^{-3} M $CdCl_2$). The characteristic variations of the positional order, however, were not accompanied with dislocations but with local orientational variations of the crystallographic axes around the mean value. Hence, the Cd arachidate films have to be considered as solids.

In contrast to these results high densities of isolated and paired dislocations were revealed by AFM (*Bourdiev et al. 1993*) in the hydrophobic surface of Ba arachidate LB films (Y-type bilayers transferred onto hydrophobic substrates from a subphase of pH =6.5 or 9). This shows that a slight change in the composition may alter the properties of the films considerably.

Crystal-like translational correlation lengths were revealed in X-ray and electron diffraction studies of monolayers of different molecules (*Jacquemain et al. 1991, 1992, Lin et al. 1990b, Möhwald 1988a*). It seems that classical crystalline states are favored in films where the intermolecular forces are governed by head group interactions whereas states between fluid and solid are frequent when the molecular interactions are governed by interchain interactions (*Möhwald 1988a*).

5.3.4 Reduction of disclination densities

Intense work towards the reproducibly perfect LB film was done by Peterson and co-workers (*Peterson 1988a*) especially on the example of 22-tricosenoic acid. Macroscopic and microscopic variations of the orientational order have been identified as a major cause of signal loss in optical LB film waveguides of this lipid (*Peterson et al. 1988b*). A method for a considerable increase of the range of orientational order was developed.

22-tricosenoic acid can be polymerized in LB films due to the vinyl end-group of the molecules (*Barraud et al. 1977*). Figure 5.8 shows an isotherm of a monomeric film. Four phases can be distinguished. While compressing the film, it passes through a liquid phase at about zero pressure followed by three smectic phases (S_I, S_F, S_B) which are thought to be hexatic (*Peterson et al. 1987b*). More details are discussed in the literature (*Peterson 1990* and references therein). Only one of the phases (S_F) can be transferred by the LB technique without considerable problems.

The isotherm of 22-tricosenoic acid is a good example to demonstrate the importance of using pure materials for Langmuir film characterizations. Due to the purity of the sample, i. e. only 1.5% contamination with 21-tricosenoic acid (*Bibo et al. 1989*) the isotherm of Figure 5.8 shows a high collapse pressure and three distinct changes of compressibility of the monofilm not corresponding to any feature of the previously published curves. Furthermore the molecular areas are significantly lower than reported previously (*Peterson 1988a*).

Peterson identified a particular low-energy variety of grain boundaries called twin boundaries in the LB films of 22-tricosenoic acid. They are connected with disclinations — well known zero dimensional defects of orientational order in

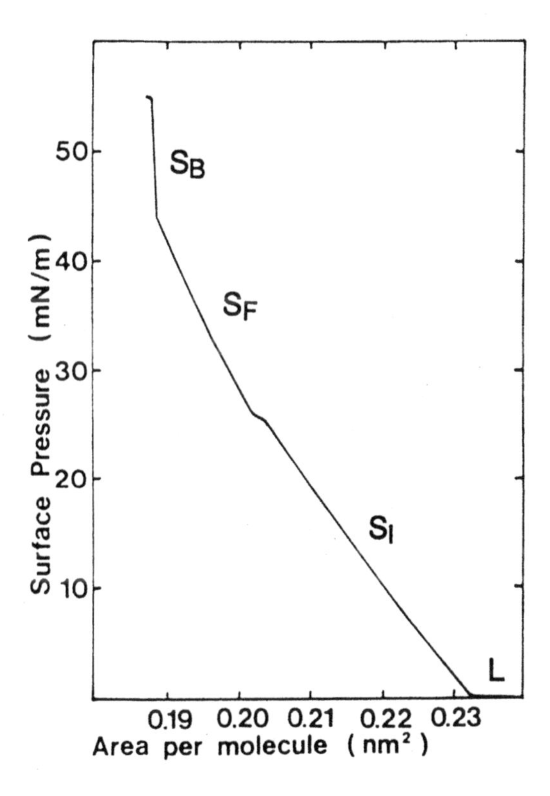

Figure 5.8 Isotherm for 22-tricosenoic acid at 20°, pH 3, compression speed 50 mNm^{-1} s^{-1} (*Peterson 1988a*). There are a liquid (L) and three smectic (S_I, S_F, S_B) phases. Reprinted by permission of Elsevier Publishers.

smectics. Figure 5.9 illustrates a monolayer lattice with two disclinations of opposite sign (Burgers angles +60°/–60°) joined by a twin boundary. It was shown that the quality of the first layer is essential for the whole LB multifilm (*Peterson 1986*). Processes similar to molecular beam epitaxy occur while depositing the following layers (*Peterson 1990*). Defects of the first monofilm are transferred to the following layers.

Indications of epitaxial processes occurring during LB deposition have been reported for various other molecules. This information is very interesting for the preparation of high quality multilayers e. g. for optical waveguiding. In particular the first layer has to be treated specially in order to reduce defect densities while the following layers can be grown conventionally.

Peterson and co-workers developed a method for the detection of disclinations and ruptures of a monolayer of 22-tricosenoic acid transferred onto a solid support. They coated the first layer epitaxially with about 200 additional layers and examined this multilayer film with a polarizing microscope. The contrast of the images is due to the in-plane anisotropy of the tilted hydrocarbon chains. Lines of discontinuity of the orientational order in the coating multifilm can be observed. The two terminating points of these lines correspond to singularities of the first

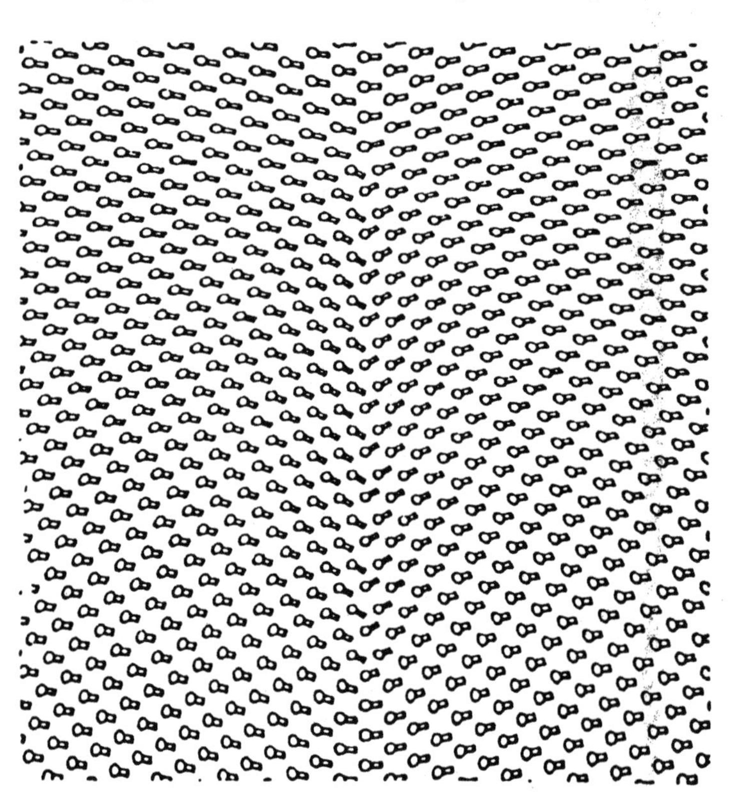

Figure 5.9 Ilustration of a twin boundary in a lattice connecting two disclinations of opposite sign (*Peterson 1990*). Reprinted by permission of IOP Publisher.

layer: disclinations with opposite signs to each other. Moreover the coating film discontinuities are bound to pass additional point defects of the first layer: disclinations, which have caused a rupture of the initial layer. Such pinholes can be identified by electroplating techniques. They limit the insulating properties of LB films (*Peterson 1990* and references therein). All the pinholes were found to lie on lines of discontinuity of the coating film (*Peterson 1986*). Hence the reduction of the density of disclinations in the first monolayer would simultaneously improve the electronic and the optical waveguiding properties of the multilayers.

It was shown that a considerable reduction of the multilayer defects latently present in the first layer can be achieved by a pressure treatment of the first film on the water surface, causing a recombination of disclinations (*Bibo et al. 1989*). The molecular mechanism of this process is still under investigation. It is not temperature activated. Rather, a compressed monolayer must be expanded in order to go through a higher order phase transition (the $S_I \leftrightarrow S_F$ transition at 25 mNm^{-1}, see Figure 5.8). Then it must be recompressed to the transfer pressure. A substantial influence of shearing forces is discussed in this context.

Applying this technique the disclination density could be reduced from about 1,000 mm^{-2} (*Peterson et al. 1987b*) to about 6 mm^{-2} (*Bibo et al. 1989*). The LB films following the careful preparation of the initial layer can be transferred relatively fast in the example of 22-tricosenoic acid (dipping speed 4 mm s^{-1}, *Bibo et al. 1989*). Another possibility for the increase of the range of orientational order in Langmuir films will be discussed below. For this purpose the typical morphologies of Langmuir films and their origin must first be described.

5.3.5 Increasing the range of orientational order

Some examples of typical textures are shown in the fluorescence micrographs of Figure 5.7 for different thermodynamic parameters and molecules. The fluid phase is stained in the examples and hence appears to be bright. The liquid phase coexists with gaseous domains for large molecular areas in the example of DMPE. The corresponding micrograph and the molecular area are marked with an a in the figure. Liquid crystalline domains coexisting with the fluid phase increase in size due to a film compression within the series of micrographs from Figure 5.7.b to e. Their number remains almost constant during growth (*Helm et al. 1988*). The circular shapes are an equilibrium feature of DMPE domains. Dendritic and fractal patterns of liquid crystalline domains occur when a film is quickly compressed. They originate from diffusion limited aggregation (*Miller et al. 1986, 1987*). An example of elongated domains in a state close to thermodynamical equilibrium (*Flörsheimer et al. 1989a,b*) is depicted in Figure 5.7.f (L-α-dipalmitoylphosphatidylcholine, DPPC). The uniform sense of curvature is caused by the chirality of the molecules.

The number of liquid crystalline or solid domains in a Langmuir film depends on the nucleation conditions, i. e on kinetic parameters (e. g. compression speed during nucleation), on the number and kind of present seeds and on the ionic strength of the subphase. These parameters can be controlled in some degree (*Helm et al. 1986, 1988, Flörsheimer 1989a, Möhwald et al. 1988b*).

The splitting of phases into large numbers of domains and unusual equilibrium domain shapes in Langmuir films are special interface phenomena. There are long range electrostatic forces originating from the organization of polar molecules on an interface. A crystalline or liquid crystalline phase has a normal excess dipole density in comparison to a coexisting fluid phase as illustrated e. g. in Figure 5.2. Hence there are repulsive forces between the molecules of a domain and between the domains of a monofilm. This causes superlattices in some materials (*Möhwald 1988a*). The line tensions of the domains are antagonists of the long range repulsive forces. The interplay of electrostatic forces and line tensions determines the domain shapes and maximum sizes to a large extent. A maximum diameter cannot be exceeded in the example of compact (e. g. circular) domains (*Keller et al. 1987, Flörsheimer et al. 1990*) since the electrostatic energy of a growing domain increases faster than the line energy. A further increase of the total area of the domains in a Langmuir film is possible, however, when the electrostatic energy is reduced by a fission or an elongation of the domains. The length of elongated straight domains is not limited thermodynamically. Their width, however, is an equilibrium parameter determined by the interplay of electrostatic forces and line tensions. The equilibrium width can be controlled systematically by the environmental conditions (*Keller et al. 1986, McConnell et al. 1988, Heckl et al. 1986 a, b, Flörsheimer 1989a, Flörsheimer et al. 1989b*).

Equilibrium sizes, shapes and shape transitions have been predicted and calculated or estimated (*Keller et al. 1986, 1987, McConnell et al. 1988, Andelmann et al. 1987, Flörsheimer 1989a, Flörsheimer et al. 1990, Vanderlick et al. 1990, Muller et al. 1991*). Predicted shape transitions of domains have been observed (*Flörsheimer 1989a, Flörsheimer et al. 1989b, Heckl et al. 1986a, b*) as well as fissions (*Flörsheimer 1989a, Flörsheimer et al. 1990*) by a systematic change of the experimental conditions. In some cases, the shape transitions were probably coupled with higher order phase transitions (*Heckl et al. 1986a, b*).

The maximum domain size was estimated and verified for the liquid crystalline DMPE phase to be in the mm or cm region (*Flörsheimer et al. 1990*). This size is not a limit for films of other molecules. Moreover the maximum domain sizes could be increased by decreasing the electrostatic forces using suitable additives. The dipole excess density of a liquid crystalline phase as compared to a coexisting fluid phase can be manipulated considerably by a selective contamination of the fluid phase with an amphiphile having large dipole moments with suitable orientation, as was demonstrated recently (*Flörsheimer et al. 1990*).

The growth of large domains in Langmuir films with a uniform orientation of the crystallographic axes is highly desirable for the preparation of low-loss optical waveguides. The long range orientational order of liquid crystalline phases was mentioned above. The orientation of the crystallographic axes was measured to change by not more than 5° over domains of some phospholipids (*Fischer 1985, Flörsheimer et al. 1990*) with typical diameters of a few 10 μm. One possibility to increase the domain sizes of a film is to reduce their number. As described above, the number can be controlled in some degree by the nucleation conditions.

It was shown additionally that the domains can be moved noninvasively and in a controlled and defined way on a water surface by applying variable *inhomogeneous electric fields* with arrays of special *electrodes* (*Flörsheimer et al. 1989b*). The lateral forces on the domains are due to the *excess dipole density* of the domains with respect to a coexisting fluid phase. Figure 5.10 sketches the principle. It is possible to separate single domains from all the others with this technique. The method allows the study of domains in the absence of the electrostatic forces from neighbor domains or e. g. to grow them selectively by a local change of the thermodynamical conditions (*Möhwald et al. 1988b, Flörsheimer et al. 1989b, Dietrich et al. 1991*). The electrically *induced motion of Langmuir film domains* also makes it possible to *fuse* two or more *domains.* Larger domains with an essentially uniform orientational order of the crystallographic axes were obtained by *fusion* (*Flörsheimer et al. 1990*).

A special *film balance* with a double trough was proposed (*Flörsheimer 1989a*) for the *selection of one* or a small number of *domains* and their *controlled manipulation.* The apparatus is sketched in Figure 5.11. Nucleation can be initiated in one of the film balances (the larger one in the example of the sketch) by compressing a monolayer. Fluid phase as well as liquid crystalline or solid domains can be sucked into the second film balance by an expansion movement of the corresponding barrier. However, it is possible to control the number of domains passing the connecting channel by the application of electrostatic forces. A selected domain can be observed with a microscope. The domain can be grown by a simple compression of the film in the second trough when a mechanical shutter between the film balances is closed. The apparatus may also be used for *zone refinement*: when *heterogeneous nucleation* takes place in the first trough during a compression,

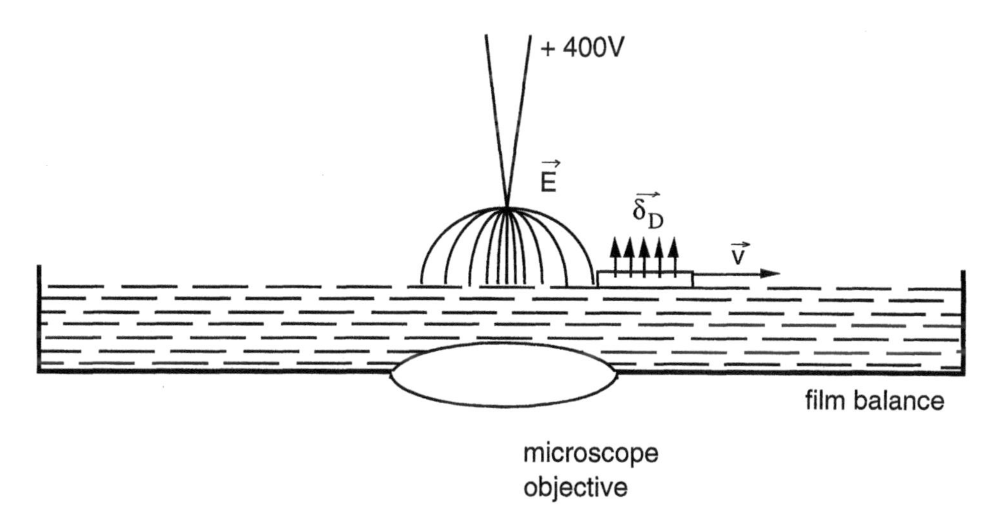

Figure 5.10 Movement of a liquid crystalline domain in an inhomogeneous electric field ($\vec{E}$) due to its dipole excess density $\vec{\rho}_D$ compared with the surrounding fluid phase. (Speed $\vec{v}$, typical voltages between electrode and water: a few 100 V, typical tip/film distance: 30 μm, fluid phase of monolayer not shown).The film is observed with a microscope from below.

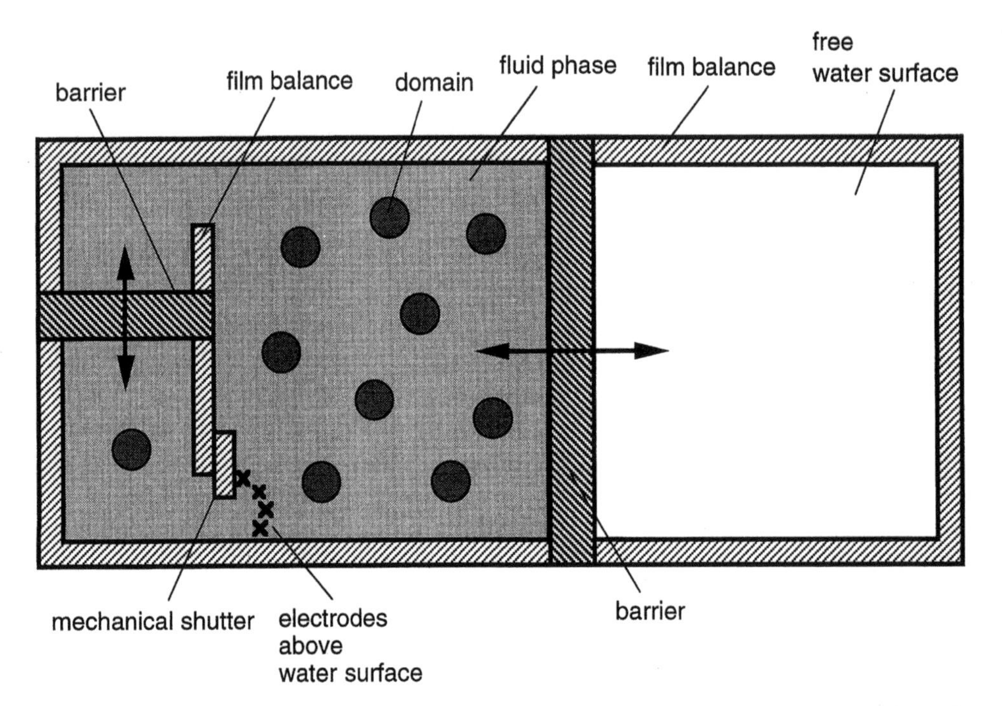

Figure 5.11 Double film balance for thermodynamical treatment of a defined small number of domains and for zone refinement. Sketch of top view. As an example, nucleation can be initiated by compressing a monolayer in the larger film balance. The fluid phase as well as liquid crystalline or solid domains can be sucked into the second film balance by an expansion movement of the corresponding barrier. However, the number of domains passing the connecting channel can be controlled applying electrostatic forces. For example a single domain can be grown in the smaller film balance after the mechanical shutter between the troughs has been closed.

contaminations which are present in the monolayer may act as seeds for domains. The concentration of contaminations in the fluid phase of the monolayer will thus be reduced.

5.3.6 Summary and outlook

Summarizing, two possibilities for the increase of the range of orientational order in Langmuir films have been described:

a) domains with uniform orientation of crystallographic axes can be grown
b) in films rich of disclinations, the recombination of disclinations with opposite sign was shown

The observation of epitaxial processes during LB film transfer suggests that the preparation of the first LB layer should receive special attention, whereas the following layers can be deposited faster and more conventionally. Another method for the increase of the uniformity of LB films utilizes preformed polymers. These polymers can be transferred from fluid-like states on the water surface which is seldom possible using monomers. LB films with a normal overall polarization can be obtained. Examples of functional preformed polymers and their LB films will be described in Chapter 8.3.

Recently various monolayer sensitive *in situ* microscopy techniques with advantages as compared to fluorescence microscopy have been developed. The Brewster angle microscope (*Hénon et al. 1991, Hönig et al. 1992*) utilizes p-polarized light which is incident on an uncoated interface under Brewster angle thus minimizing the reflected intensity. When the interface is coated with a thin film, however, the reflectivity becomes strongly dependent on the thickness, the roughness and the refractive index of this film. Ellipsometric microscopy (*Reiter et al. 1992*) is also sensitive to the refractive index or the thickness of monolayer structures. Neither technique requires fluorescent additives. However, the films are observed via an objective whose optical axis is tilted with respect to the surface. Therefore, high magnification and resolution near Abbé's limit are not obtained with simple optical means.

Birefringent monolayers can be observed utilizing a polarization microscope with slightly uncrossed polarizers where the incident light is within the absorption region of the material. The contrast has been explained recently with a microscopic model, approximating the molecules as dipole oscillators on an interface (*Flörsheimer et al. 1993b*), and with a macroscopic model based on Maxwell's equations (*Flörsheimer et al. 1994b*). The method provides high light intensities so that the films can be observed with the naked eye. The lateral resolution is near Abbé's limit. The need of an absorption band of the material in the visible or ultraviolet range of the spectrum is not a strong restriction for material sciences. All highly polarizable or hyperpolarizable molecules or e. g. intermolecular charge transfer compounds for the development of low dimensional electric conductors have electronic absorption bands. In addition to this second-order susceptibilities of monolayer structures can now be imaged *in situ* with a second-harmonic microscope developed recently (*Flörsheimer et al. 1993a*). Not only Langmuir films but also the LB transfer process was observed on the border between substrate and subphase using a fluorescence microscope (*Riegler 1988, Spratte et al. 1991*).

With all the new contrast methods which use direct imaging, important material properties without the need of staining or decoration of the sample will considerably facilitate the characterization of films of new synthetic molecules. They are well suited for the investigation of the order and the range of order in Langmuir films as a function of preparation parameters. The new contrast techniques could also be used for the *in situ* investigation of the LB film transfer process as was demonstrated with the fluorescence microscope. The new tools, together with local probe techniques and other integrating methods, will help us to understand and to control films of new classes of functional molecules, as well as some previously characterized lipid films. Nonlinear optical, electro-optic and waveguide properties of LB films with synthetic molecules will be reviewed in Chapter 8.3.

6. ELECTRO-OPTIC AND NONLINEAR OPTICAL POLYMERS

6.1 INTRODUCTION

Organic polymers have become an important class of nonlinear optical materials. They have been the subject of intense research, since they combine the nonlinear optical properties of conjugated π-electron systems with the feasibility of molecular engineering, i.e. creating new materials with appropriate optical, structural and mechanical properties (*Williams 1983*). Some of them are summarized in Table 6.1.

Table 6.1 Properties of electro-optic polymers.

Optical	Structural	Mechanical
— large nonresonant non–linearities	— low cost materials	— resistance to radiation, heat and shock
— low dc dielectric constants	— molecular engineering	
— low switching energy	— integrated optics	
— subpicosecond response times	— room temperature operation	
— broadband transparency	— chemical stability	
— high optical damage threshold	— ease of processing and synthesis modifications	
— low absorption		

Low dielectric constants, which are of the order of 3, make travelling wave devices possible since matching of optical and microwave velocities can be achieved (see Chapter 9.2). Therefore lower driving voltages and power requirements are needed, and render polymers ideal for high frequency electro-optic modulator materials.

However several restrictions to the promising features of polymers also have to be made. To exploit the optical properties in electro-optic and second-order nonlinear optical experiments fully, one has to break the isotropic structure of polymer films. Poling under an electric field leads to the desired noncentrosymmetry through the alignment of the nonlinear optical active moieties. Unfortunately, this electric field induced order tends to decay when the poling field is turned off, especially at higher temperatures expected in future device applications. The stabilization of the noncentrosymmetric state is still a major problem in developing polymer electro-optic devices.

Another polymer class attracted researcher's interest; third order nonlinear optical polymers with their large nonresonant susceptibilities and subpicosecond response times are very promising candidates in the field of optical signal processing, since the principle of operation of an e.g. all optical waveguide device is basically a third order process: the intensity dependence of the refractive index.

6.2 SECOND-ORDER NONLINEAR OPTICAL POLYMERS

6.2.1 Material aspects

The packing density of nonlinear optical molecules in a macroscopic sample of matter determines the expected value for the second order susceptibilities. Crystal growth is one possibility of incorporating nonlinear optical molecules in a bulk material. Most organics crystallize in a centrosymmetric form and are therefore not useful for second-order nonlinear effects. In addition the preparation and the processing of organic crystals demand a great deal of know-how.

The incorporation of nonlinear optical molecules in glassy polymers is comparatively easy and can be done in different ways: by admixing of active carriers in a polymeric matrix forming a guest-host system, by covalently linking them to a polymer backbone in the form of a side chain or by incorporating them into the main chain itself. Figure 6.1 depicts schematically these three kinds of second-order nonlinear optical polymers.

One of the major problems of guest-host systems is the solubility of guest molecules in the host matrix which allows only moderate doping levels. Side-chain polymers avoid this problem.

One of the potential advantages of polymeric materials is the possibility of easy film formation. Viscous polymer solutions can be spread as thin films on different substrates by a variety of techniques (spin coating, dipping, or doctor blade methods). A model of the spin coating process, accounting for variations of concentration, viscosity and diffusivity across the thickness of the spin coated film is given by Bornside et al. (*1989*).

Second-order nonlinear optical effects require noncentrosymmetric molecules ($\beta \neq 0$) with macroscopic polar order. Unfortunately, most polymers belong to the class of isotropic materials. To break this symmetry one has to apply an external electric field that aligns the nonlinear optical chromophores by coupling to their dipole moment. This poling procedure imposes an ∞mm point group symmetry on the polymer (*Mortazavi et al. 1989*). It can be shown (*Nye 1967*) that the susceptibility tensor for second harmonic generation (Eq.(3.10)) has only three independent elements d_{33}, d_{31} and d_{15} where the 3-axis is along the poling field.

Systems like liquid crystalline (mesogenic) polymers are very interesting because of the expected enhancement of polar order due to spontaneous alignment of the dipoles (see below). Figure 6.2 shows possible arrangements of the mesogenic units and the nonlinear optical groups. The mesogenic groups themselves may also be nonlinear (*Boyd et al. 1988*). However recent calculations (*Van der Vorst et al. 1990*) suggest that initial liquid crystallinity is not required to obtain large polar

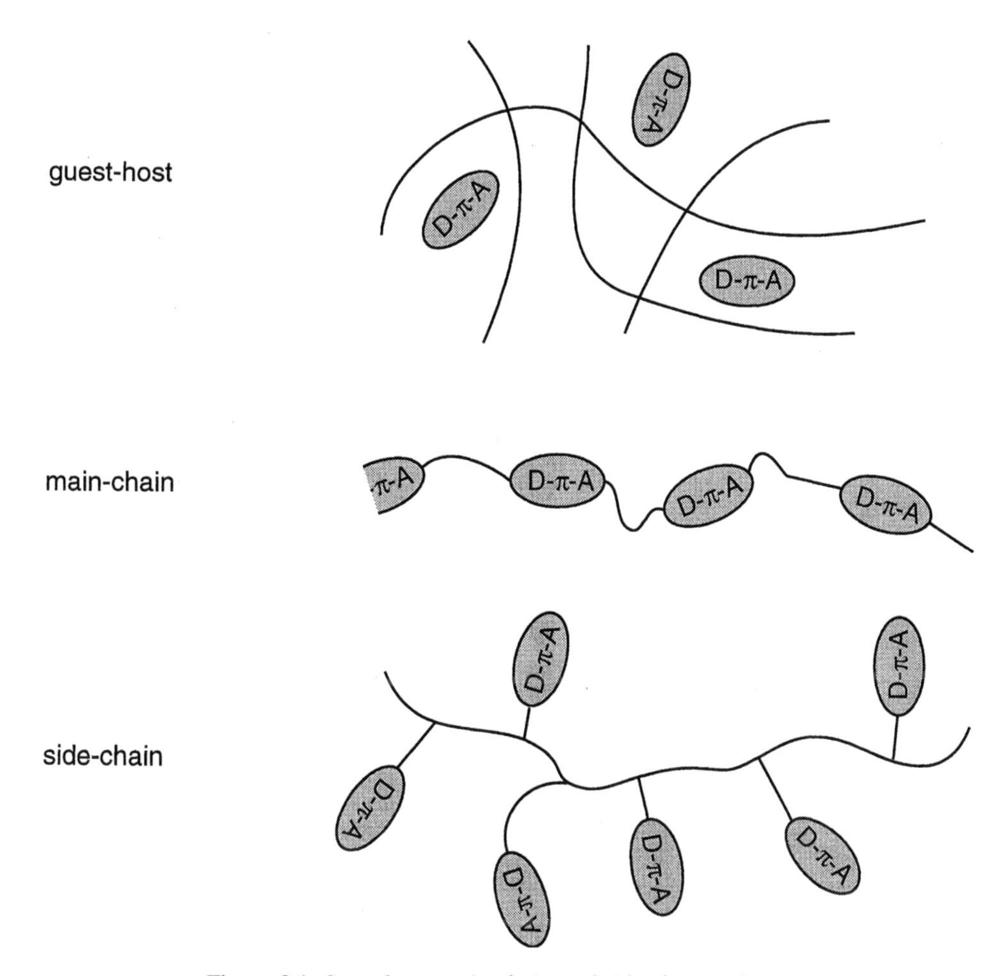

Figure 6.1 Guest-host, main-chain and side-chain polymers.

order in the strong field regime (E ≈ 10^8 V/m). In addition liquid crystalline polymer systems often show light scattering by crystalline domain boundaries.

6.2.2 Poling process

6.2.2.1 Theoretical aspects. As discussed in Chapter 3, the main component of the microscopic hyperpolarizability tensor β is β_{zzz}. For poled films, the relation between the macroscopic value of the second-order nonlinearity and the microscopic hyperpolarizability then follows from (the stars refer to values corrected for local fields, c.f. (3.25)):

$$\chi^{(2)}_{333} = N \cdot \beta^*_{zzz} \left\langle \cos^3 \theta \right\rangle \tag{6.1}$$

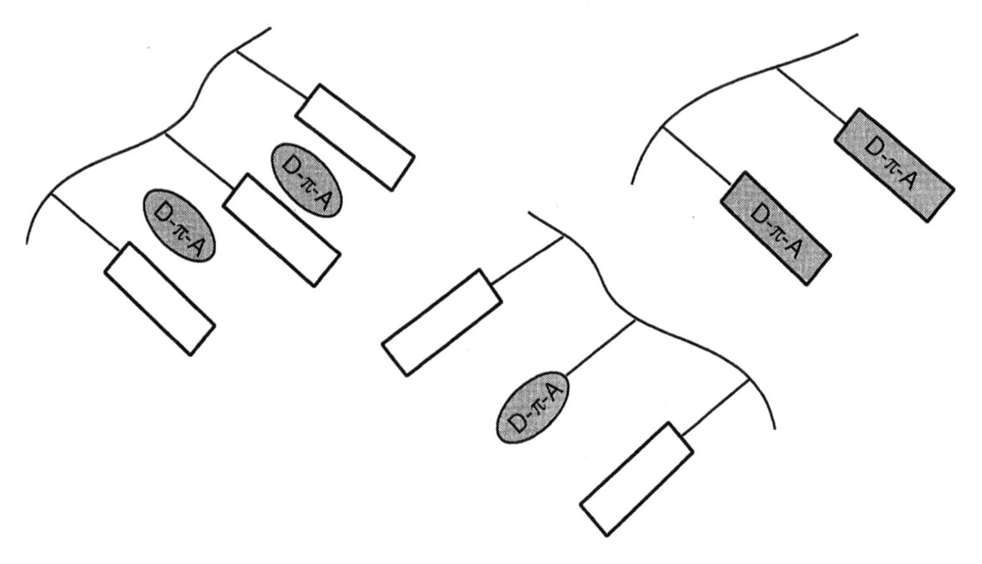

Figure 6.2 Arrangements for mesogenic polymer systems. Rectangles refer to mesogenic units.

and
$$\chi^{(2)}_{311} = N \cdot \beta^{*}_{zzz} \left\langle \frac{1}{2} \cos\theta \sin^{2}\theta \right\rangle \tag{6.2}$$

where θ is the angle between the molecular dipole moment and the electric poling field in the 3-direction and the bracket $\langle \ldots \rangle$ stands for averaging over all molecular orientations weighted by an order distribution function G. Writing Eqs. (6.1) and (6.2), it is assumed that polymeric films possess azimuthal symmetry. In this case G depends on θ only. At thermal equilibrium G equals a Maxwell-Boltzmann distribution. Therefore

$$G(\theta) = \frac{1}{Z} \cdot e^{-U(\theta)/k_B T} \tag{6.3}$$

where $k_B T$ is the Boltzmann factor and Z the partition function

$$Z = \int_{-1}^{1} d(cos\theta)\, e^{-U(\theta)/k_B T} \tag{6.4}$$

Hence the average value of any function $F(\theta)$ becomes

$$\langle F(\theta) \rangle = \int_{-1}^{1} d(cos\theta)\, F(\theta)\, G(\theta) \tag{6.5}$$

The potential $U(\theta)$ is expanded in terms of the poling field strength E_p:

$$U(\theta) = U_0(\theta) + U_1(\theta) + U_2(\theta) \tag{6.6}$$

where the subscripts denote the order in the field.

Four statistical models for $U(\theta)$ are treated in the literature: the isotropic and the Ising model (*Meredith et al. 1983*), the model of Singer, Kuzyk and Sohn (SKS) (*Singer et al. 1987*) and the model of Van der Vorst and Picken (VVP) (*Van der Vorst et al. 1990*). Common to all models is the term

$$U_1(\theta) = \vec{\mu}_g^* \cdot \vec{E}_p \tag{6.7}$$

where $\vec{\mu}_g^*$ is corrected for the local static poling field by the Onsager expression

$$\mu_g^* = f^0 \mu_g = \frac{\varepsilon(n_\infty^2 + 2)}{n_\infty^2 + 2\varepsilon} \mu_g \tag{6.8}$$

with ε the static dielectric constant and n_∞ the optical index of refraction.

In the Ising model ($\theta = 0$ or π) and in the isotropic model $U_0(\theta)$ and $U_2(\theta)$ are both zero. The averaged value $\langle cos^3\theta \rangle$ can be evaluated exactly (*Van der Vorst et al. 1990*). The linear approximation of the original isotropic and Ising models yields

$$\left\langle \cos^3\theta \right\rangle_{Ising} \approx \frac{\mu_{g,z}^* E_p}{k_B T} \tag{6.9}$$

and

$$\left\langle \cos^3\theta \right\rangle_{isotropic} \approx \frac{1}{5} \cdot \frac{\mu_{g,z}^* E_p}{k_B T} \tag{6.10}$$

that is, the two limiting distribution functions differ by a factor of 5.

In the SKS model the distribution function is expanded in terms of Legendre polynomials. The $\langle P_2 \rangle$ and $\langle P_4 \rangle$ order parameters, where P_l is the l-th order Legendre polynomial, give the zero field energy $U_0(\theta)$ which quantifies the order induced by intermolecular forces. No $U_2(\theta)$ term is included. For the macroscopic coefficients one gets (*Singer et al. 1987*)

$$\chi_{333} = \frac{N\beta_{zzz}^* \mu_{g,z}^* E_p}{k_B T} \left(\frac{1}{5} + \frac{4}{7}\langle P_2 \rangle + \frac{8}{35}\langle P_4 \rangle \right) \tag{6.11}$$

$$\chi_{311} = \frac{N\beta_{zzz}^* \mu_{g,z}^* E_p}{k_B T} \left(\frac{1}{15} + \frac{1}{21}\langle P_2 \rangle - \frac{4}{35}\langle P_4 \rangle \right) \tag{6.12}$$

In the isotropic phase of liquid crystals or in glassy polymers $\langle P_2 \rangle$ and $\langle P_4 \rangle$ are zero which leads to the same results as in the isotropic model.

These three foregoing linear models apply only for moderate poling fields where saturation effects do not yet occur.

The VVP model is intended for strong poling fields. The term $U_2(\theta) = -(1/3)\Delta\alpha E_p^2 P_2(cos\theta)$, where $\Delta\alpha = \alpha_{\|} - \alpha_{\perp}$ is the difference in the linear polarizability parallel and perpendicular to the main axis of the molecule, takes the energy of the linearly induced dipole moment into account. It leads to a field induced axial order since $P_2(cos\theta)$ is minimal for $\theta = 0$ or π. Axial order means that the molecules are aligned along a certain axis with conserved inversion symmetry. This up-down symmetry is broken in the case of a polar order where the molecules are aligned in a certain direction along a vector.

As suggested by Eqs. (6.9) and (6.10), a spontaneously present axial order without any field may enhance the field induced polar order up to a factor 5 over the maximum achievable value in isotropic media by use of a liquid crystal host (*Meredith et al. 1983*). The VVP model accounts for this mutual axial alignment by introducing an effective single particle potential

$$U_0(\theta) = -\varepsilon\langle P_2\rangle P_2(cos\theta) \tag{6.13}$$

where the parameter ε determines the absolute strength of the potential. The order parameter $\langle P_2\rangle$ describes the influence of the environment on the alignment of the molecules. Since for an isotropic medium $\langle P_2\rangle = 0$ at zero field strength, no axial order is induced by the $U_0(\theta)$ term whereas $\langle P_2\rangle = 1$ manifests perfect alignment. The θ-dependence is reflected by the factor $P_2(\theta) = (3cos\theta - 1)/2$.

Nevertheless, the most often used expressions for the macroscopic susceptibilities are the linear approximations of the isotropic model in the weak field regime:

$$\chi_{333} \approx \frac{1}{5} \cdot \frac{N\beta^*_{zzz}\mu^*_{g,z}E_p}{k_BT} \tag{6.14}$$

$$\chi_{311} \approx \frac{1}{3} \cdot \chi_{333} \tag{6.15}$$

One should note the factor 2 difference between Eq.(3.25) and Eqs.(6.1,6.2): β-values from EFISH-measurements (see Chapter 7.2) are usually calibrated with some crystal d-values (e.g. quartz, see Eq.(7.18)). Therefore

$$\beta(Eq.3.25) = \frac{1}{2}\beta(Eq.6.1)$$

since $d = 1/2\chi$.

One of the prerequisites for future application of poled polymers is the stabilization of the alignment of the nonlinear optical moieties even when the poling field is turned off. This is quite a severe problem. A variety of experiments have shown that the relaxation mechanisms for the molecular dipole orientational distribution function are correlated to the specific volume of the glass forming liquid. We make the assumption that this is approximately the case, although it is expected that the specific volume behavior describes best that of a guest-host type polymer system.

The following phase diagram illustrates most of the general properties of the glass transition phenomena using the specific volume and temperature as the thermodynamic variables. At high temperatures (above T_o) the material is in the liquid state, with a thermal expansion coefficient of α_l. Below T_{eq}, the material is a glass with a thermal expansion coefficient $\alpha_g < \alpha_l$. T_{eq} represents a theoretical temperature where the glass is in a true equilibrium state. Due to kinetic and structural constraints, however, the actual transition occurs at a temperature T_g which can be defined as the intersection of the glass and liquid lines. The glass transition temperature T_g is a kinetic phenomena that depends on the rate of cooling, q, as indicated in the diagram (*Blythe 1979*).

In the temperature range above T_g the concept of free volume adequately describes the relaxation behavior of most polymers. The total volume per gramm, v, is the sum of the free and the occupied volumes v_f and v_0:

$$v_f = v - v_0 \tag{6.16}$$

v_0 is given by the van der Waals radii including the volume associated with vibrational motions.

In order to allow moves of molecular segments from one site to another, there is a critical free volume v_{cf} needed. The rate of segmental motion depends exponentially on the ratio $-v_{cf}/v_f$ (*Bueche 1962*). The ratio of the rates at two different temperatures T_1 and T_2 can be expressed as

$$ln\frac{r_2}{r_1} = v_{cf}\left(\frac{1}{v_{f1}} - \frac{1}{v_{f2}}\right) \tag{6.17}$$

Assuming that the excess rate of the expansion of the liquid over that of the glass state is caused by an increase in thermal expansion coefficient and the free volume at the glass transition temperature, one may write

$$v_{f2} = v_{f1} + \Delta\alpha . v_1 (T_2 - T_1) \tag{6.18}$$

where $\Delta\alpha = \alpha_l - \alpha_g$ is the difference in the thermal expansion coefficient above and below the transition temperature ($\alpha = (1/v)\cdot(dv/dT)$, which implies an exponential dependence of v on temperature and not just a linear relation (as Eq.(6.18) suggests)).

Equation (6.17) together with Eq. (6.18) provides

$$log\frac{r_2}{r_1} = \frac{2.303\cdot (v_{cf} / v_{f1})(T_2 - T_1)}{(v_{f1} / \Delta\alpha v_1) + (T_2 - T_1)} \tag{6.19}$$

Taking the dipolar relaxation time τ as a measure for segmental mobility

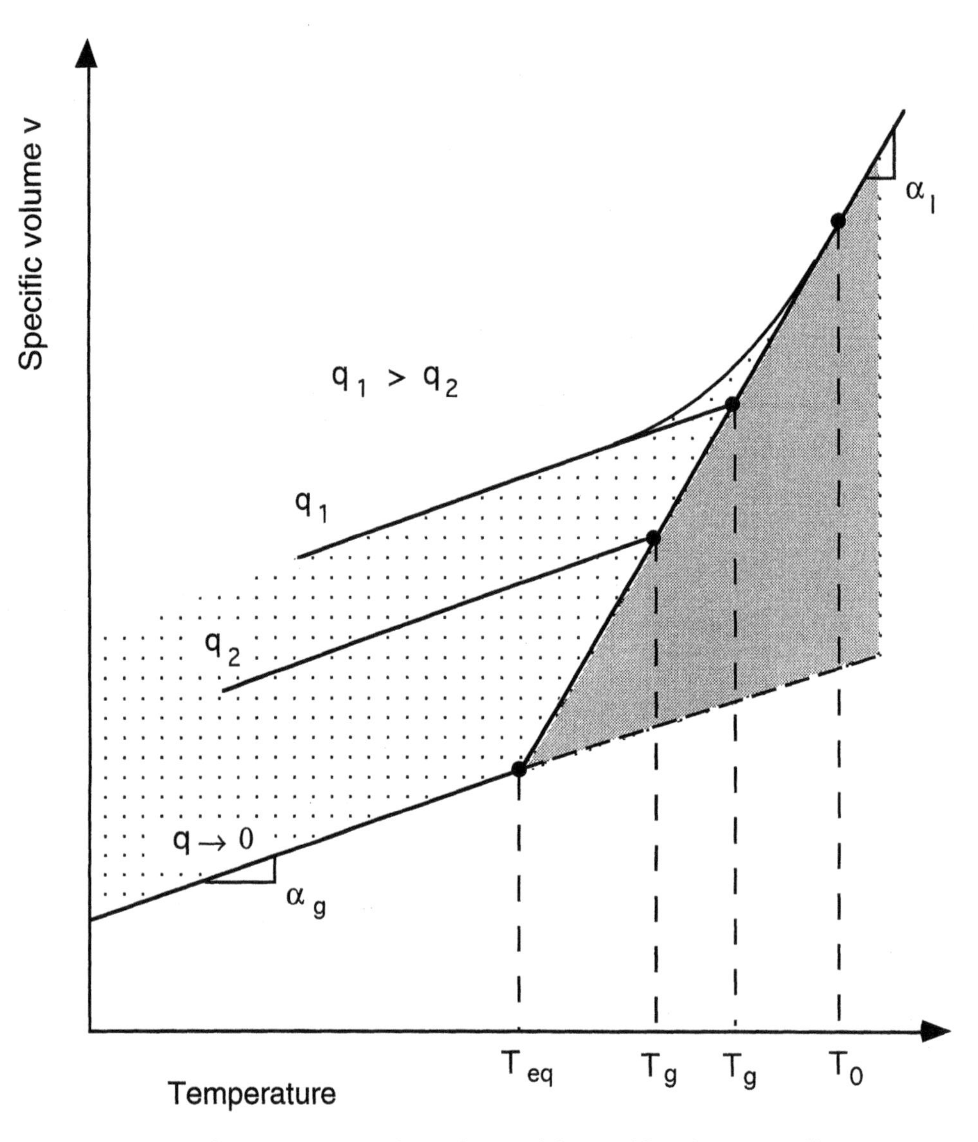

Figure 6.3 Temperature dependence of the specific volume near T_g.

$$r \propto \frac{1}{\tau} \tag{6.20}$$

the temperature dependence of τ above an arbitrary reference temperature T_{ref} (often taken to be the glass transition temperature) is well described by the WLF-equation (*Williams, Landel, Ferry 1955*)

$$\log a_T \equiv \log \frac{\tau_T}{\tau_{ref}} = \frac{-C_1 (T - T_{ref})}{C_2 + (T - T_{ref})} \tag{6.21}$$

This functional form also describes the behavior of the viscosity and time-temperature shift factors. 'Universal' behavior of polymers can often be described when the constants C_1 and C_2 are chosen to be equal to 17 and 52 K, respectively. The product of C_1 and C_2 is supposed to be independent of the reference temperature, T_{ref}. The WLF equation can be rearranged into the Vogel-Fulcher formula given by

$$\tau = A^* \exp\left(\frac{B}{T - T^*}\right) \tag{6.22}$$

where $T^* = (T_{ref} - C_2)$. Although phenomenologically correct above the glass transition temperature, these formulations are unable to describe relaxation processes in the glassy state. Both of these equations have a singularity at $T = (T_{ref} - C_2)$, which is an unphysical description of the nonequilibrium properties of a slowly relaxing glass. In the range of $T_g < T < T_g + 100$ K, Eq. (6.21) applies not only to a variety of polymers but also to organic glass forming liquids and inorganic glasses (*Eich et al. 1989a*).

It is therefore clear how one can achieve the desired stability. First the polymer film is heated up slightly above T_g. One expects maximum alignment under a poling field in the vicinity of T_g because of the high mobility of the nonlinear optical guest molecules. To fix this anisotropy in the polymeric film it has to be cooled down below T_g with the poling field on. The shrinking of the specific volume reduces the free space for relaxations of the aligned moieties.

Nevertheless, there still occur relaxations with time when the field is turned off. These residual relaxations are only partly attributable to the phenomenon described above since large scale reorientations of the chains should be suppressed far below T_g. Other processes (conventionally labeled α, β, γ,..., beginning at high temperatures) such as rotation around c-c bonds may be responsible for the residual thermal decay of the induced order. Such relaxations often follow an Arrhenius type of law since they fit quite accurately models of thermal activation over a potential energy barrier.

6.2.2.2 Experimental aspects. The most commonly used poling techniques are electrode and corona poling. The former has the advantage of well controllable field strength (*Eich et al. 1989b*), whereas the latter allows higher fields because of the absence of electrical breakdown due to imperfections in the film. A typical set-up for electrode and needle to plane poling is illustrated in Figure 6.4.

Corona discharge is a partial breakdown of air because of high inhomogeneous electric fields ($V \approx 10$ kV) (*Mortazavi et al. 1989*). A sharp tungsten needle is placed a few centimeters above the polymeric film. Ionized molecules of the surrounding atmosphere are deposited on the film surface and generate the poling field depending on the potential V as well as on the corresponding conductivities. Studies of the influence of positive or negative poling polarity and the composition of the atmosphere are discussed by Hampsch et al. (*1990*). A typical poling cycle is shown in Figure 6.5 (*Eich et al. 1989c*).

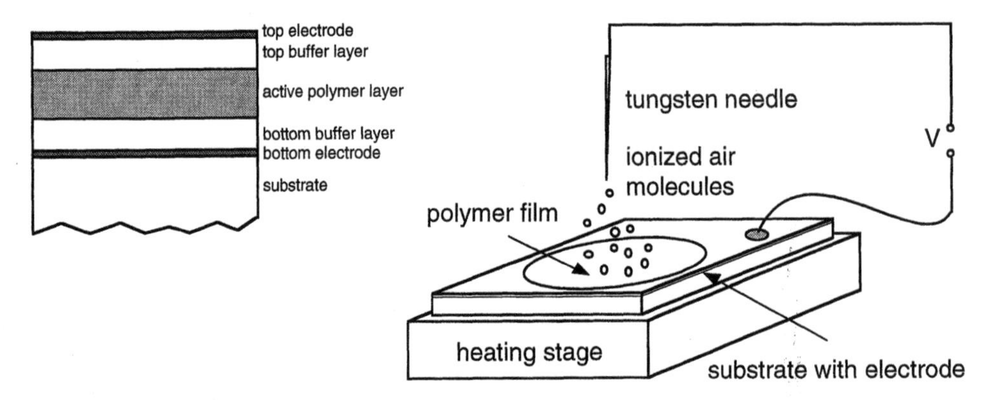

Figure 6.4 Electrode and needle corona poling.

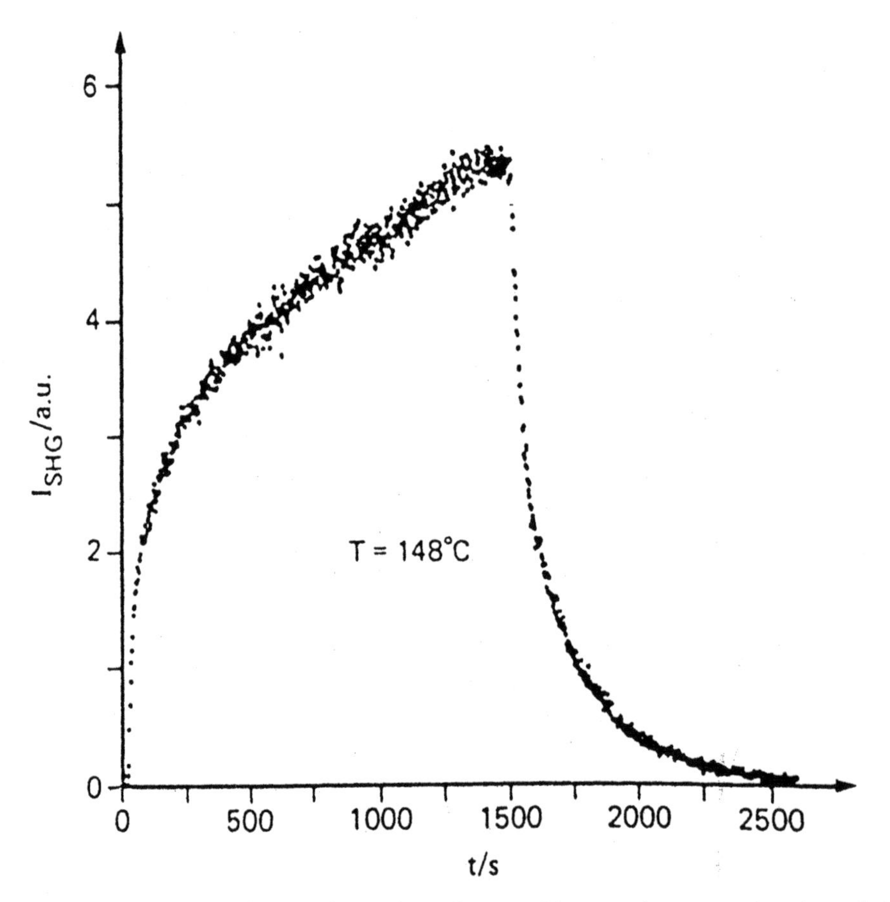

Figure 6.5 Time dependence of monitored second-harmonic generation in poly(p-nitroaniline) (PPNA) as the corona potential is switched on and off (T_g = 125°C) (*Eich et al. 1989c*). Reprinted by permission of AIP Publisher.

Observation of the second-harmonic response gives direct information about the long time stability of the probe polymer. Möhlmann et al. (*1990*) estimated 50%-decay-lifetimes for a polymer with a T_g of 140°C to be 10^{11} years (sic!) at ambient room temperature (20°C) or still 19 years for an operation temperature of 80°C. Via chemical cross linking induced vitrification of nonlinear optical polymers Eich et al. (*1989b*) obtained nonlinear optical susceptibilities in a bisphenol-A-4 nitro-1,2 phenylenediamine (Bis A-NPDA) polymer stable over more than 500 hours even at elevated temperatures (Figure 6.6).

Other efforts tend to make use of the intrinsic ordering of some hosts such as nematic polymeric liquid crystals (*Yitzchaik et al. 1990*), or polyimide films (*Wu et al. 1991*). Hill *et al.* (*1989*) use a ferroelectric host material which is partially crystalline. The poling field aligns the molecular chains in the crystalline regions. Guest molecules residing in the amorphous parts of the host also align under the influence of the poling field. When the latter is removed one expects a remaining coercive field from the crystalline region which could maintain the ordering of the guests. Fields as high as 125 MV/m have been recorded.

6.2.3 Second-harmonic generation as a probe of poled polymer systems

Usually second harmonic coefficients d_{IJK} of poled polymer films are measured using a Maker-fringe technique (*Jerphagnon et al. 1970*). The film spun on a substrate is turned around an axis perpendicular to the laser beam as illustrated in Figure 6.7.

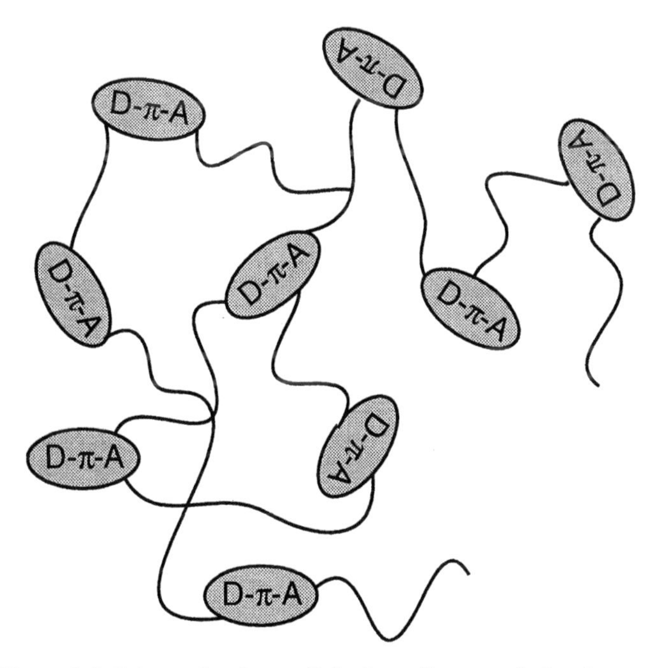

Figure 6.6 Schematic of cross-linked nonlinear optical polymers.

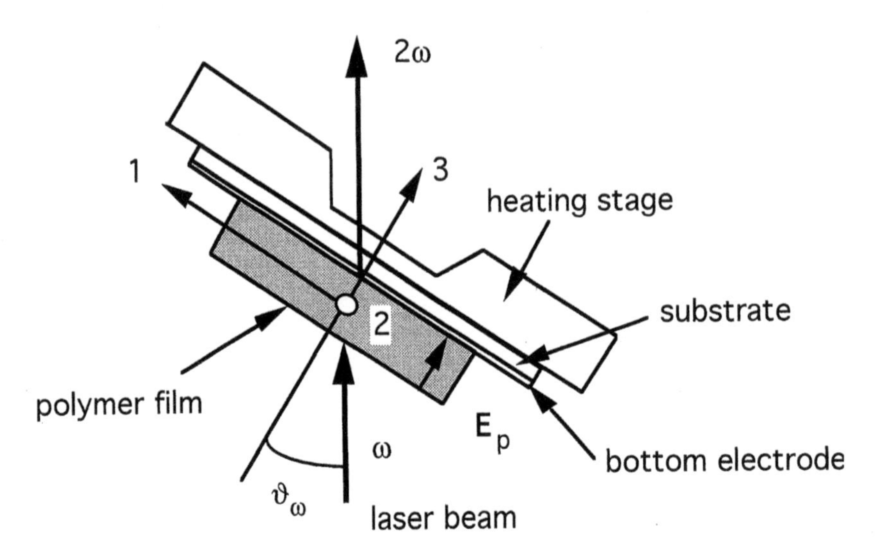

Figure 6.7 Maker-fringe measurement geometry.

The field induced optical axis is parallel to the 3 axis. The nonlinear optical susceptibility tensor d_{IJK} for the point group ∞mm is given by

$$d_{IJK} = \begin{pmatrix} 0 & 0 & 0 & 0 & d_{15} & 0 \\ 0 & 0 & 0 & d_{15} & 0 & 0 \\ d_{31} & d_{31} & d_{33} & 0 & 0 & 0 \end{pmatrix} \tag{6.23}$$

Assuming Kleinman symmetry ($d_{15} = d_{31}$) in the wavelength range of negligible absorption (*Kleinman 1962*) the nonlinear optical polarization $\boldsymbol{P}_{NL}$ at frequency 2ω then takes the form

$$P_{NL} = \varepsilon_0 \cdot \begin{pmatrix} 2d_{31}E_1^{\omega}E_3^{\omega} \\ 2d_{31}E_2^{\omega}E_3^{\omega} \\ d_{31}\left((E_1^{\omega})^2 + (E_2^{\omega})^2\right) + d_{33}(E_3^{\omega})^2 \end{pmatrix} \tag{6.24}$$

where $\boldsymbol{E}^{\omega} = (E_1^{\omega}, E_2^{\omega}, E_3^{\omega})$ denotes the optical field inside the film. The intensity of the generated second harmonic signal $\boldsymbol{I}^{2\omega}$ is proportional to $|P_{NL}|^2$. Therefore d_{31} can be measured directly with a fundamental beam polarized along the 2-direction (s-polarization) whereas for p-polarized light (polarization in the plane of incidence) d_{33} can be measured assuming $d_{33} = 3 \cdot d_{31}$. Figure 6.8 shows a typical second-harmonic response of a poled polymer film. For further details on this method, the reader is referred to Chapters 7.3.2 and 7.3.3.

According to Eqs. (3.28, 3.29) the electronic contributions to the electro-optic effect r_{IJK}^{e} can be directly calculated from the measured nonlinear optical coefficients d_{IJK}.

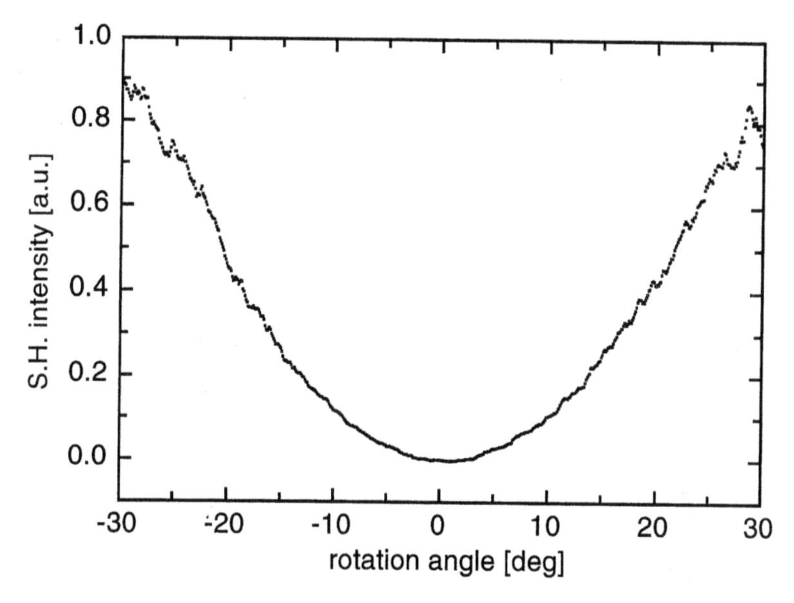

Figure 6.8 Typical second-harmonic response of a poled polymer film.

The electro-optic coefficients can be measured in waveguiding experiments either by an interferometric or a polarization rotation method (*Hayden et al. 1990*). A modulating voltage applied across two electrodes shifts the phase between travelling TE and TM modes. This technique has the disadvantage that it requires the polymer to be fabricated into a waveguide. A simple reflection method for measuring the electro-optic coefficient of poled polymers is reported by Teng et al. (*1990*). Its analysis has recently been improved (*Levy et al. 1993*).

6.2.4 Electro-optic devices

The properties of nonlinear optical polymers discussed in Table 8.5 make them promising candidates for integrated electro-optic devices. Provided that the aforementioned poling and stabilization procedures are transferable to the level of integrated optic devices such as channel waveguides, phase and intensity modulators or optical switches are realizable.

Multilayer structures including electrodes, buffer layers and the optically nonlinear polymer have to be tailored to an adequate form. Lytel and co-workers (*1989a*) solved this problem with their so-called selective poling procedure, SPP, as illustrated in Figure 6.9. Since the poled regions become birefringent, with the optical axis along the poling field, propagating waves polarized in this direction experience an index of refraction higher than the one of the unpoled regions. Therefore they can be confined in the lateral dimension. To complete the device the upper poling electrode is etched off and an upper buffer layer, including switching electrodes on top, is deposited.

Other authors use a UV-bleaching technique to lower the refractive index of the already poled polymer layer beside the desired waveguide patterns. Figure 6.10 depicts the sandwich construction of an integrated phase modulator, Figure 6.11 an intensity modulator. The top electrode that has been used for poling the polymer was removed before bleaching. The driving electrodes are processed by photolithographic techniques and wet chemical etching.

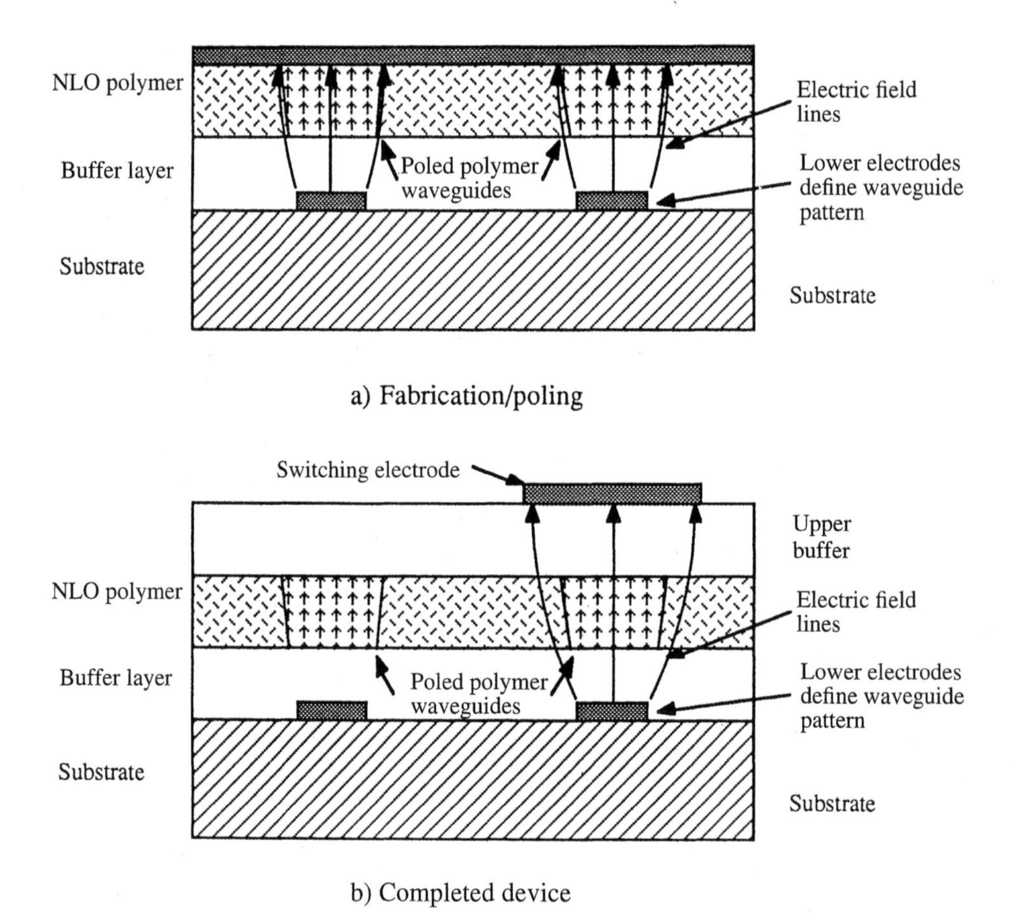

Figure 6.9 Selective poling procedures for active channel waveguides. *(Lytel et al, 1989a)* Reprinted by permission of Kluwer Academic Publishers.

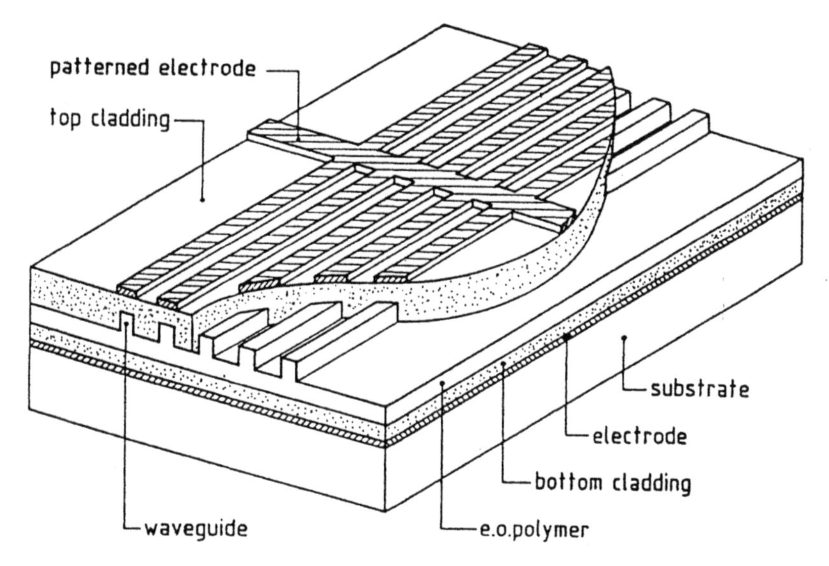

Figure 6.10 Schematic representation of an array of polymeric integrated phase modulators (from Möhlmann et al. 1990). Reprinted by permission of SPIE Publisher.

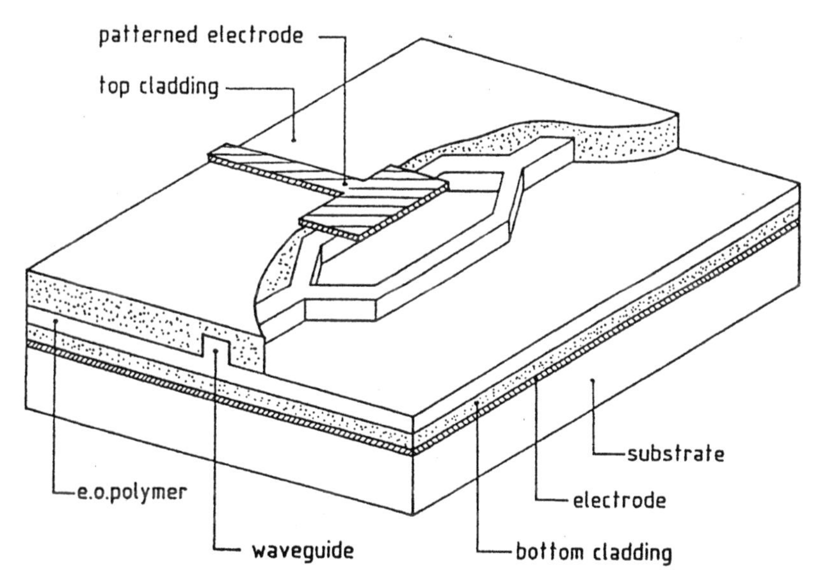

Figure 6.11 Schematic representation of an integrated Mach-Zehnder interferometer (from Möhlmann et al. 1990). Reprinted by permission of SPIE Publisher.

7. EXPERIMENTAL METHODS

This chapter presents a discussion on experimental methods for the optical, electro-optic and nonlinear optical characterization of molecular crystals. Corresponding techniques for LB films and polymers can be found in Chapters 5 and 6.

7.1 LINEAR OPTICAL EXPERIMENTS

The knowledge of the linear optical properties is essential for the characterization of materials for any kind of optical application. Whereas absorption measurements can be performed quite easily, the determination of refractive indices can become very troublesome, especially for organic crystals where, at the initial stage of research, the size of crystals with sufficient optical quality may be limited.

7.1.1 Absorption

Polarized transmission experiments are of fundamental importance for the characterization of materials. The absorption can be obtained from the transmission measurements using Eq. (7.1) (see e.g. *Schoenes 1984*). Multiple reflection from both surfaces are taken into account. Equation (7.1) is valid for the case of small absorption with $k/n < 0.1$ (complex refractive index $n_{\text{complex}} = n + ik$). Interference effects between reflected beams do not have to be considered in this evaluation because incoherent light is generally used for these measurements.

$$a(\lambda) = -\frac{1}{L} ln \left(\frac{-\frac{(1-R(\lambda))^2}{T(\lambda)} + \sqrt{\left(\frac{(1-R(\lambda))^2}{T(\lambda)}\right)^2 + 4R(\lambda)^2}}{2R(\lambda)^2} \right) \tag{7.1}$$

$$R(\lambda) = \frac{\left(n(\lambda) - 1\right)^2}{\left(n(\lambda) + 1\right)^2} \tag{7.2}$$

L: crystal thickness
$T(\lambda)$: measured transmission

Figure 7.1 shows an example of absorption for COANP (using a UV/VIS/NIR spectrometer (Perkin Elmer, Lambda 9)). The incoming beam propagating along the a-axis was polarized either along the b- or c-axis of the crystal. Measurements were carried out from 450 to 2,000 nm. The transmission range extends from 480 nm to 1,510 nm (50% points for crystals with $L = 1.37$ mm) and

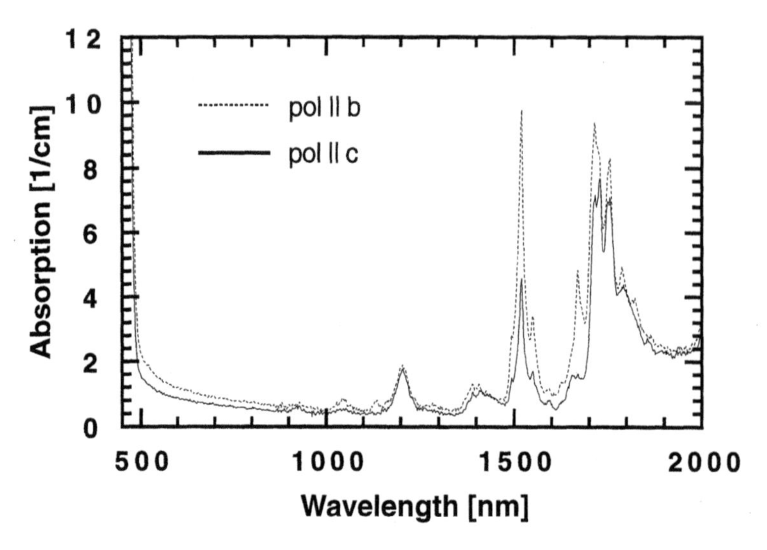

Figure 7.1 Absorption of COANP for light polarized either parallel to the b- or c-axis of the crystal. A polarization dependence is seen for wavelengths larger than 1,500 nm.

is practically the same for the two polarizations, with some differences above 1,500 nm due to vibrational excitations. The absorption coefficient is lowest at $\lambda \sim$ 1,350 nm where $\alpha_{min} \leq 0.5\ \text{cm}^{-1}$ (note that α is generally an upper limit because of scattering losses). Weak absorption bands appear near 1,200 nm and 1,400 nm.

Like many of the other potential materials with high second-order nonlinearities, COANP also absorbs in the blue part of the spectrum (see transparency-efficiency trade-off in Chapter 2).

7.1.2 Refractive indices

The knowledge of the refractive indices is most important for the evaluation of the suitability of a given material for nonlinear optical applications. These data can be used for the determination of phase-matching configurations for efficient frequency-doubling, sum- and difference-frequency generation and parametric amplification. Since most organic crystals are optically biaxial, three independent refractive indices generally have to be determined. Because of difficult sample preparation of the soft organic materials, many of the commonly used methods are not applicable. Table 7.1 lists some of the most often used methods for the determination of refractive indices applicable to organic crystals.

Depending on the circumstances (crystal size, solubility of crystals, range of refractive indices) different methods have to be used. Some of the techniques applied to molecular crystals will be described in the following discussion.

7.1.2.1 Minimum deviation method. The minimum deviation method is one of the most accurate techniques for determining refractive indices of transparent

Table 7.1 Methods to determine refractive indices of organic crystals. The references refer to articles describing the method and to published work where these techniques were applied to molecular crystals.

technique	advantages	disadvantages	references
• minimum deviation method	• can give very accurate results	• due to inherent cleavage planes, cutting and polishing of organic crystals (especially wedges) is often very difficult	• see e.g. *Hecht 1987a* • *Oudar et al. 1977b* • *Zyss et al. 1981* • *Ledoux et al. 1990* • *Bierlein et al. 1990*
• ellipsometry	• experimentally not very difficult	• crystal surfaces often not perfect enough rather difficult analysis	• *Azzam et al. 1977* • *Morita et al. 1987*
• Brewster angle	• experimentally not very difficult	• crystal surfaces often not perfect enough	• see e.g. *Hecht 1987b*
• method by Duc de Chaulnes	• simple experiment	• not very accurate in the case of birefringent materials • crystal surfaces often not perfect enough	• *Lawless et al. 1964* • *Deshpande et al. 1980*
• Abbe refractometer	• simple experiment	• need of high-quality surfaces • need of immersion fluids with high refractive indices • limited by the refractive indices of the prisms used	• *Born 1985*
• Becke line	• applicable to small samples	• solubility of crystal in immersion fluid • need of immersion fluids with high refractive indices • crystal alignment difficult	• see e.g. *Hartshorne et al. 1970, Wahlstrom 1969* • *Sutter et al. 1988a* • *Bosshard et al. 1989* • *Kerkoc et al. 1989* • *Kondo et al. 1989a*
• Jamin Lebedeff interferometer	• no immersion fluids required • applicable to small samples	• rather time consuming • crystal orientation difficult	• *Gahm 1962, 1963* • *Bosshard et al. 1989* • *Sutter et al. 1988a*
• interferometric	• easy method • also applicable for high refractive indices • often as grown crystal platelets can be used	• need of parallel plates (see also advantages)	• *Shumate 1966* • *Sutter et al. 1988b* • *Bosshard et al. 1989* • *Kerkoc et al. 1989* • *Bierlein et al. 1990* • *Sutter et al. 1991a* • *Knöpfle et al. 1993*

Table 7.1 Continued.

technique	advantages	disadvantages	references
• conoscopy	• surface quality less important	• critical optical adjustment • extensive analysis	• see e.g *Wahlstrom 1969* • *Bosshard et al. 1989* • *Sutter et al. 1991a*
• birefringence measurements	• easy measurement	• fails if the birefringence is strongly wavelength dependent	• *Sutter et al. 1988a*
• from minima of Maker-fringe or wedge measurements	• applicable for wavelengths in the near infrared (1–2 μm)	• refractive indices at one wavelength must be known • accuracy depends on combination of refractive indices used	• *Bosshard 1991a* • *Sutter et al. 1991a*

substances (see e.g *Hecht 1987a*). An appropriately cut prism is fixed on a rotation stage. A light beam incident on the prism will be deflected from its original direction by an angle δ, the angular deviation. The prism is then rotated leading to a variation of the deflection angle. At a certain angle, the angle of minimum deviation δ_m, the change in direction of the deflection is reversed. The knowledge of δ_m and the apex angle α of the prism immediately allows the determination of the refractive index n.

This technique has been applied to various molecular crystals such as methyl-(2,4-dinitrophenyl)-aminopropanoate (MAP, *Oudar et al. 1977b*), 3-methyl-4-nitropyridine-1-oxide (POM, *Zyss et al. 1981*), N-(4-nitrophenyl)-(L)-prolinol (NPP, *Ledoux et al. 1990*), 3-Methyl-4-methoxy-4'-nitrostilbene (MMONS, *Bierlein et al. 1990*). As was mentioned above, the cutting and polishing of organic crystals (especially wedges) is often very difficult due to inherent cleavage planes and is a difficult problem to overcome.

7.1.2.2 Becke line method. The Becke line method is based on the comparison of the refractive index of a crystal with the index of the fluid in which it is immersed using a polarizing microscope (see e.g *Hartshorne et al. 1970, Wahlstrom 1969*). A polarizer determines the refractive index to be measured, whereas the analyzer is removed from the light path. First the crystal is put into focus. Subsequently the distance between the microscope objective and the crystal is increased. In doing so a bright line, the Becke line, leaves the sample edge. The line either moves towards the crystal region (if $n_{crystal} > n_{immersion\ fluid}$) or away from it (if $n_{crystal} < n_{immersion\ fluid}$). The aim is now to find a fluid with the same refractive index as the crystal. In this case the Becke line is not moving at all and the refractive index of the sample is thus determined.

A problem may arise from the solubility of the crystals in the immersion fluids. In addition a whole set of such fluids with slightly varying refractive indices is needed. The Becke line method has been successful applied e.g. for the refractive indices n_c of 2-cyclooctylamino-5-nitropyridine (COANP, *Bosshard et al. 1989*), n_c of

2-methyl-4-nitro-N-methylaniline (MNMA, *Sutter et al. 1988a*), and n_y of 4-(N,N-dimethylamino)-3-acetamidonitrobenzene (DAN, *Kerkoc et al. 1989*).

7.1.2.3 Interferometric set-up after Jamin-Lebedeff. This interferometric set-up allows a measurement of the optical path length (refractive index × thickness) of crystals under the microscope (*Gahm 1962, 1963*). With known thickness the refractive index can be determined.

This method has also been applied to molecular crystals. Because of the high value of n_b of COANP and the lack of appropriate immersion fluids, this refractive index had to be measured with a modified interferometric technique of Jamin-Lebedeff. The difference in thickness of a slightly wedged crystal plate (difference in thickness Δd smaller than a few μm) was determined for a polarization of the incoming light for which the refractive index (here n_c) is known. The optical path difference Γ is measured by counting the interference fringes and by using a Senarmont compensator. Then the rotation of the polarization by 90° allows the determination of n_b. The same method has also been used with another organic material, 2-methyl-4-nitro-N-methylaniline (MNMA) (*Sutter 1988a*).

7.1.2.4 Birefringence measurements. High dispersion and absorption in the interesting spectral range can produce unexpected interference effects and can cause problems for measurements with polychromatic light. Polychromatic light may be used in e.g. birefringence measurements with a compensator between crossed polarizers in a microscope, or measurements of the delay time of short pulses in the material. In such measurements, where the group velocity instead of the phase velocity is determined, or where the zeroth order black interference fringe is used, the error Δ in the measured birefringence or refractive index is in first approximation (*Ehringhaus 1951, Hariharan 1985*)

$$\Delta = -\lambda_o \cdot \frac{\partial n}{\partial \lambda} \tag{7.3}$$

where λ_o is the central wavelength of the quasi-monochromatic light used and n the refractive index or birefringence to be determined. In organic materials the index of refraction is usually highly dispersive at wavelengths near the absorption edge and can vary by 10^{-3}/nm (for COANP (*Bosshard et al. 1989*)) or even by $3 \cdot 10^{-3}$/nm as was reported for MNMA (*Sutter et al. 1988a*). Thus the error Δ for a refractive index measured at 600 nm can be up to 0.6 or 1.8 if no special precautions are taken. Furthermore, Eq. (7.3) only accounts for a linear change of n in λ and not for higher order derivatives. They will not only cause a shift of the zero-order fringe, but also a decrease in contrast because the positions of order zero depend on the wavelength. Thus such birefringence measurements cannot be used for materials where the birefringence is strongly wavelength dependent.

7.1.2.5 Interferometric refractive index measurements. Shumate (1966) developed an interferometric method to determine large refractive indices of materials: a crystal plate with two parallel surfaces is mounted in one arm of a Michelson interferometer. It can be rotated around an axis perpendicular to the laser beam. The light is polarized parallel to this axis of rotation. When rotating the crystal away from its position of normal incidence by an angle θ, the change of the optical path length can be determined by counting the number m of interference fringes at the interferometer output (Figure 7.2). From m and θ the refractive index n can be determined using the formula

$$n=\frac{\alpha^2+2(1-\cos\theta)(1-\alpha)}{2[1-\cos\theta-\alpha]} \tag{7.4}$$

where $\alpha = (m \cdot \lambda)/(2 \cdot L_s)$, λ is the wavelength of incident radiation, L_s is the sample thickness, θ is the rotation angle away from normal incidence, and n is the refractive index of the sample.

Such experiments have been applied to various organic crystals (COANP, *Bosshard 1989*; PNP, *Sutter et al. 1988b*; DAN, *Kerkoc et al. 1989*; MMONS, *Bierlein et al. 1990*; N-(4-nitro-2-pyridinyl)-phenylalaninol (NPPA), *Sutter et al. 1991a*; 4'-nitrobenzylidene-3-acetamino-4-methoxyaniline (MNBA), *Knöpfle et al. 1994*).

The light can also be polarized perpendicularly to the rotation axis. In the case where the rotation axis is parallel to a dielectric axis but where the surface normal does not coincide with such an axis (Figure 7.3), the relation between m and θ is given by

$$m=\frac{2}{\lambda}\cdot L_s\left\{\sqrt{n^2(\phi_z-\theta')-n_o^2\sin^2(\theta)}-n_o\cos(\theta)-[n(0)-n_o]\right\} \tag{7.5}$$

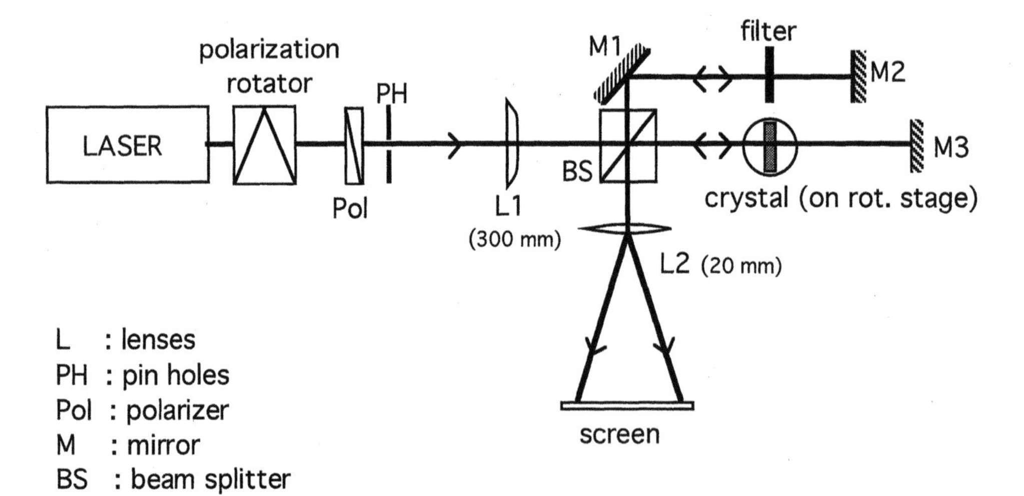

Figure 7.2 Experimental set-up for the determination of refractive indices in a Michelson interferometer.

$$n^2(\phi_z - \theta') = \frac{1}{\dfrac{\cos^2(\phi_z - \theta')}{n_x^2} + \dfrac{\sin^2(\phi_z - \theta')}{n_z^2}}$$

$$\theta' = arctg(1/x),\ x = \frac{1}{2a}\left(-b \pm \sqrt{b^2 - 4ac}\right) \qquad + \text{ for } \theta' > 0,\ - \text{ for } \theta' < 0$$

$$a = n_z^2 \cos^2(\phi_z) + n_x^2 \sin^2(\phi_z)$$

$$b = \sin(2\phi_z) \cdot (n_z^2 - n_x^2)$$

$$c = n_z^2 \sin^2(\phi_z) + n_x^2 \cos^2(\phi_z) - \frac{n_x^2 n_z^2}{n_o^2 \sin^2(\theta)}$$

n_o is the refractive index of the surrounding medium (usually air: $n_o = 1$). All parameters are shown in Figure 7.3. A theoretical fit of the number of counts m versus angle of incidence θ then yields ϕ_z, n_x and n_z (Figure 7.4).

The accuracy of the results is limited by the misorientation of the crystals, by the error in the parallelism of the surfaces of the samples, and by the crystal thickness. Fringe counts m up to 300 have been measured. As light sources several lines of Ar^+-, HeNe-, dye (Rhodamine 6G), Ti:Sapphire, and Nd:YAG-lasers within the spectral range from 488 to 1064 nm were used (see references in this section).

7.1.2.6 Conoscopic measurements. With conoscopy, crystals can be oriented and some crystal parameters (optic axis, position of index ellipsoid with respect to the crystallographic axes) can be determined. In the usual orthoscopic observation under a microscope one point in the object plane is projected onto one point of the retina of the observer. In the conoscopic set-up, however, all beams that are parallel in the object plane are projected onto one point of the retina. Therefore a projection from the wave vector space in the object plane on the direct space of

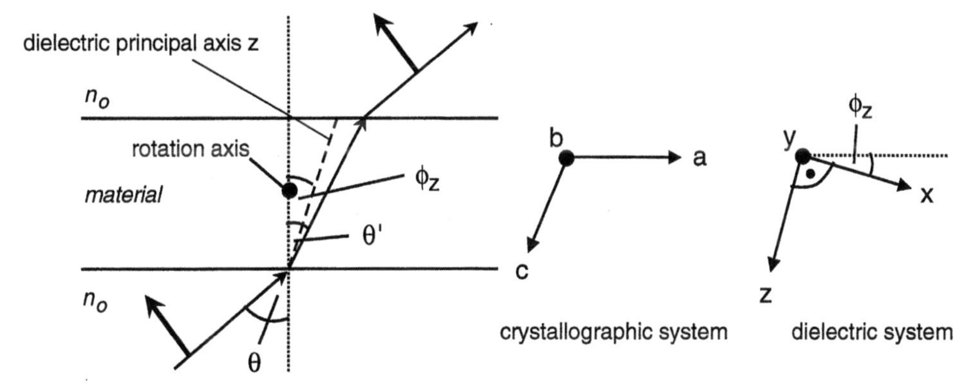

Figure 7.3 Example of possible crystallographic and dielectric systems in an interferometric experiment to determine refractive indices.

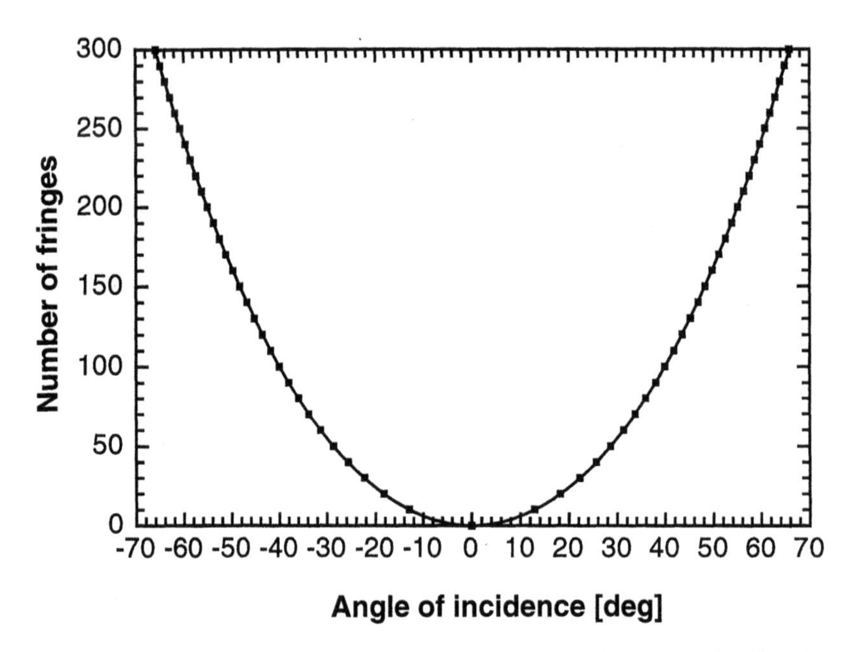

Figure 7.4 Number of interference fringes m versus angle of incidence for MNMA at λ = 1,064 nm (crystal thickness L = 965 μm, rotation axis, a, polarization of incoming beam in bc-plane (b parallel surface normal)). The light passed the crystal once only. With the knowledge of n_c, n_b could be obtained from a fit using Eq. (7.5) and $\phi_z = 0$ (unpublished).

the retina takes place (*Wahlstrom 1969*). An experimental conoscopic set-up is shown in Figure 7.5.

Conoscopy was applied in the case of e.g. COANP to determine the refractive index n_a because thin crystals with good surfaces perpendicular to the b- or c-axis could not be prepared (*Bosshard et al. 1989*). The number of isochromatic lines between the optical axis and the main axis (b) determined $n_a - n_b$. Knowing n_b we could calculate n_a. This method is more accurate than the determination measuring the optic angle.

Conoscopic methods were also used in the case of NPPA (*Sutter et al. 1991a*). The measurement of the angles corresponding to the optic and the principal axis of the index ellipsoid and the determination of the number of interference fringes between these two positions allowed the calculation of the refractive indices n_x and n_z. The authors also describe the additional difficulties in the evaluations if the main axis of the index ellipsoid does not coincide with the surface normal.

7.1.2.7 Dispersion relation for refractive indices. In order to describe the dispersion of the refractive indices, a model using several harmonic oscillators is used. These oscillators are more or less strongly excited by the electric light field. If one oscillator with frequency ω_o dominates, $n(\omega)$ is given by (*Born 1980a*)

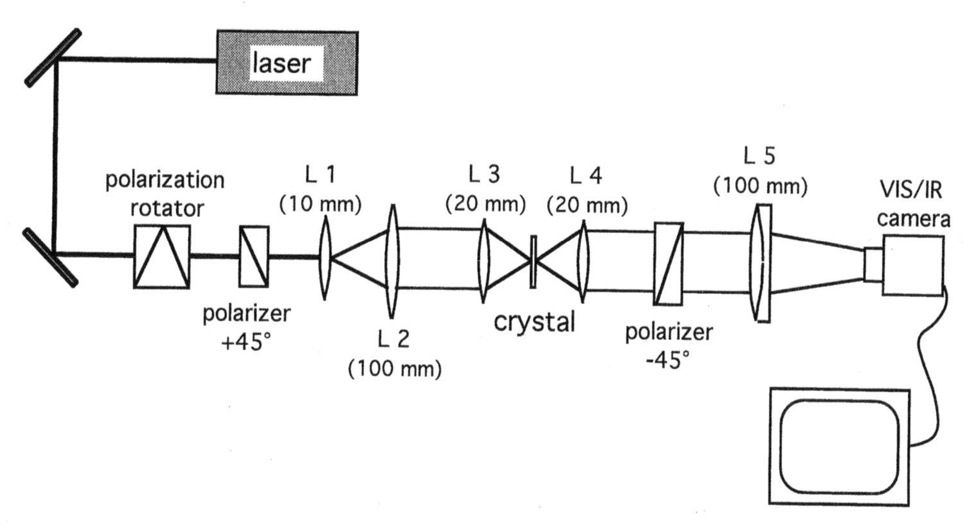

Figure 7.5 Experimental set-up for a conoscopic determination of the refractive indices.

$$n^2(\omega) - 1 = \frac{\omega_p^2 \gamma_o}{\omega_{eg}^2 - \omega^2} + A = \frac{p}{\omega_{eg}^2 - \omega^2} + A = \frac{q}{\lambda_{eg}^{-2} - \lambda^{-2}} + A \tag{7.6}$$

where ω_p is the plasma frequency ($\omega_p^2 = Ne^2/m\varepsilon_o$, N: oscillator density, e: oscillator charge, m: oscillator mass, ε_o: vacuum permittivity), γ_o is a local field parameter, $q = \omega_p^2\gamma_o/(2\pi c)^2$, ω_{eg} is the eigen frequency of the dominant oscillator, $\lambda_{eg} = 2\pi c/\omega_{eg}$, and A is a constant containing the contributions of all the other oscillators. p, ω_{eg}, q, λ_{eg}, and A are Sellmeier coefficients.

It should be noted that other authors use a different formula to describe the dispersion of the refractive indices (see e.g *Oudar et al. 1977b*):

$$n^2(\omega) = A + \frac{B \cdot \lambda^2}{\lambda^2 - C} - D \cdot \lambda^2 \tag{7.7}$$

where A, B, C, and D are also Sellmeier coefficients. For the case $\omega^2 \ll \omega_{eg}^2$ Eq. (7.6) can be written as

$$n^2(\omega) - 1 = \frac{\omega_p^2\gamma_o + \omega_{eg}^2 A - \omega^2 A}{\omega_{eg}^2 - \omega^2} \approx \frac{\omega_p^2\gamma_o + \omega_{eg}^2 A}{\omega_{eg}^2 - \omega^2} \equiv \frac{S_o^2\lambda_{eg}^2}{1 - \lambda_{eg}^2/\lambda^2} \tag{7.8}$$

which is the well-known Sellmeier formula. A plot $1/(n^2(\omega) - 1)$ vs $1/\lambda^2$ should result in a straight line. As can be seen from Figure 7.6 for the case of n_c of COANP this is not the case for most organic materials. We see that Eq. (7.8) is only a good approximation for wavelengths far enough away from the absorption edge. Most organics, however, have the absorption edges around 450 nm to 500 nm (e.g. λ = 480 nm for COANP (50% point for crystals with L = 1.37 mm)). Therefore one

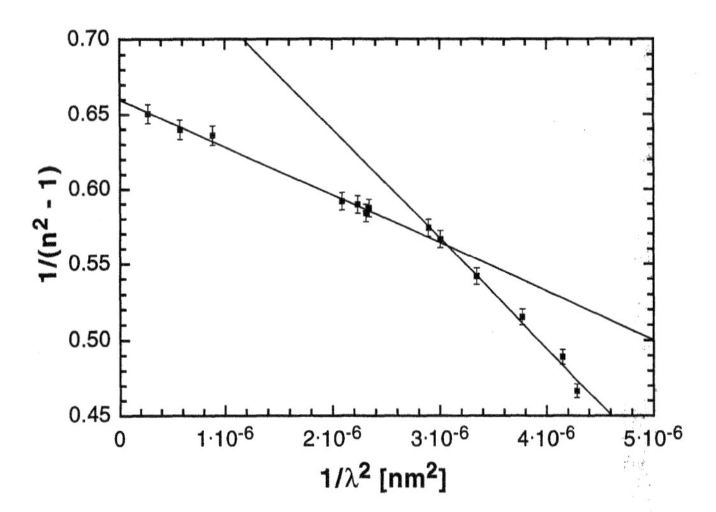

Figure 7.6 Plot of $1/(n^2(\omega) - 1)$ vs $1/\lambda^2$ for n_c of COANP. It is clearly seen that one straight line is not appropriate to describe the dispersion.

should use Eq. (7.6) to describe the dispersion of the refractive indices of many molecular crystals appropriately.

Figures 7.7a, b show examples of the dispersion for the refractive indices of COANP (*Bosshard et al. 1991a*) and NPPA (*Sutter et al. 1991a*). Sellmeier parameters for several crystals from fits using Eq. (7.6) are listed in Table 7.2.

7.2 DETERMINATION OF MOLECULAR NONLINEAR OPTICAL PROPERTIES

7.2.1 General remarks

There are several techniques to investigate microscopic nonlinear optical properties of molecules. Table 7.3 lists some of them with their advantages and disadvantages.

A detailed description of the solvatochromic method can be found (e.g. *Bosshard et al. 1992b*). Hyper-Rayleigh scattering is a very new method. Due to its simplicity and versatility it may well become the standard method for measuring second-order molecular susceptibilities (see section 7.2.3). Since, so far, electric field-induced second-harmonic generation (EFISH) is the standard technique, it will be described in more detail in the next section.

7.2.2 Nonlinear optics in solution: electric field-induced second-harmonic generation (EFISH)

The knowledge of the nonlinear optical second-order polarizabidlities β provides insight into molecular structures and functional origins of second-order optical

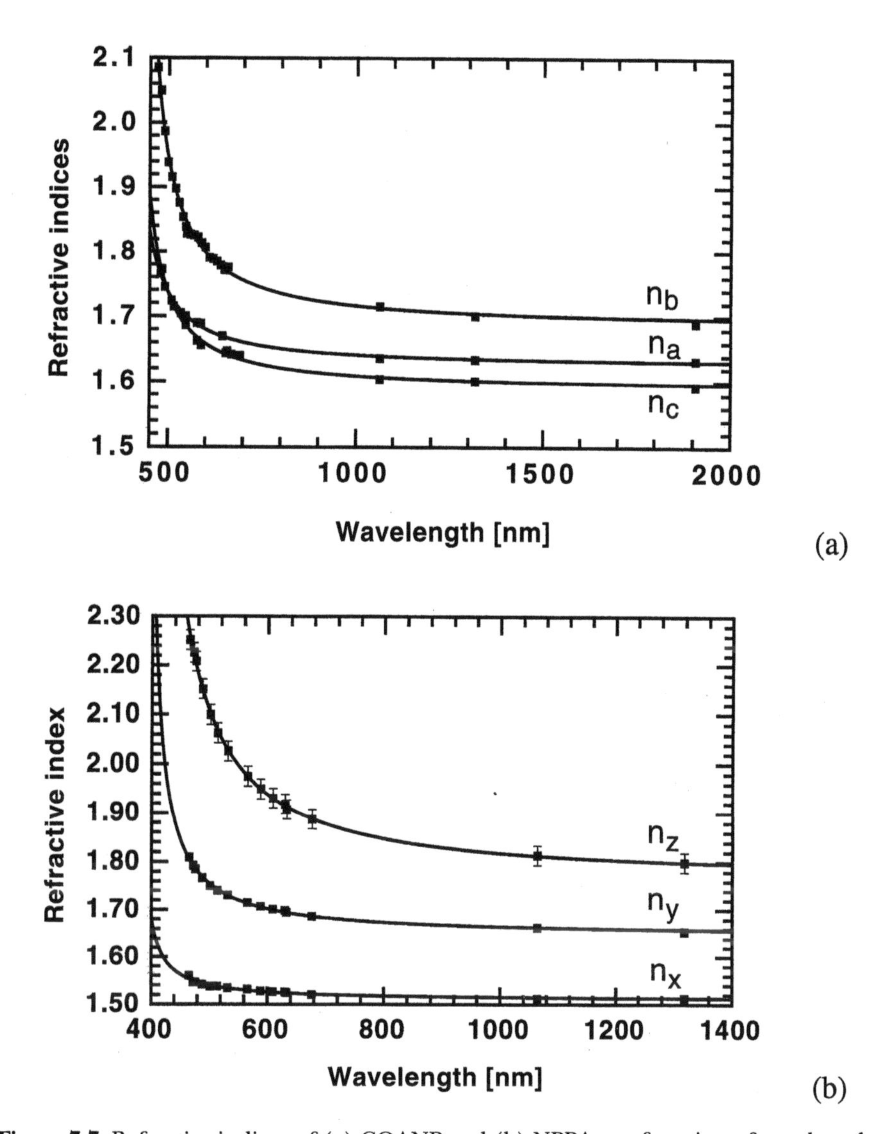

Figure 7.7 Refractive indices of (a) COANP and (b) NPPA as a function of wavelength.

effects. It is important for the optimization of organic crystalline nonlinear optical properties. Comparison of molecular and crystalline second-order susceptibilities allows the determination of the influence of intermolecular interactions on the nonlinearity.

The most frequently used method to measure nonlinear optical properties of molecules is the electric field-induced second-harmonic generation (EFISH) (*Levine et al. 1975*). As one is interested in the determination of molecular properties, one tries to exclude intermolecular contributions to the second-harmonic generating process. Therefore the measurement is carried out in dilute

Table 7.2 Dispersion parameters p, ω_{eg}^2, q, λ_{eg} and A of the refractive indices for different organic crystals. Data of refractice indices at selected wavelengths are also listed. x, y and z always refer to the dielectric axes (COANP and MNMA: dielectric system (x, y, z) = crystallographic system (a, b, c). The corresponding references can be found in the text.

material	refractive index	p [10^{30}s^{-2}]	ω_{eg}^2 [10^{30}s^{-2}]	q [10^{12}m^{-2}]	λ_{eg} [nm]	A	632.8 [nm]	1064 [nm]
COANP	n_x	5.6340	23.4856	1.5879	388.68	1.4090	1.672	1.639
	n_y	7.7778	20.1303	2.1921	419.83	1.4801	1.781	1.714
	n_z	5.1875	21.5238	1.4620	406.01	1.3038	1.647	1.608
MNMA	n_x	40.352	22.616	11.373	396.09	0.6797	2.148	1.936
	n_z	1.4886	20.2199	.41955	418.90	1.1798	1.520	1.506
PNP	n_x	24.908	24.188	7.0200	383.0	1.3454	1.990	1.880
	n_y	8.2990	22.110	2.3391	400.6	1.5658	1.788	1.732
	n_z	.64646	21.100	0.1822	401.6	1.0961	1.467	1.456
DAN	n_x	3.8600	26.186	1.0879	368.1	1.1390	1.539	1.517
	n_y	7.0458	22.938	1.9858	393.3	1.3290	1.682	1.636
	n_z	14.515	20.172	4.0908	419.4	1.5379	1.949	1.843
NPPA	n_x	1.6963	24.9629	0.47808	377.01	1.2151	1.524	1.511
	n_y	5.3450	23.4985	1.5064	388.58	1.5015	1.694	1.663
	n_z	18.5341	23.0861	5.2236	392.04	1.3548	1.907	1.813

solution, and a static electric field is applied in order to break the isotropic symmetry of the liquid.

The EFISH experiment is usually performed using a wedge Maker-fringe technique. By translating the wedged liquid cell across the beam Maker-fringe amplitude oscillations of the generated second-harmonic are obtained. The liquid cell consists of two glass windows (e.g. BK7) positioned between two stainless steel electrodes (Figure 7.8). As was pointed out by Levine and Bethea (*1975*), liquid cell considerations are important. The static field has to be uniform over the whole region of the liquid, in order to ensure that the full nonlinearity Γ_L is achieved at the glass-liquid boundary. If the electrodes were immersed in a liquid without glass walls almost no second-harmonic signal would be produced.

The macroscopic polarization $P^{2\omega}$ induced in a solution by an incident laser field E^{ω} is described by

$$P_I^{2\omega} = \varepsilon_o d_{IJK}(E^o) E_J^{\omega} E_K^{\omega} \tag{7.9}$$

where the components d_{IJK} of the nonlinear optical susceptibility tensor are dependent on the strength of the applied dc field E^o. ε_o is the vacuum permittivity. Assuming both the dc field and the polarization of the fundamental laser field to be parallel to the 3-axis (Figure 7.8), the only susceptibility component producing a second-harmonic signal is $d_{333}(E^o)$.

Table 7.3 Methods to determine microscopic nonlinear optical properties of molecules.

technique	advantages	disadvantages	references
electric field-induced second-harmonic generation (EFISH)	• widely used • the standard method • much data available • can be done quite fast	• solvents/solute effects not accounted for • sometimes problems with solubility	• *Levine et al. 1975* • *Oudar 1977a* • *Teng et al. 1983* • *Singer et al. 1981* • *Bosshard et al. 1992b*
method of solvatochromism	• no expensive laser sources needed	• many delicate assumptions necessary • gives mostly trends of the strengths of the nonlinearity	• see e.g *Amos et al. 1973* • *Paley et al. 1989* • *Bosshard et al. 1992b*
Langmuir-Blodgett films	• on water and substrate • fast scans of materials possible	• only useful for molecules with long aliphatic chains	• see e.g. *Shen 1984*
Hyper-Rayleigh scattering in solution	• easy measurement, no electric fields needed • ionic species measurable • several tensor components can be measured • no correction of local fields necessary • no knowledge of μ_g required • no knowledge of γ required	• relatively weak signals to be detected (incoherent scattering) • very good beam quality required	• *Clays et al. 1991* • *Clays et al. 1992* • *Verbiest et al. 1993*
vapor phase experiments	• no interaction between molecules	• experimentally very difficult and time consuming	• *to the authors' knowledge not yet performed with organic molecules*

For weak fields $d_{333}(E^o)$ becomes proportional to the external field. If the molecular z-axis is chosen to lie parallel to the ground state dipole moment μ_g, then $P_3^{2\omega}$ is written as

$$P_3^{2\omega} = \varepsilon_o \Gamma_L (E_3^{\omega})^2 E_3^o \tag{7.10}$$

with
$$\Gamma_L = \frac{d_{333}(E_3^o)}{E_3^o} = Nf^o (f^{\omega})^2 f^{2\omega}\left(\gamma + \frac{\mu_g \beta_z}{5kT}\right) \tag{7.11}$$

where N is the number density of the molecules and f^o, f^{ω} and $f^{2\omega}$ are local field factors evaluated at the indicated frequency. If Kleinman symmetry is assumed, the microscopic quantities γ and β_z are given by

$$\gamma = \frac{1}{5}\gamma_{iijj} \tag{7.12}$$

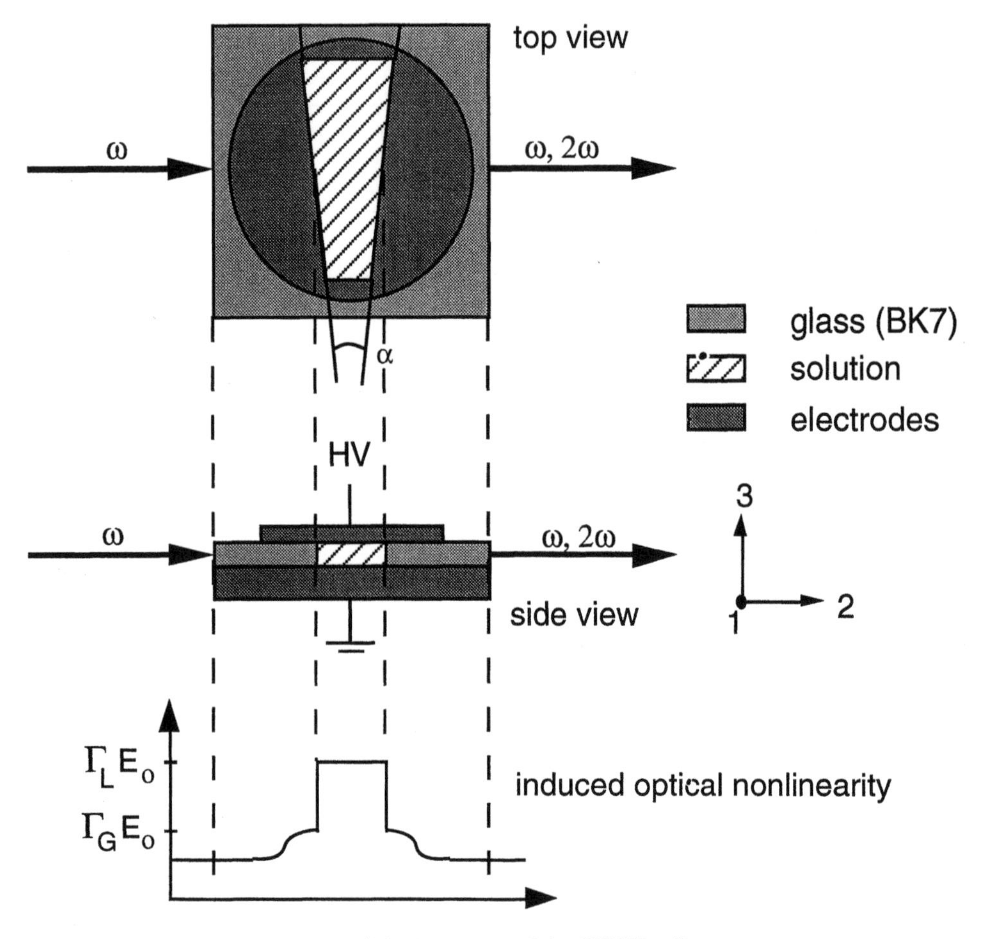

Figure 7.8 Geometry of the EFISH-cell.

$$\beta_z = \beta_{zzz} + \beta_{xxz} + \beta_{yyz}$$

where β_z is the vector part of the hyperpolarizability tensor β_{ijk} along the direction of the permanent dipole moment. Equation (7.11) means that besides β_z, the permanent dipole moment μ_g also has to be determined. This can be obtained by measuring the dielectric constant of the substance of interest dissolved in an apolar solvent and using the Debye-Guggenheim equation (*Guggenheim 1949*). The purely electronic effect described by the third-order polarizability $\gamma(-2\omega,\omega,\omega,0)$, however, can usually be neglected for strongly conjugated molecules with regard to the molecular reorientation contribution given by $\mu_g\beta_z/5kT$ (see e.g. *Singer et al. 1981*)

The intensity of the generated second-harmonic signal after the liquid cell is given by several authors (*Levine et al. 1975, Oudar 1977a, Teng et al. 1983*). The formula given here can be found in Oudar (1977a) and Bosshard et al. (1992b):

$$I_L^{2\omega}(L) = (I_L^{2\omega})_E \cdot f(L) = K\eta_L\,[T_{G/L}\Gamma_G l_c^G - T_L\Gamma_L l_c^L]^2\,(E^0)^2\,(I^\omega)^2 f(L) \qquad (7.13)$$

I^ω and $I_L^{2\omega}(L)$ are the intensities of the fundamental and of the generated second-harmonic wave, respectively. $(I_L^{2\omega})_E$ is the envelope of the interference fringes (if no absorption is present), and E^0 is the static field strength. The factor $\eta_L = (t^\omega)^4 \cdot (t^{2\omega})^2$ contains Fresnel transmission factors of the fundamental and the second-harmonic wave at the air/glass and the glass/air interface, respectively. The factors $T_{G/L}$ and T_L result from the electromagnetic boundary conditions at the glass/fluid and liquid/glass interfaces. l_c^G and l_c^L are the coherence lengths of the glass and the liquid, respectively, and are given by

$$l_c^G = \frac{\lambda}{4(n_G^{2\omega} - n_G^\omega)} \quad \text{and} \quad l_c^L = \frac{\lambda}{4(n_L^{2\omega} - n_L^\omega)} \qquad (7.14)$$

The constant K is given by

$$K = \frac{128}{\varepsilon_0 c\lambda^2}\ \text{(SI)} \qquad K = \frac{8192\pi^3}{c\lambda^2}\ \text{(esu)}$$

Generally $f(L)$ is given by

$$f(L) = \frac{1}{2}\,exp\left[-\left(\alpha^\omega + \frac{\alpha^{2\omega}}{2}\right)L\right]\cdot\left\{cosh\left[\left(\alpha^\omega - \frac{\alpha^{2\omega}}{2}\right)L\right] - cos\left(\frac{\pi L}{l_c^L}\right)\right\} \qquad (7.15)$$

describing the well-known oscillations as L is varied. α^ω and $\alpha^{2\omega}$ are the absorption coefficients at the corresponding wavelengths. In the case of negligible absorption $f(L)$ is reduced to

$$f(L) = sin^2\left(\frac{\pi L}{2l_c^L}\right) \qquad (7.16)$$

Figure 7.9 shows an example of an EFISH measurement performed with a COANP solution at λ = 1,318 nm (10 weight percent) (*Bosshard et al. 1992b*). The curve has been analyzed using a least squares fit of the form

$$I^{2\omega} = a_1\,sin^2\left(\frac{\pi L}{2a_3} + \frac{a_4}{2}\right) + a_2 \qquad (7.17)$$

where the parameters a_3 and a_4 are the coherence length and the phase offset, respectively. a_2 is the fringe minimum and a_1 is the fringe amplitude.

A quartz wedge is often used as reference material. In order to obtain the same intensity I^ω and the same confocal parameters in the quartz crystal as in the EFISH cell, the crystal is usually immersed in an index matching liquid (*Levine et al.*

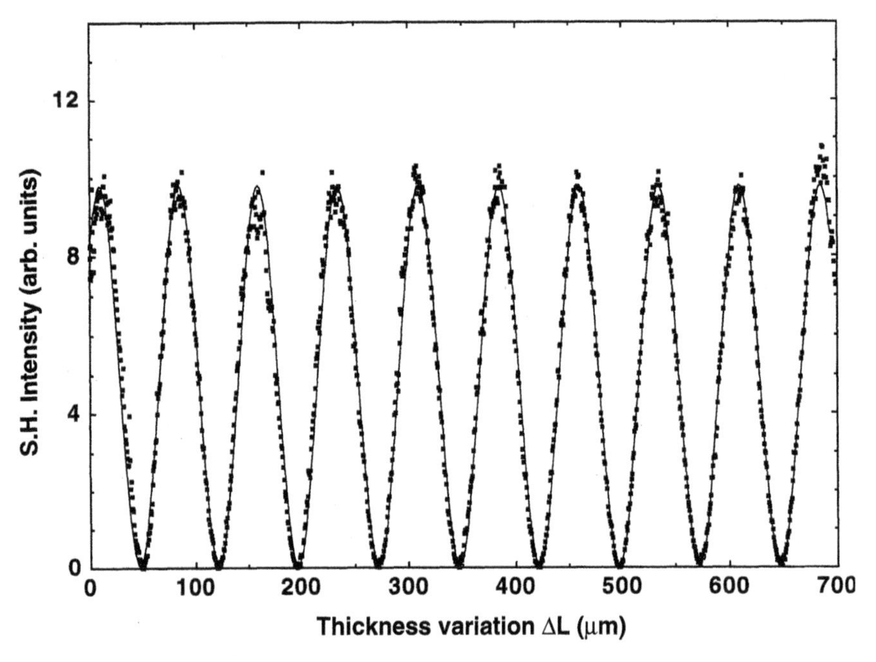

Figure 7.9 Example of an EFISH measurement performed with COANP at λ = 1,318 nm (10 weight percent solution). The curve has been analyzed using a least squares fit to the theoretical function Eq. (7.17).

1975). In a recent publication (*Bosshard et al. 1992b*) we proposed a slightly different configuration: the crystal was placed between two glass windows, identical to the ones used for the liquid cell. The quantitative description of this type of configuration is almost the same as for the one used by Levine (where the crystal is immersed in an index matching liquid inside a glass cell). The intensity of the generated second-harmonic signal in quartz is thus given by (*Levine et al. 1975*)

$$I_Q^{2\omega} = K\eta_Q \frac{d_{11}^2 (l_c^Q)^2}{(n_Q^\omega + n_Q^{2\omega})^2} (I^\omega)^2 \cdot \sin^2\left(\frac{\pi L}{2 l_c^Q}\right) \tag{7.18}$$

$$\text{where} \quad \eta_Q = C(T^\omega)^4 (T^{2\omega})^2$$

T^ω and $T^{2\omega}$ are products of Fresnel transmission factors for the fundamental and second-harmonic waves and C is a factor resulting from the electromagnetic boundary conditions. l_c^Q is the coherence length of quartz and K is the same constant as in Eq. (7.13). n_Q^ω and $n_Q^{2\omega}$ are the refractive indices of quartz at the fundamental and second-harmonic frequency. d_{11} is the nonlinear optical susceptibility of quartz (= 0.4 pm/V (*Landolt-Börnstein 1979b*)).

For a two-component solution, the macroscopic third-order nonlinearity Γ_L is the sum of solute (index 1) and solvent (o) contributions expressed as (*Singer et al. 1981*)

$$\Gamma_L = N_o f_o^o (f_o^\omega)^2 f_o^{2\omega} \gamma_o' + N_1 f_1^o (f_1^\omega)^2 f_1^{2\omega} \gamma_1' \quad 7.19$$

where the microscopic third-order nonlinearity γ' is defined as

$$\gamma' = \left(\gamma + \frac{\mu_g \beta_z}{5kT} \right) \quad (7.20)$$

Mostly pure Lorenz-Lorentz type local field factors with $n^\omega = n^{2\omega}$ are applied.

In order to minimize solvent-solvent and solute-solute interactions, Singer and Garito (*1981*) have developed an extrapolation procedure to infinite dilution. The nonlinearity Γ is measured for different concentrations (again using a reference for calibration). Using this extrapolation procedure the quantity γ_1' is expressed in terms of the concentration dependences of the nonlinearity, the dielectric constant, the specific volume and the refractive index of the solution. The terms containing the concentration dependence of the specific volume and of the refractive index can often be neglected because of their minor influence on γ_1' as compared to the other terms. γ_1' is then given by

$$\gamma_1' = \frac{81 M_1}{N_A (\varepsilon^o + 2)(n_o^2 + 2)^3} \left\{ v_o \left(\frac{\partial \Gamma_L}{\partial w} \right)_o + v_o \Gamma_o \left[1 - \frac{1}{\varepsilon^o + 2} \left(\frac{\partial \varepsilon}{\partial w} \right)_o \right] \right\} \quad (7.21)$$

where M_1 is the molar weight of the solute and N_A is Avogadro's number, w represents the weight fraction of the solute ($w = m/(m + m_o)$, m : mass of solute, m_o: mass of solvent) and v_o, n_o, and ε^o are the specific volume (volume/mass), the refractive index, and the dielectric constant of the pure solvent.

Fig. 7.10 shows an example of the concentration dependence of Γ_L for COANP and PNP at $\lambda = 1318$ nm.

If $\partial\Gamma/\partial\omega)_o$ and $\partial\varepsilon/\partial\omega)_o$ (and therefore μ_g (*Guggenheim 1949*)) are known and if γ can be neglected β_z can easily be obtained from Eq. (7.20). Data on β_z of different compounds can be found in Chapter 8.

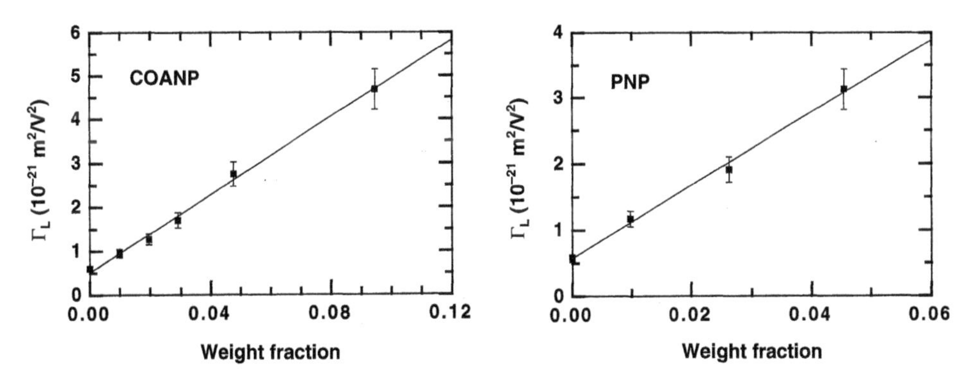

Figure 7.10 Concentration dependence of the liquid nonlinearity Γ_L for COANP and PNP measured at $\lambda = 1{,}318$ nm.

7.2.3 Hyper-Rayleigh scattering in solution

Hyper-Rayleigh scattering in solution (*Clays et al. 1991*) is a new technique to determine the hyperpolarizability of nonlinear optical molecules. It has several advantages over the EFISH method and will be discussed here briefly.

Incoherent second-order light scattering in a solution is mainly due to orientation fluctuations of the dissolved nonlinear optical molecules. The intensity of the second-harmonic scattered light depends on the orientational correlation

$$\langle \beta_{uvw,1}(-2\omega,\omega,\omega),\ \beta_{xyz,2}(-2\omega,\omega,\omega)\rangle_{av} \tag{7.22}$$

for molecules at two different volumes 1 and 2. The subscript $_{av}$ indicates time averaging of the motion of the particles over one cycle of the optical field. Assuming that there is no coupling between neighboring volume elements and the rotational motion of the dissolved molecules is slow compared to the optical frequencies, the intensity of the second harmonic scattered light is proportional to the number density N and the orientational correlation in Eq. (7.22). As mentioned above, for molecules with strong nonlinearities along a single charge transfer axis, by far the largest contribution to Eq. (7.22) is the term β^2_{zzz}. The total intensity of the second harmonic scattered light can then be written as

$$I_{sc}^{2\omega} = g\sum_s N_s\beta^2_{zzz,s}I_0^2 \tag{7.23a}$$

The proportionality factor g contains the conversion from the molecular to the laboratory coordinate frame (averages of direction cosines over all directions) and local field factors at optical frequencies similar to Eq. (3.25). The summation is performed over all species s. Since no external static field has to be applied as in the EFISH configuration, only local field factors at optical frequencies are involved, which are most often taken to be the same as in Eq. (3.27). In addition no dipole moments and third-order hyperpolarizabilities have to be known compared to Eq. (7.20).

For the solvent-solute system used in the experiment, Eq. (7.23a) takes the form

$$I_{sc}^{2\omega} = g\left(N_{solvent}\beta^2_{zzz,solvent} + N_{solute}\beta^2_{zzz,solute}\right)I_0^2 \tag{7.23b}$$

For low concentrations of solute molecules $N_{solvent}$ is nearly constant. Measurements of the scattered light intensity at different solute concentrations therefore show a linear dependence of $g(N_{solvent}\beta^2_{zzz,solvent} + N_{solute}\beta^2_{zzz,solute})$ on N_{solute}. From the slope and the intercept $\beta_{zzz,solute}$ can then be calculated if the solvent hyperpolarizability $\beta_{zzz,solvent}$ is known.

Experimental data for para-nitroaniline (p-NA) dissolved in methanol are shown in Figure 7.11 (*Clays et al. 1991*).

The β-value for p-NA extracted with this method is in good agreement with previously reported values from EFISH experiments in different solvents, cf. Table 8.1.

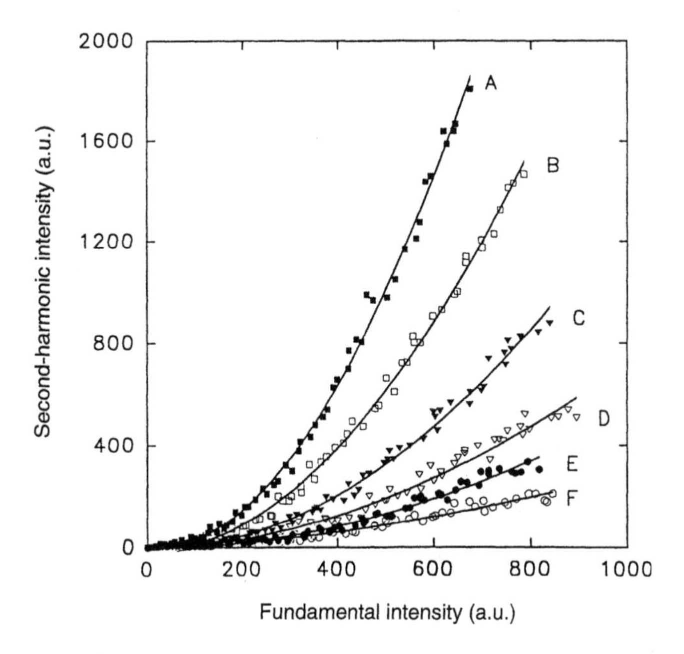

Figure 7.11a Second-harmonic scattered light intensity for p-NA in methanol at 293 K at different number densities in units of $10^{18}cm^{-3}$: curve A(92), B(46), C(23), D(9.2), E(4.6), F(1.8). The solid lines are fitted curves (Eq. (7.23b)). Reprinted by permission of APS Publishers.

The experimental set-up is quite simple. Besides the laser source, all one needs is a properly designed cell and a highly efficient condensor system (*Clays et al. 1992*).

7.3 DETERMINATION OF NONLINEAR OPTICAL PROPERTIES OF MOLECULAR CRYSTALS

7.3.1 General remarks

As in the case of molecules there are several techniques to investigate macroscopic nonlinear optical properties. Table 7.4 lists some of them with their advantages and disadvantages. Note that recently the powder test technique was improved (*Kiguchi et al. 1992*) to get rather accurate results also for non-phase-matchable compounds.

The most often used method applied to molecular crystals is the Maker-fringe technique. A detailed description is therefore given in section 7.3.2.

7.3.2 Maker-fringe method

The nonlinear optical coefficients of organic materials are most often measured with the Maker-fringe technique, which was introduced in 1962 (*Maker et al. 1962*).

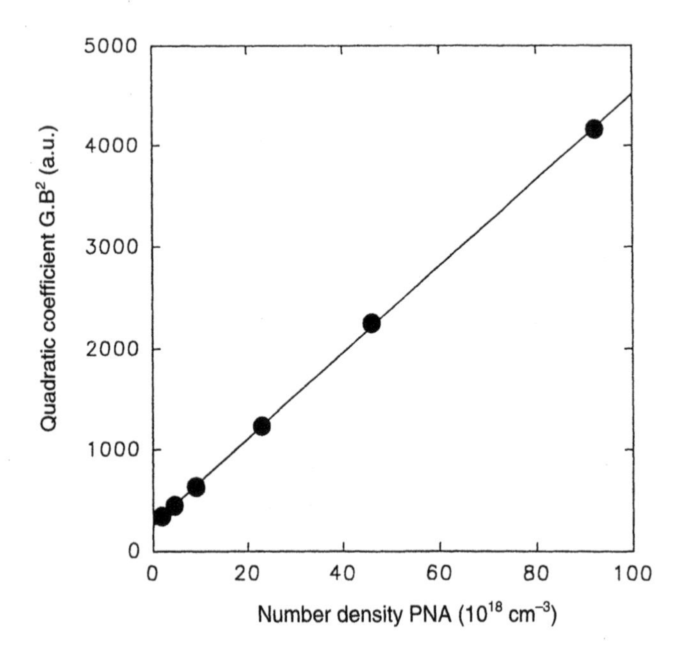

Figure 7.11b Quadratic coefficient $GB^2 = g(N_{solvent}\cdot\beta^2_{solvent} + N_{solute}\cdot\beta^2_{solute})$, obtained from the curves in Figure 7.11a vs N_{solute}. Reprinted by permission of APS Publishers.

This method is well suited for many organic materials because as grown crystals often possess parallel faces due to cleavage planes, which also facilitates the polishing of the samples. Using the Maker-fringe technique a plane parallel sample is rotated around an axis perpendicular to the incoming laser beam (Figure 7.12). As can be seen from Eq. (3.13) the intensity of the second-harmonic wave $I^{2\omega}$ generated in the sample shows oscillations due to different phase velocities of the fundamental and frequency-doubled beams in the material. For the exact evaluation of the nonlinear optical susceptibilities, transmission factors have to be taken into account in Eq. (3.13) (see below). The envelope of such curves is evaluated at zero-degree incidence and compared to the Maker-fringes of a reference crystal, quartz in most cases, whose nonlinear optical coefficients are known. An experimental set-up is shown in Figure 7.12 (*Bosshard 1991a*).

It should be noted that there is still no reliable absolute scale for nonlinear optical coefficients, although new attempts have been made very recently (*Kurtz 1990, Eckhardt et al. 1990*). One of the best choices for a reference crystal is quartz because its growth, handling and stability are under control. Its absolute nonlinear optical susceptibilities are, however, still not well established. In our laboratory we use a value for quartz of $d_{11} = 0.4$ pm/V (*Landolt-Börnstein 1979b*).

The application for optically isotropic and uniaxial crystals was discussed in detail by Jerphagnon and Kurtz (*1970*). The complete theory for optically biaxial crystals as in the case of many organics was developed by Bechthold (*1976*).

The intensity of the second-harmonic signal is of the form (see e.g. *Kurtz 1972*)

Table 7.4 Methods to determine macroscopic nonlinear optical properties.

technique	advantages	disadvantages	remarks	references
powder test	• widely used • allows fast screening	• gives only order of magnitude	• very useful to decide if a material is centrosymmetric or not	• *Kurtz et al. 1968* • *Kiguchi et al. 1992*
Maker-fringe	• widely used • most versatile • no absolute power measurement required • the standard method for organic crystals	• works best with thin plates which can sometimes be a disadvantage	• yields results relative to a reference crystal • works best with thin plates	• *Maker et al. 1962* • see e.g. *Kurtz 1972* • *Bechthold 1976* • see also text below
wedge method	• no absolute power measurement required	• wedge-shaped crystal required • not very useful for organics	• yields results relative to a reference crystal	• see e.g. *Kurtz 1972* • *Boyd et al. 1971*
phase-matching	• provides absolute results	• exact knowledge on laser beam parameters and quality needed • absolute power measurements • crystalline optical quality critically important	• only to be used for phase-matchable materials	• see e.g. *Kurtz 1972*
spontaneous parametric emission	• provides absolute results with no absolute power measurements • laser beam quality not critical	• experimentally the most difficult method	• seldom used	• see e.g. *Kurtz 1972*

$$I^{2\omega}(\theta) = 4AB\sin^2\psi + (A-B)^2 \tag{7.24}$$

where $\psi = \pi L/2l_c(\theta)$ and A and B are constant for a specific angle of incidence θ. The first part of the equation describes the well-known interference pattern originating from different phase velocities of the fundamental and frequency-doubled beam in the material, whereas the second one is independent of the crystal thickness L and is usually some orders of magnitude smaller than the first one in the case of nonabsorbing materials. It is therefore usually neglected.

For weakly absorbing materials ($\alpha \leq 10^3$ cm^{-1}) Eq. (7.24) changes to (for perpendicular incidence)

$$I^{2\omega}(0) = 4ABe^{-(\alpha_\omega+\alpha_{2\omega}/2)L}\sin^2\psi + (Ae^{-\alpha_\omega L} - Be^{-\alpha_{2\omega}L/2})^2 \tag{7.25}$$

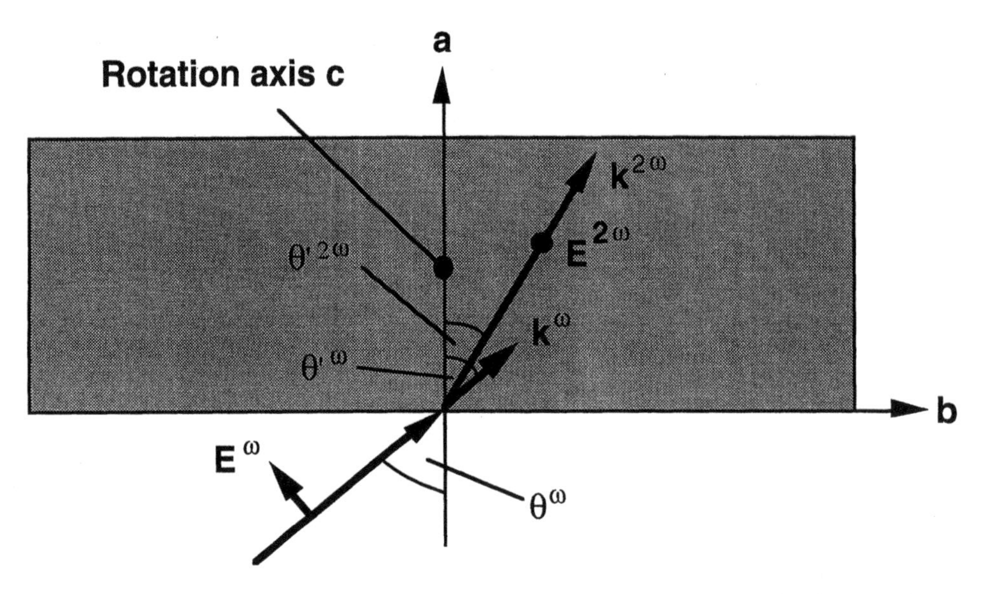

Figure 7.12 Configuration for nonlinear optical experiments using the Maker-fringe technique. The geometry for the determination of the nonlinear optical coefficient d_{32} of COANP (rotation axis c) is shown.

If $\alpha^{\omega}L \ll 1$ and $\alpha^{2\omega}L \ll 1$ the non-oscillating term in Eq. (7.25) can again be neglected. A thorough analysis of measurements in absorbing materials is given by Chemla et al. (*1971a, b*).

Thus for nonabsorbing materials we have (see also Eq. (3.13))

$$I^{2\omega}(\theta) = I_E(\theta)\sin^2\left(\frac{\pi L}{2l_c(\theta)}\right) \tag{7.26}$$

$$I_E(\theta) = \frac{2^5}{\varepsilon_o c\lambda^2} d_{eff}^2(\theta)\, l_c^2(\theta)\eta^{-1}(\theta)\,(I^{\omega})^2 \tag{7.27}$$

I_E describes the envelope of the Maker interferences. It contains the effective nonlinear optical coefficient, the coherence length $l_c(\theta)$, the fundamental intensity and transmission factors of fundamental and second-harmonic waves $\eta(\theta)$. Multiple reflection $R(\theta)$ and Gaussian beam corrections $B(\theta)$ are also included in $\eta(\theta)$. In many experiments we have $B(0) = 1$. Figure 7.14 shows examples of Maker-fringe curves for COANP and quartz (*Bosshard 1991a*).

Multiple reflection losses for $n \leq 2$ can be approximated by

$$R(0) \approx 1 + \left(\frac{n_{2\omega} - 1}{n_{2\omega} + 1}\right)^4 \quad (= 1.01 \text{ for } n = 2) \tag{7.28}$$

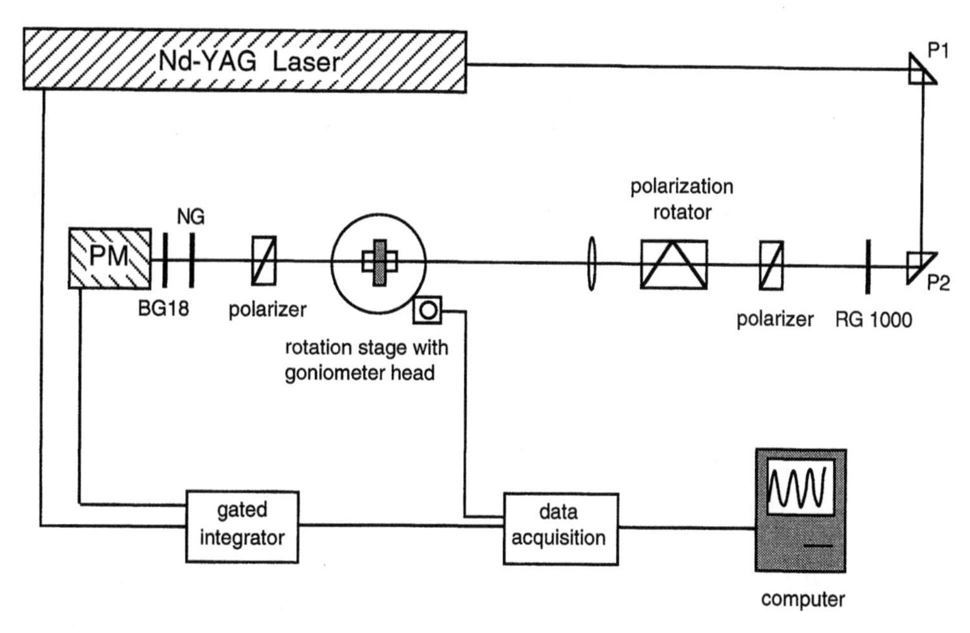

Figure 7.13 Experimental set-up for Maker-fringe measurements.

One can see that $R(0) \approx 1$ for most organics (see this chapter for refractive indices). The formula holds as long as the thickness of the crystal varies more than $\lambda/(8n)$ (= 67 nm for $n = 2$, $\lambda = 1{,}064$ nm) but less than $l_c(0)/2$ (~1.4–17 μm for our crystals) across the illuminated area to still allow oscillations of $I^{2\omega}$ (*Jerphagnon et al. 1970*). This means that the crystal should be a plane-parallel slab but not an interferometer.

As mentioned above, the reference crystal is often an α-quartz (point group 32, nonlinear optical coefficient d_{11} = 0.4 pm/V (*Landolt-Börnstein, 1979b*)). The corresponding formula for $I^{2\omega}$ of quartz is (*Jerphagnon et al. 1970*, SI units)

$$I^{2\omega}(\theta) = \frac{8}{\varepsilon_o c}\frac{1}{\left((n^{\omega})^2 - (n^{2\omega})^2\right)^2} d_{eff}^2 (I^{\omega})^2 \left[t^{\omega}(\theta)\right]^4 T^{2\omega}(\theta) sin^2\left(\frac{\pi L}{2l_c(\theta)}\right) = I_E(\theta) sin^2\left(\frac{\pi L}{2l_c(\theta)}\right) \tag{7.29}$$

$$I_E(\theta) = \frac{2^5}{\varepsilon_o c\lambda^2} d_{eff}^2(\theta)\, l_c^2(\theta) \eta^{-1}(\theta)\, (I^{\omega})^2 \tag{7.30}$$

where the Gaussian beam correction and multiple reflection factors have again been neglected. $t^{\omega}(\theta)$ and $T^{2\omega}(\theta)$ are transmission factors at the fundamental and the second-harmonic frequency, respectively. $d_{eff} = d_{11}\, cos^3(\theta'^{\omega})$ is the effective nonlinear optical coefficient for the geometry used in the experiments. θ'^{ω} is the internal angle at frequency ω. Combining Eq. (7.26) and Eq. (7.29) then leads to

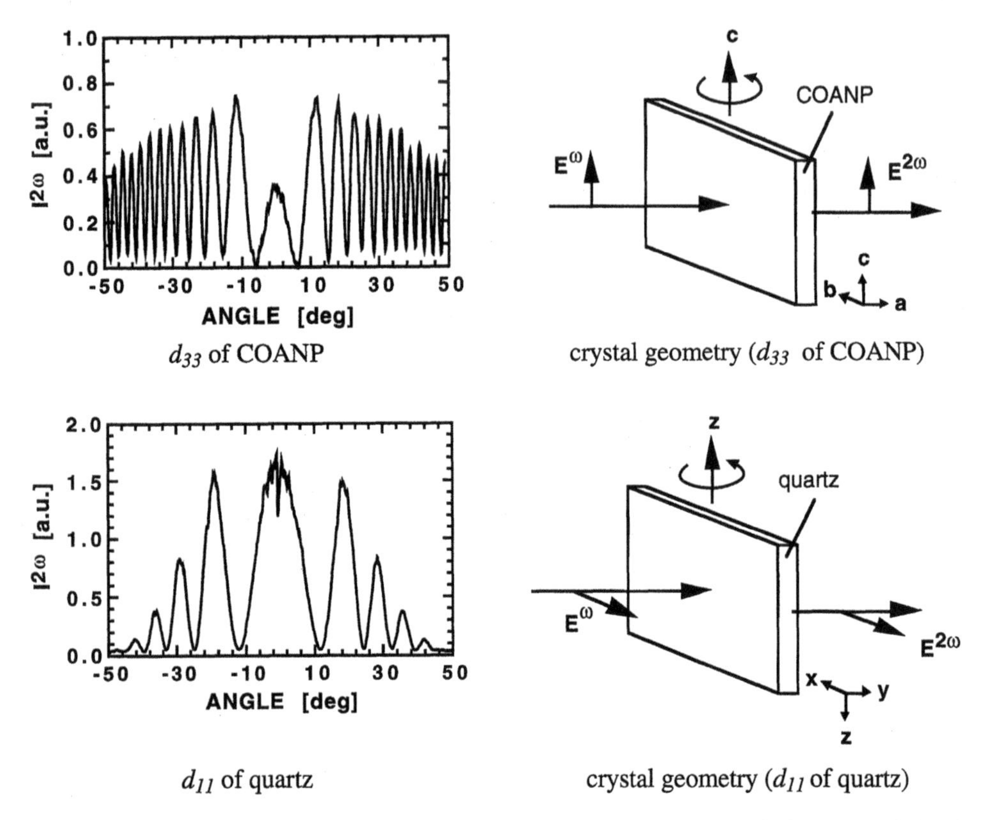

Figure 7.14 Maker-fringe curves of d_{33} of COANP (crystal thickness $L = 1.37$ mm) and d_{11} of quartz (crystal thickness $L = 2.008$ mm) at $\lambda = 1{,}318$ nm.

$$d_{IJ} = \sqrt{\frac{I_E(0)\eta(0)}{I_E^Q(0)\eta^Q(0)}\frac{l_c^Q(0)}{l_c(0)}}\,d_{11}^Q \tag{7.31}$$

$$l_c = \frac{\lambda}{4\left|n^{2\omega} - n^{\omega}\right|} \tag{7.32}$$

for the determination of the nonlinear optical coefficients d_{IJ} for perpendicular incidence. $I_E(0)$ and $I_E^Q(0)$ are the envelopes of the Maker-fringes, $\eta(0)$ and $\eta^Q(0)$ include the correction for reflection losses at the interfaces, and $l_c(0)$ and $l_c^Q(0)$ are the coherence lengths, all at zero-degree incidence. In the case of absorption at either the fundamental or second-harmonic wave, $I_E(0)$ has to be multiplied by

$$exp\left[\left(\alpha^{\omega} + \frac{1}{2}\alpha^{2\omega}\right)L\right] \tag{7.33}$$

where α^{ω} and $\alpha^{2\omega}$ are the absorption coefficients at the corresponding wavelengths and $\alpha^{\omega}L \ll 1$, $\alpha^{2\omega}L \ll 1$ (L: sample thickness).

It should be noted that often scattering from the surface or from impurity centers within the crystal leads to a reduced transmission in some parts of the visible spectrum. More accurate values can often be obtained by correcting the transmission factors T (fresnel factors), by multiplying Eq. (7.31) by

$$\frac{T^{\omega}}{T^{\omega}_{meas}} \cdot \sqrt{\frac{T^{2\omega}}{T^{2\omega}_{meas}}}, \qquad T = \frac{4n}{(n+1)^2} \tag{7.34}$$

where T_{meas} is the measured transmission.

Maker-fringe curves can be evaluated at any angle of incidence. The theoretical analysis just becomes more elaborate since the angle of dependence has to be explicitly taken into account, resulting in complicated formulae. For different orientations between dielectric and crystallographic axes the analysis may have to be carried out in that way (*Bechthold 1976, Sutter et al. 1991a*).

7.3.3 Maker-fringe method for thin polymeric films

As long as the change in the optical path length during the rotation of a thin film is small compared to one coherence length the oscillating term in Eq. (7.29) can be expanded to (*Singer 1986*)

$$sin^2\left(\frac{\pi L}{2l_c(\theta)}\right) = \left(\frac{\pi L}{4l_c} \frac{(n_{\omega} + n_{2\omega})}{\left(\frac{1}{2}(n_{\omega}^2 + n_{2\omega}^2) - sin^2\theta\right)^{1/2}}\right)^2 \tag{7.35}$$

Figure 6.8 is an illustrative example: no fringes are visible for a thin film. The main advantage of this expansion is the cancellation of all terms containing differences of refractive indices in the expression for the second-harmonic intensity. Inserting Eq. (7.32) for the coherence length into Eq. (7.35) the term in the nominator cancels exactly with the prefactor in Eq. (7.29), that is

$$I_{2\omega}(\theta) = \frac{8}{\varepsilon_0 c} d_{eff}^2 \left[t^{\omega}(\theta)\right]^4 T^{2\omega}(\theta)\left[t^{2\omega}(\theta)\right]^2 I_{\omega}^2 \left(\frac{1}{n_{2\omega}^2 - n_{\omega}^2}\right)^2 sin^2\psi$$

$$\approx \frac{8}{\varepsilon_0 c} d_{eff}^2 \left[t^{\omega}(\theta)\right]^4 T^{2\omega}(\theta)\left[t^{2\omega}(\theta)\right]^2 I_{\omega}^2 \cdot \left(\frac{1}{n_{2\omega}^2 - n_{\omega}^2}\right)^2 \cdot \left(\frac{\pi L}{\lambda}\right)^2 \cdot \frac{(n_{\omega} + n_{2\omega})^2 (n_{\omega} - n_{2\omega})^2}{\frac{1}{2}(n_{\omega}^2 + n_{2\omega}^2) - sin^2\theta}$$

$$\approx \frac{8}{\varepsilon_0 c} d_{eff}^2 \left[t^{\omega}(\theta)\right]^4 T^{2\omega}(\theta)\left[t^{2\omega}(\theta)\right]^2 I_{\omega}^2 \cdot \left(\frac{\pi L}{\lambda}\right)^2 \cdot \frac{1}{\frac{1}{2}(n_{\omega}^2 + n_{2\omega}^2) - sin^2\theta} \tag{7.36}$$

Therefore the refractive indices do not have to be known very accurately for the determination of the nonlinear optical coefficients. Since polymer films are coated on a substrate an extra factor $t^{2\omega}(\theta)$ for the transmission of the second-harmonic wave from the substrate to air has to be taken into account.

Data on nonlinear optical coefficients of various molecular crystals and polymers are summarized in Chapter 8.

7.4 ELECTRO-OPTICS ON MOLECULAR CRYSTALS

These are basically two experimental techniques to investigate electro-optic effects in crystals. Both of them will be discussed in the following sections.

7.4.1 Interferometric measurement of the electro-optic coefficients

A phase modulation technique using Michelson and Mach Zehnder interferometers (Figure 7.15, see e.g. *Sigelle et al. 1981*) is often used to measure electro-optic coefficients of crystals. In the experiment shown in Figure 7.15 a lens (f = 500 mm) focuses the laser beam onto the crystal. A second lens is used to widen up the laser beam in front of the detector. The laser beam passes through the crystal once (Mach Zehnder) or twice (Michelson interferometer). For the detection of the intensity changes in the interference pattern, photomultipliers and photodiodes can be used. A lock-in amplifier is used for phase-sensitive detection of the field-induced phase shift. Electrodes can either be made of silver paste, or evaporated metals, or the sample can be mounted between two parallel metal plates with areas exceeding those of the samples. With both arrangements the same results can be obtained (*Bosshard et al. 1993*).

Two monochromatic plane waves of the same polarization with intensities I_1 and I_2 are combined at the output of a Michelson interferometer where they interfere

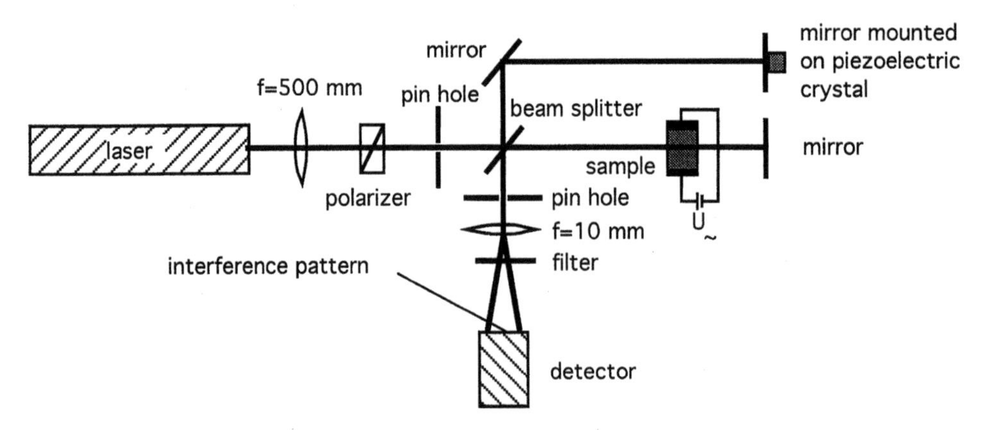

Figure 7.15 Michelson interferometer used for the determination of electro-optic coefficients.

(Figure 7.15) (electric field along J, polarization along I). If one of the beams passes the crystal, the resulting intensity of the interference pattern at the output is given by

$$I = I_1 + I_2 + 2\sqrt{I_1 \cdot I_2}\, cos\Delta\Phi \tag{7.37}$$

where

$$\Delta\Phi = \Delta\Phi_o + \Delta\Phi(E) = \frac{2\pi}{\lambda}\left(\Delta L + (n_I - 1)L\right) - \frac{\pi}{\lambda} n_I^3\, r_{IJ} L E_J. \tag{7.38}$$

n_I is the appropriate refractive index, L is the length of the beam path in the crystal (equal to one or two times the crystal thickness for perpendicular incidence) and ΔL is the difference in length between the two interferometer arms. ΔL can be adjusted with the piezoelectric mirror. For a preset ΔL of $2\pi/\lambda(\Delta L + (n_I - 1)L) = (2m - 1/2)\pi$, $m = 0, \pm 1, \pm 2$, the application for a weak sinusoidal modulation voltage U

$$U = U_o\, sin(2\pi\nu t) \quad \text{with} \quad U_o \ll \frac{d\lambda}{\pi n_I^3\, r_{IJ} L} = \frac{V_\pi}{\pi} \tag{7.39}$$

d: electrode distance $\quad$ V_π: half-wave voltage

leads to

$$I = I_1 + I_2 - 2\sqrt{I_1 \cdot I_2} \cdot \frac{\pi}{\lambda} n_I^3\, r_{IJ} \frac{L}{d} U_o\, sin(2\pi\nu t) \tag{7.40}$$

With the two following definitions (Figure 7.16)

$$\begin{aligned} &\Delta I \equiv I_{\max} - I_{\min} = 4\sqrt{I_1 \cdot I_2} && \text{intensity contrast} \\ &I_{eff} = \frac{\Delta I}{2}\frac{\pi}{\lambda} n_I^3 \left|r_{IJ}\right| \frac{L}{d} U_{eff}, \quad U_{eff} \equiv \sqrt{\langle U^2 \rangle_t} = \frac{U_o}{\sqrt{2}} && \text{modulation signal} \end{aligned} \tag{7.41}$$

and the measurement of the contrast ΔI and the effective modulation signal I_{eff} (Figure 7.16) the electro-optic coefficients can be determined:

$$\left|r_{IJ}\right| = \frac{2\lambda}{\pi}\frac{I_{eff}}{\Delta I}\frac{d}{L}\frac{1}{n_I^3}\frac{1}{U_{eff}} \tag{7.42}$$

Typically the two beams should have approximately equal intensities $I_1 \approx I_2$. The electric field is applied along a dielectric axis. The polarization of the light beam propagating in the crystal is chosen to select a particular electro-optic coefficient. The dimensions of the crystals are preferred to be $d/L < 1$. Often , however, this cannot be achieved (*Bosshard et al. 1993*).

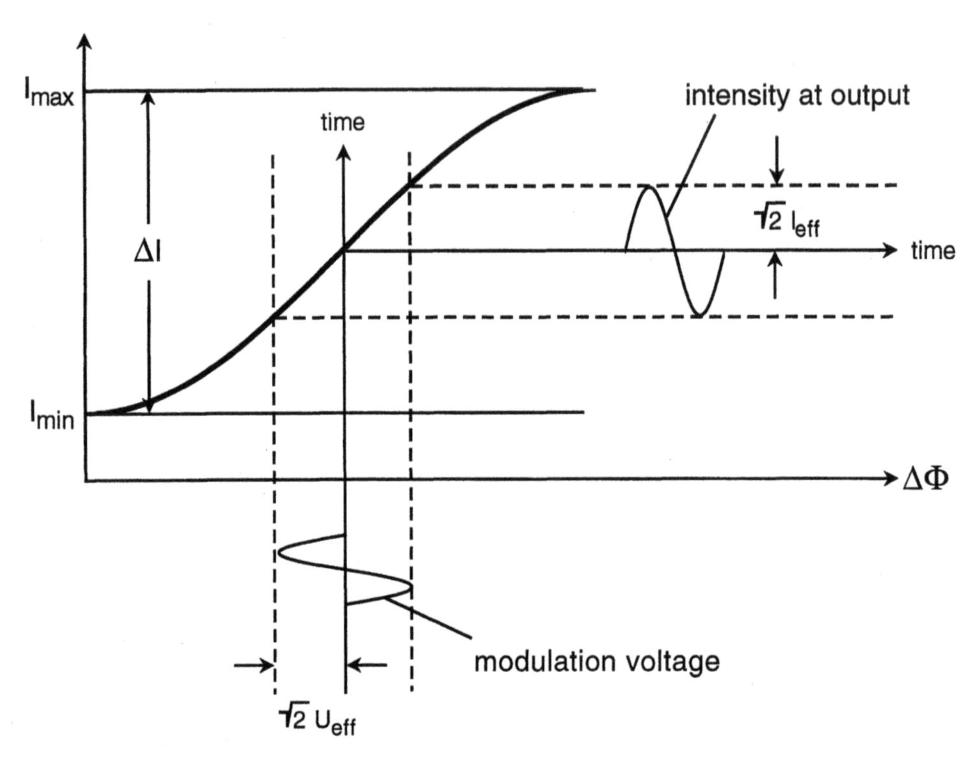

Figure 7.16 Intensity at the output of the interferometer. The modulator is biased to the point $I_1 + I_2$ with a mirror mounted on a piezoelectric crystal. A small applied sinusoidal voltage modulates the light intensity about the bias point.

The electro-optic effect is usually measured at several local spots of the optical end faces. Typical modulation voltages vary between 5 and 200 V rms, the frequency is typical around $\nu = 1$ kHz. Examples of the dispersion of electro-optic coefficients are shown in Figure 7.17. A theoretical fit using Eq. (3.30) is overlaid. The resonance frequencies ω_{eg} were determined from the values obtained from the measurement of the dispersion of the refractive indices (*Sutter et al. 1988b, Bosshard et al. 1989*).

The absolute signs of the electro-optic coefficients can be determined as follows. In a first step we define a positive field direction in our crystals using a pyroelectric measurement. In a second step we compare the sign of the refractive index change Δn with the sign of the applied field, which allows a determination of the sign of the electro-optic coefficients using Eq. (3.22).

First the sign of the electric field E has to be known. A positive sign for E can be defined by a heating experiment. At room temperature the voltage across the crystal is 0 (compensated surface charge). By heating up the crystal, pyroelectric charges are generated leading to positive and negative surface charges S_+ and S_-. We can then define a positive sign for an electric field E with a field vector pointing from S_+ to S_-. (Figure 7.18).

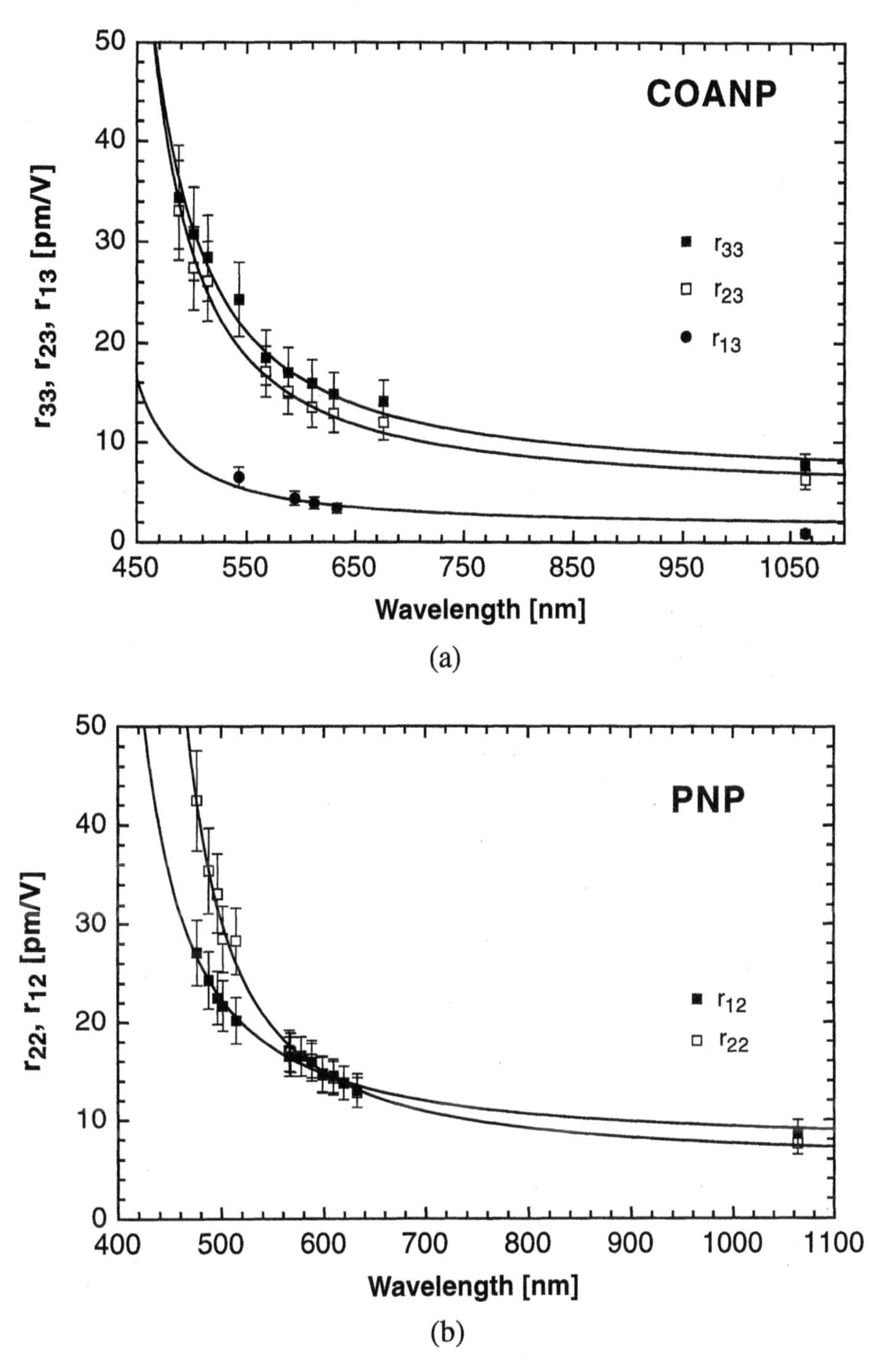

Figure 7.17 Wavelength dispersion of the electro-optic coefficients of (a) COANP and (b) PNP. A theoretical fit using Eq. (3.30) is overlaid.

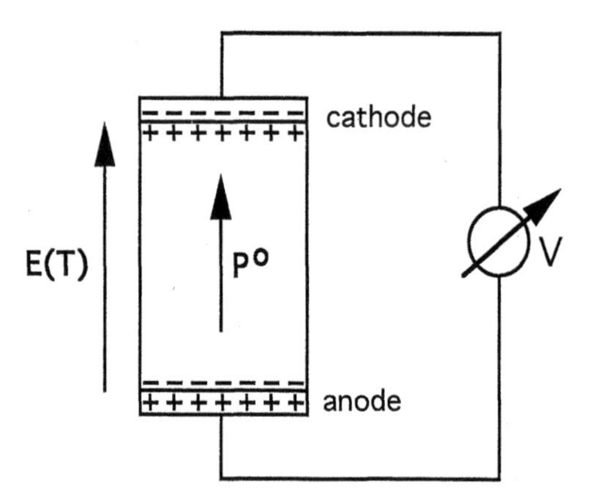

Figure 7.18 Determination of the direction of the spontaneous polarization by the pyroelectric method.

This measurement allowed us to define a positive field direction along the polar axis of COANP and PNP crystals.

In the second step a thin glass plate (thickness L ~ 0.1 mm) is mounted in the path of the reference beam of the Michelson interferometer and is then rotated in such a way that *U* and *I* are in phase. Then an increase of *U* also increases the intensity measured by the detector. A further rotation of the glass plate leads to an increase of the optical path length in this arm of the interferometer and to an increase in average output intensity. This corresponds to a decrease of the optical path of the probe beam and therefore $\Delta n < 0$ for an increase of *U*. Using Eq. (3.22) $\Delta n < 0$ leads to $r > 0$. The sign of the electro-optic coefficients could be determined in this way for different configurations of COANP and PNP (*Bosshard et al. 1993*). The wavelength used in this experiment was $\lambda = 632.8$ nm.

7.4.2 Field-induced birefringence with crossed polarizers

The experimental set-up for field-induced birefringence measurements with crossed polarizers is shown in Figure 7.19.

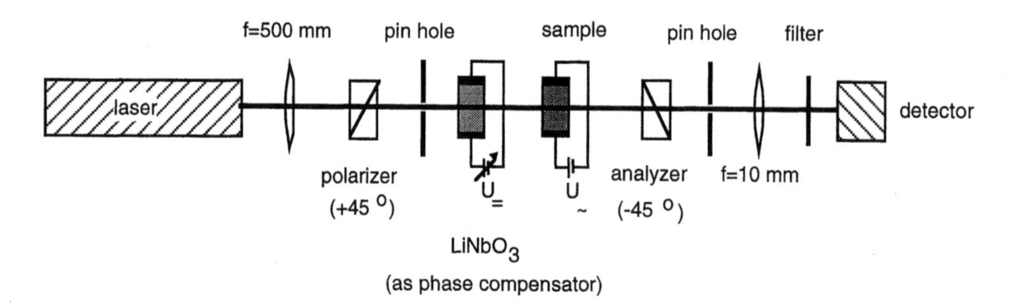

Figure 7.19 Experimental set-up for the determination of the field-induced birefringence.

The light transmission T of such a set-up is well known (see e.g. *Yariv 1975*)

$$T = sin^2\left(\frac{\Delta\phi}{2}\right) \tag{7.43}$$

$$\Delta\phi = \frac{2\pi}{\lambda}(n_2 - n_3)\cdot L - \frac{\pi}{\lambda}(n_2^3 r_{23} - n_3^3 r_{33})\frac{U}{d}L + \Delta\Phi \tag{7.44}$$

The term $\Delta\phi$ is induced by the $LiNbO_3$ crystal. Regulated by a dc voltage it allows one to compensate for the natural birefringence $(n_2 - n_3)$ and to set the working point properly. This leads to

$$\Delta\phi = \frac{\pi}{2} - \frac{\pi}{\lambda}(n^3 r)_{eff}\frac{U}{d}L \tag{7.45}$$

$$(n^3 r)_{eff} = (n_2^3 r_{23} - n_3^3 r_{33}) \tag{7.46}$$

This gives the transmitted intensity $I = T{\cdot}I_0$. For small modulations $U \ll V_\pi$ we obtain

$$\left|n_I^3 r_{II} - n_J^3 r_{JI}\right| = \frac{2\lambda}{\pi}\frac{I_{eff}}{I_{max} - I_{min}}\frac{d}{L}\frac{1}{U_{eff}} \tag{7.47}$$

I_{max} and I_{min} are the beam intensities for field-dependent maximum and minimum transmission. The other parameters have been defined above. Comparisons between interferometric and field-induced birefringence measurements of COANP and PNP are listed in Table 7.5. The agreement is good for PNP and for $|n_3^3 r_{33} - n_1^3 r_{13}|$ of COANP and satisfactory for $|n_3^3 r_{33} - n_2^3 r_{23}|$ of COANP. In the latter case the large error for the calculated values arises from inconvenient

Table 7.5 Comparison of the $|n_I^3 r_{II} - n_J^3 r_{IJ}|$ (at 1 kHz) with the results calculated from the known r_{IJ} (interferometric measurements) given in pm/V.

wavelength [nm]	$\|n_3^3 r_{33} - n_2^3 r_{23}\|$ measured	$\|n_3^3 r_{33} - n_2^3 r_{23}\|$ calculated	$\|n_3^3 r_{33} - n_1^3 r_{13}\|$ measured	$\|n_3^3 r_{33} - n_1^3 r_{13}\|$ calculated
		COANP		
514.5	37 ± 6	38 ± 6	—	98 ± 20
632.8	8.6 ± 1.3	6 ± 20	54 ± 11	52 ± 14
		PNP		
wavelength [nm]	$\|n_2^3 r_{22} - n_1^3 r_{12}\|$ measured	$\|n_2^3 r_{22} - n_1^3 r_{12}\|$ calculated	—	—
514.5	28 ± 4	16 ± 46	—	—
632.8	27 ± 4	31 ± 17	—	—

combinations of the r_{IJ}. In addition this experiment showed that both electro-optic coefficients have the same sign in agreement with the above measurements where the sign of the electro-optic coefficients was determined (*Bosshard et al. 1993*).

Data on electro-optic coefficients are compiled in Chapters 3 and 8.

7.4.3 Contribution of acoustic phonons to the electro-optic coefficients

With experiments using field-induced birefringence with crossed polarizers (±45°) the contribution of acoustic phonons to the electro-optic coefficients can easily be observed (*Bosshard et al. 1993*, see also Chapter 3). Figure 7.20 shows the dependence of the effective electro-optic coefficient $|n_3^3 r_{33} - n_2^3 r_{23}|$ on the modulation frequency ν (from 1 kHz to 1 MHz) measured at λ = 514.5 nm for COANP. The dotted line shows the value of $|n_3^3 r_{33} - n_2^3 r_{23}|$ calculated from r_{33} and r_{23} (interferometric measurement at 1kHz). The crystal was fixed between two parallel metal plates where care was taken not to clamp the crystal too strongly. We see that there are several resonances starting at about 150 kHz with enhancements of $|n_3^3 r_{33} - n_2^3 r_{23}|$ up to a factor of 25.

Bosshard et al. (*1993*) identified the resonances coming from longitudinal waves in the crystallographic b-direction. A rough estimation of the velocity of

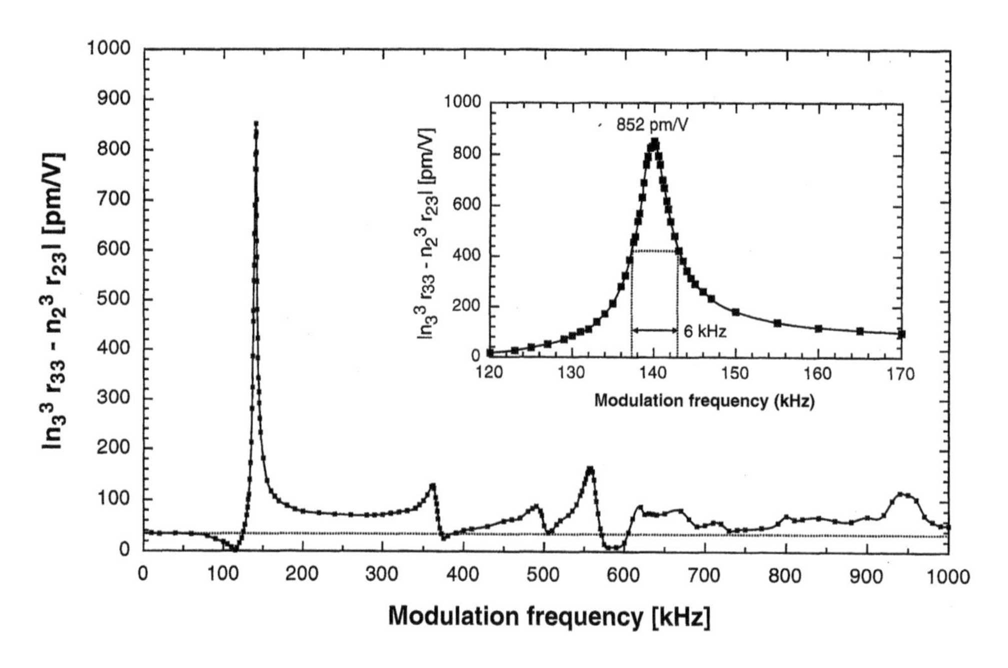

Figure 7.20 Dependence of the effective electro-optic coefficient $|n_3^3 r_{33} - n_2^3 r_{23}|$ on the modulation frequency ν measured at λ = 514.5 nm (applied voltage: 7 V (rms), electric field $|E|_{rms}$ = 4.1 kV/m) for COANP. The dashed line shows the value of $|n_3^3 r_{33} - n_2^3 r_{23}|$ calculated from r_{33} and r_{23} (interferometric measurement at 1kHz). The crystal was mounted between metal plates. Inset: first resonance with a better resolution.

sound for longitudinal waves along the b-direction yielded $c_s = 2{,}100$ m/s. From c_s also an approximate value for the elastic constant $c_{33} = c_s^2\,\rho$ could be deduced: $c_{33} = 5.5 \cdot 10^9 \mathrm{Nm}^{-2}$.

In Chapter 8.5.3 the correspondence between calculated (from second-order polarizabilities) and measured electro-optic coefficients COANP, PNP, MMONS, and MNA will be discussed. There is a disagreement for r_{33} of COANP and the values of PNP which may arise from two effects. On the one hand, two tensor components of β may have to be used since the nitrogen atom in the ring pulls electron density out of the ring. On the other hand, we believe that the discrepancies are at least partially due to non-negligible optic phonon contributions to the electro-optic coefficients. In crystals with almost parallel alignment of the molecules, and therefore large electronic values of r, such effects are not as important (see e.g. MNA) because the electronic values are dominant. Contributions from acoustic phonons to the static electro-optic coefficients can be ruled out as can be seen e.g. from Figure 7.20 (similar values of n^3r for 1 kHz and 1 MHz are observed).

8. NONLINEAR OPTICAL AND ELECTRO-OPTIC PROPERTIES OF ORGANIC COMPOUNDS

8.1 OVERVIEW ON PRESENTLY KNOWN NONLINEAR OPTICAL MOLECULES

Since the first investigation of nonlinear optical properties of organic molecules many new substances have been found, most of which have been analyzed by the EFISH method (Chapter 7). The data thus obtained provide valuable insight into the origin of the hyperpolarizability and help to synthesize molecules with optimized properties. Table 8.1 displays some well known molecules together with the recent approaches which tend towards the use of two ring systems. In addition Figure 8.1 displays the dispersion of β_z for some of the molecules from Table 8.1. Theoretical fits using the two-level model are also shown.

The more one of the wavelengths of the interacting beams approaches the wavelength of maximum absorption, the more resonance effects come into play. When characterizing nonlinear optical molecules, we can eliminate this effect by introducing the hyperpolarizability β_o at zero frequency (see Eq. (3.3), Chapter 3). It characterizes the strength of the molecules in a more reliable way.

$$\beta_o = \frac{3}{2\varepsilon_o \hbar^2} \frac{\Delta\mu\, \mu_{eg}^2}{\omega_{eg}^2} \tag{8.1}$$

β_o is also marked in Figure 8.1.

It is important to realize that the solvents used in the EFISH measurements are of great importance since polar solvents will lead to an increase of β_z and β_0 (*Levine et al. 1976, Stähelin et al. 1992*). A recent work has clearly demonstrated this solvent dependence of the second-order polarizability for the case of p-NA (*Stähelin et al. 1992*).

It is clearly seen from Table 8.1 that there already exist many molecules with high second-order polarizabilities. Values from 500 to 700 $\times$ 10^{-40} m^4/V (and possibly even larger values) for β_o are large enough for most nonlinear optical and electro-optic applications (see e.g. Chapter 3). The greatest problem associated with nonlinear optical organic compounds is, however, not the search for ideal molecules but more the incorporation of these molecules to form ideal macroscopic samples (crystals, polymers, LB films) for electro-optics and nonlinear optics.

Note that so-called octupolar molecules have also been recently investigated (see e.g. *Zyss 1991, 1993*).

Section 8.2 will deal with known electro-optic and nonlinear optical properties of crystals, Section 8.3 with these properties in LB films, and Section 8.4 with these properties in polymers.

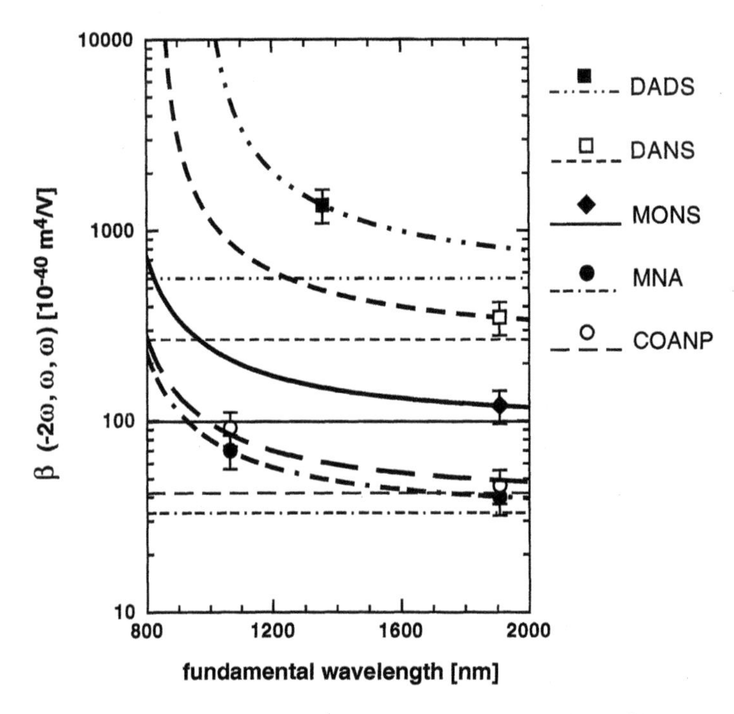

Figure 8.1 Dispersion of the molecular second-order polarizabilities $\beta(-2\omega;\omega,\omega)$ for some of the molecules listed in Table 8.1. The data were fitted using Eqs. (3.4) and (8.1) where the absorption wavelength was taken from Table 8.1 and only the constant had to be adjusted. In addition the molecular hyperpolarizability β_0 (at zero frequency, Eqs. (3.4) and (8.1)) is also plotted (horizontal lines).

8.2 OVERVIEW ON PRESENTLY KNOWN ELECTRO-OPTIC AND NONLINEAR OPTICAL CRYSTALS

Since the discovery of the first good organic nonlinear optical crystals such as nitroaniline and nitropyridine derivatives many new materials have been synthesized. Most of the research now concentrates on materials with two or more π-electron rings. They are superior with regard to the molecular hyperpolarizability β_z. However, there still exists the efficiency-transparency trade-off, that is, one observes a red-shift of the absorption edge associated with a higher optical nonlinearity. Table 8.2 summarizes the nonlinear optical properties of the best presently known organic crystals together with the two well characterized inorganic materials $LiNbO_3$ and $KNbO_3$. The electro-optic properties already compiled in Table 3.1 (Chapter 3) will also be discussed here.

An important parameter for the characterization of nonlinear optical materials is the figure of merit

$$f = \frac{d^2}{n^3} \tag{8.2}$$

Table 8.1 Nonlinear optical second-order polarizabilities β_z and β_o of organic molecules and wavelength of maximum absorption λ_{eg}.

molecule	λ_{eg} [nm]	β_z [10^{-40}m^4/V]	β_0 [10^{-40}m^4/V]	reference
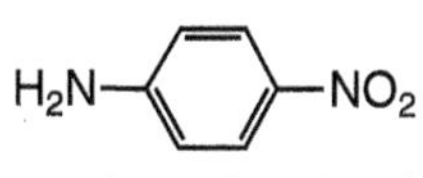 para-nitroaniline (p-NA)	358 [1] 284 [6]	71.0 [a,B,1] 40.2 [d,B,1] 96.6 [a,E,2]	35 [1]	*Teng et al. 1983* *Clays et al. 1992*
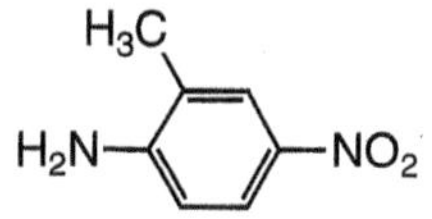 2-methyl-4-nitroaniline (MNA)	361 [1] 288 [6]	69.9 [a,B,1] 39.8 [d, B, 1]	33 [1]	*Teng et al. 1983*
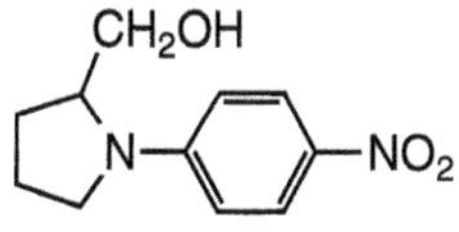 N-(4-nitrophenyl)-(L)-prolinol (NPP)	397 [4] 391 [5]	176 [a,B,4]	67 [4]	*Barzoukas et al. 1987* *Twieg et al. 1986*
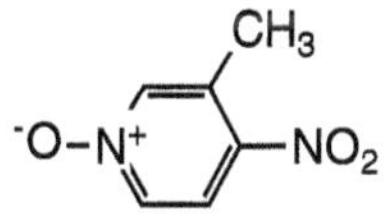 3-methyl-4-nitropyridine-1-oxide (POM)	320 [8]	36 [a,B,8]	21 [8]	*Zyss et al. 1981, 1982*
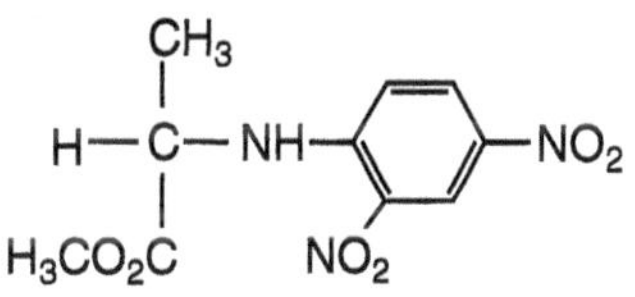 methyl-(2,4-dinitrophenyl)-aminopropanoate MAP)	360 [8]	92 [a,B,4]	44 [4]	*Oudar et al. 1977b, c*
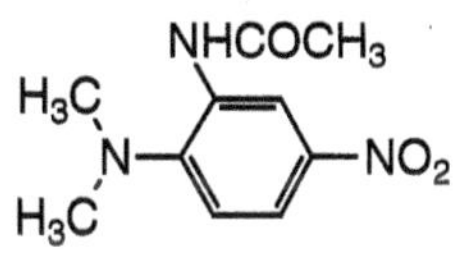 4-(N,N-dimethylamino)-3 acetamidonitrobenzene (DAN)	360 [1]	55 [d,C,1]	45 [1]	*Paley et al. 1989*
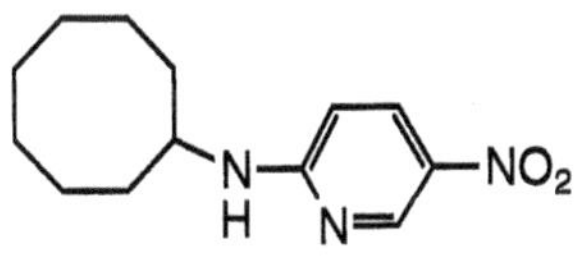 2-cyclooctylamino-5-nitropyridine (COANP)	361 [1] 292 [6]	92 [a,A,1] 46 [d,A,1]	42 [1]	*Bosshard et al. 1992b*

Table 8.1 Continued.

molecule	λ_{eg} [nm]	β_z [$10^{-40}m^4/V$]	β_0 [$10^{-40}m^4/V$]	reference
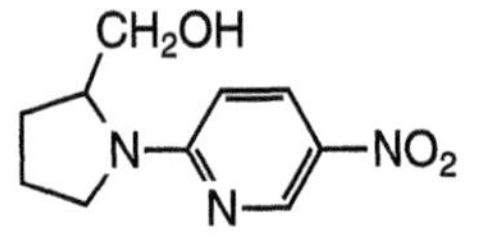 2-(N-prolinol)-5-nitropyridine (PNP)	376[1] 308[6] 370[5]	76 [a,A,1] 47 [d,A,1]	38 [1]	*Bosshard et al. 1992b* *Twieg et al. 1986*
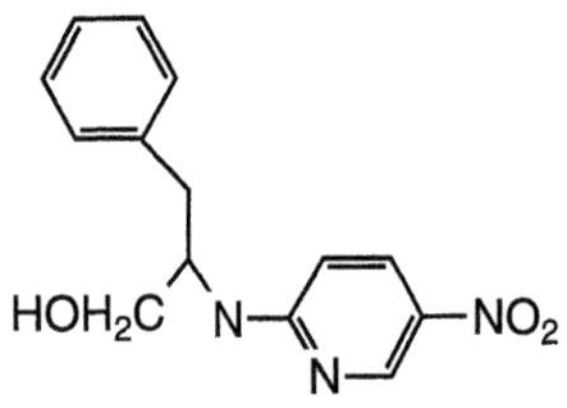 N-(4-nitro-2-pyridinyl)-phenylaninol (NPPA)	345[8]	113 [a,A,8] 76 [b1, A, 8]	56 [8]	*Sutter et al. 1991a*
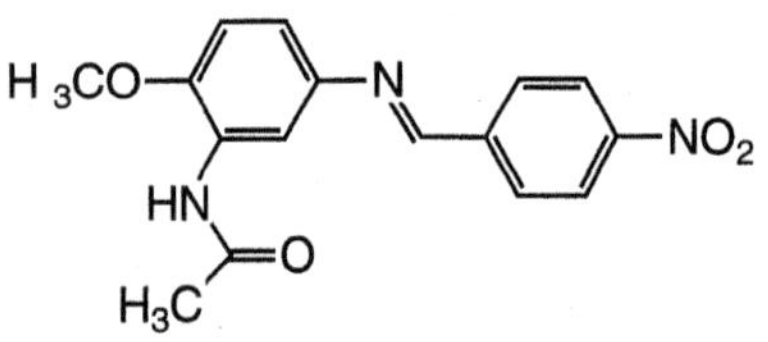 4′-nitrobenzylidene-3-acetamino-4-methoxyaniline (MNBA)	380[9]	126 [a,A,1] 64 [d,A,1]	53 [1]	*Knöpfle et al. 1994*
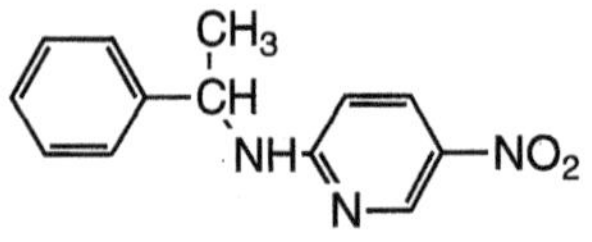 (-) 2-(α-methylbenzylamino)-5-nitropyridine (MBANP)	359[5]	63 [a, F, 1]	30 [1]	*Kondo et al. 1989b*
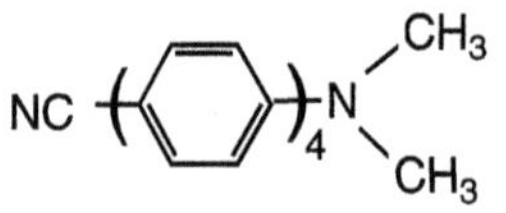 polyphenyl derivative	330[2]	424 [a,B,10] 424 [b2,B,10]	235 [2]	*Ledoux et al. 1991*
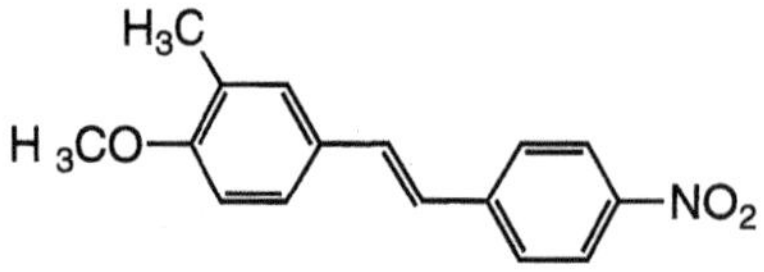 3-Methyl-4-methoxy-4′ nitrostilbene (MMONS)	366[1]	109 [d,D,1]	90 [1]	*Cheng et al. 1989*

Table 8.1 Continued.

molecule	λ_{eg} [nm]	β_z [$10^{-40}m^4/V$]	β_0 [$10^{-40}m^4/V$]	reference
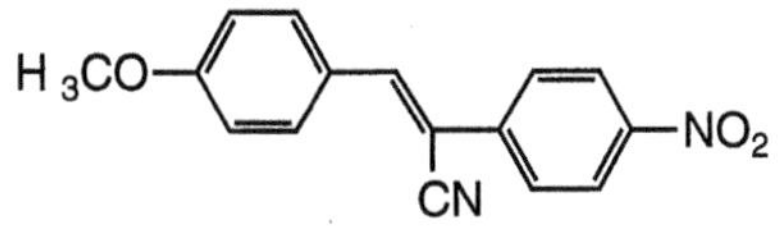 1-(4-nitrophenyl)-2-(4-methoxyphenyl)-1-cyanoethylene (CMONS)	361[1]	88 [d, D, 1]	73 [1]	*Cheng et al. 1989*
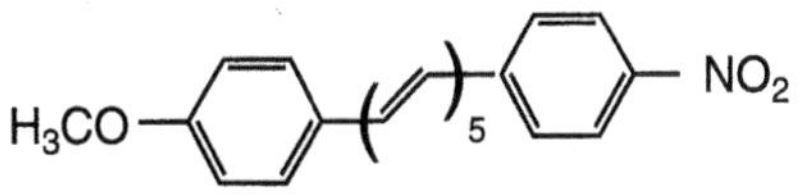 MPNP5	435 [1]	2617 [a,B,1]	722 [1]	*Huijts et al. 1989*
4-dimethylamino-4′-nitro-stilbene (DANS)	—	1134 [a]	—	*Möhlmann et al. 1990*
	424 [2]	307 [d,D,2]	234 [2]	*Cheng et al. 1991*
4-[N-ethyl-N-(2-hydroxyethyl)] amino-4′-nitroazobenzene (disperse red 1 (DR1))	509 [7]	525 [b3,B,7]	197 [7]	*Singer et al. 1989*
	455 [1]	206 [d,D,1]	150 [1]	*Cheng et al. 1991*
4-metoxy-4′-nitro-stilbene (MONS)	—	340[a]	—	*Möhlmann et al. 1990*
	364[1]	120 [d,D,1]	99 [1]	*Cheng et al. 1989*
	360 [1]	340 [a,B,1]	163 [1]	*Huijts et al. 1989*
		441[a,E,2]		*Clay et al. 1992*
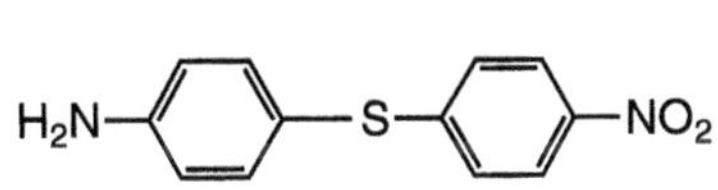 4-amino-4′-nitrodiphenyl sulfide (ANDS)	341 [4]	1200 [a,b,4]	634 [4]	*Barzoukas et al. 1989*
	400 [3]			
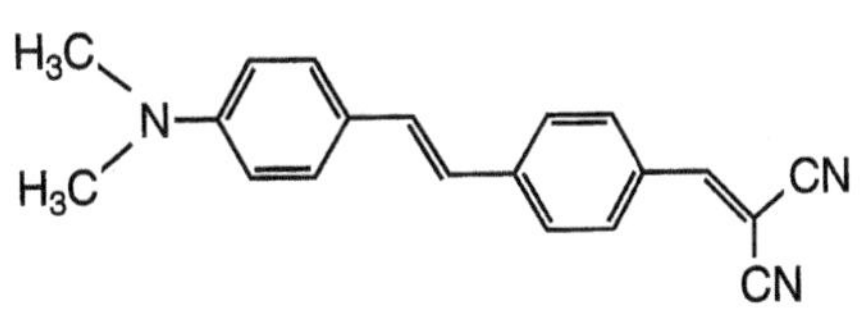 4-(N,N-dimethylamino)-4′-(2,2-dicyanovinyl)-stilbene (DADS)	492 [7]	1360 [b3,B,7]	559 [7]	*Singer et al. 1989*

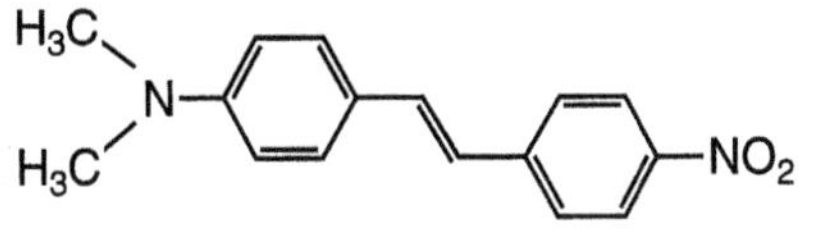

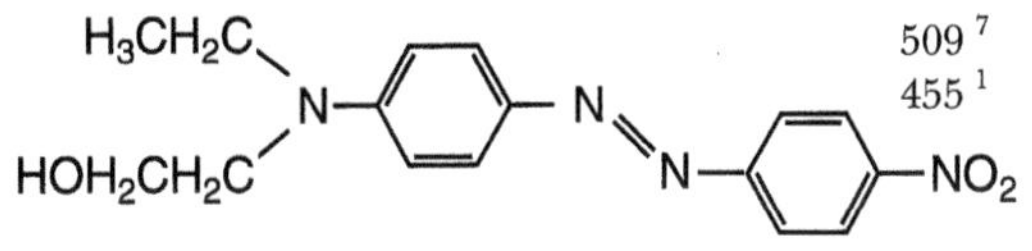

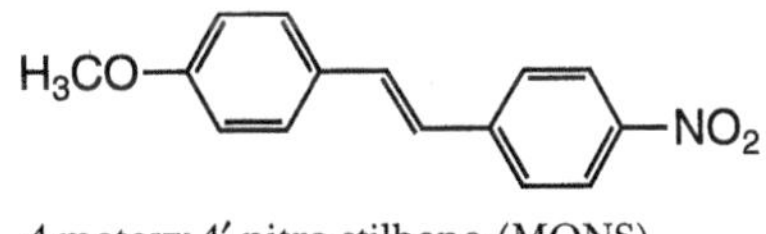

Table 8.1 Concluded.

molecule	λ_{eg} [nm]	β_z [10^{-40}m^4/V]	β_0 [10^{-40}m^4/V]	reference
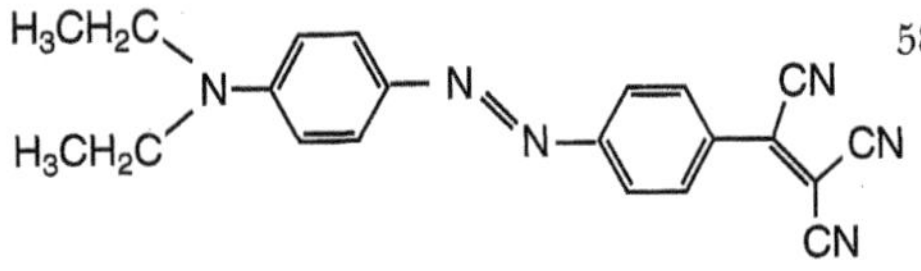 4-(N,N-diethylamino)-4'-tricyanovinylazobenzene (DADAB)	582 [7]	1640 [c,B,7]	648 [7]	*Singer et al. 1989*
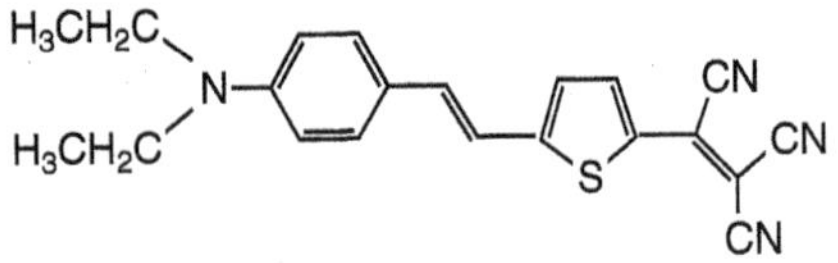 (E)-benzeneamine N, N-diethyl-4-[2-(2-tricyanovinylthiophene) ethyl]	640 [1]	86860 [d,F,1] (*)	42351 [1] (*)	*Rao et al. 1993b*
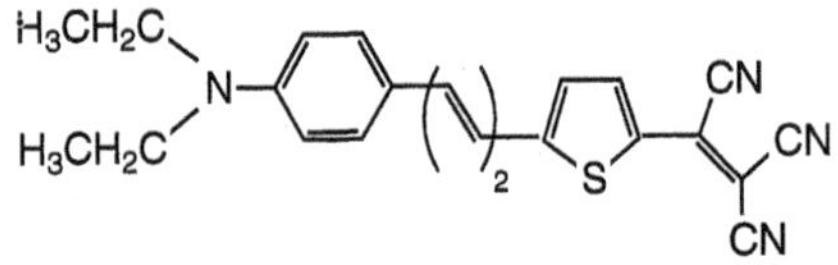 (E)-benzeneamine N, N-diethyl-4-[2-(2-tricyanovinylthiophene) butadienyl]	662 [1]	127488 [d,F,1] (*)	58084 [1] (*)	*Rao et al. 1993b*
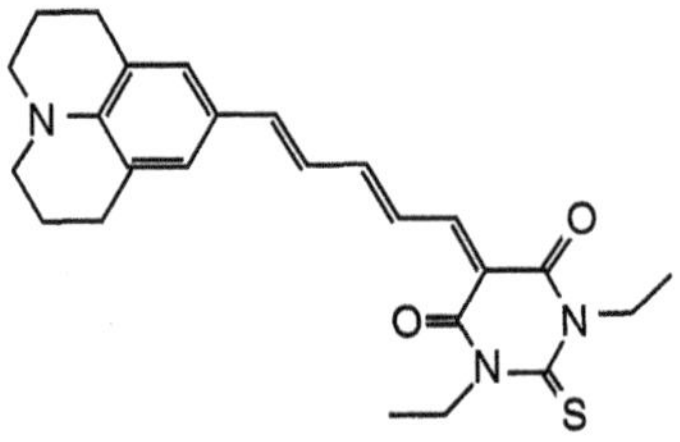 (E,E) 1-(benzeneamine 4-(di-2-propenylamino)) pentadieneylidene thiobarbituric	—	4792 [d,F,2] 100200 [d,F,2] (*)	—	*Bourhill et al. 1993*

[1] measured in 1,4-dioxane
[2] measured in $CHCl_3$
[3] measured in methanol
[4] measured in acetone
[5] measured in ethanol
[a] measured at λ = 1,064 nm
[b1] measured at λ = 1,318 nm
[b2] measured at λ = 1,340 nm
[b3] measured at λ = 1,356 nm
[c] measured at λ = 1,580 nm
[d] measured at λ = 1,907 nm
[6] calculated in gas phase (p-NA, MNA: measured in gas-phase)
[7] measured in DMSO
[8] from crystal data
[9] from dispersion of β_z
reference: [A] d_{11} = 0.4 pm/V (quartz)
[B] d_{11} = 0.5 pm/V (quartz)
[C] $LiIO_3$ crystal as reference
[D] d_{11} = 0.44 pm/V (quartz)
[E] $\beta_z = 2.01 \times 10^{-40}$m^4/V of $CHCl_3$
[F] not specified
(*) product $\beta\mu$ in units of 10^{-70} m^{5}C/V (compare to e.g. DANS with $\beta\mu = 8120 \times 10^{-70}$ m^{5}C/V (at λ = 1.907 μm (*Rao et al. 1993a*))

which originates from Eq. (3.13). The figure of merit is also listed in Table 8.2. It is important to notice that different reference materials (nonlinear optical coefficient d_{ref}) are used for the determination of d_{IJK} and that even for the same reference different values for d_{ref} are applied. Therefore given values of d_{IJK} have to be carefully examined before different materials can be compared.

Two aspects leading to large nonlinear optical coefficients should be mentioned:

- The angle between the charge transfer axis and the polar crystal axis θ_p is of great importance for an optimized material (see Chapter 3). θ_p should either be zero for electro-optic or nonlinear optics in waveguides, or it should be 54.7° for nonlinear optical effects in bulk crystals.
- The number of molecules per unit volume is another important factor to be considered. An unfavorable crystal packing can immediately lead to lower values of r_{IJK} or d_{IJK}.

Some of the compounds in Table 8.2 will be discussed in the following.

The largest nonlinear optical coefficients published so far are found in DAST (4′-dimethylamino-N-methyl-4-stilbazolium tosylate, *Perry et al. 1991, Lawrence 1992*). This compound crystallizes in the monoclinic point group m and has a cut-off wavelength of around 700 nm. The charge transfer axis of the cations is inclined under $\theta_p = 20°$ towards the symmetry axis. This arrangement is almost ideal for electro-optics or for nonlinear optical applications in waveguides since the corresponding coefficient d_{11} or r_{11} is only decreased by a factor $1/\cos^3\theta_p \approx 1.21$ compared to the ideal configuration with $\theta_p = 0$ (in this estimation local field effects are neglected).

Other interesting stilbene type salts have been proposed by Okada et al. (*1990*). They have measured a nonlinear optical coefficient $d_{11} = 500$ pm/V for 4′-hydroxy-N-methyl-4-stilbazolium tosylate. In both this compound and DAST the largest nonlinear optical susceptibilities are not phase-matchable in the bulk crystalline form. These materials should also possess higher optical nonlinearities when compared to the upper limits given in Table 8.2. The main reason is probably the strong Coulombic interaction present in such salts. Therefore the simple model used for approximate upper limits is not applicable to these molecular systems.

When looking at one ring systems, large phase-matchable nonlinear optical coefficients have been obtained in 3,5-dimethyl-1-(4-nitro-phenyl) pyrazole (DMNP, $d_{32} = 90$ pm/V, *Harada et al. 1991*), N-(4-nitrophenyl)-(L)-prolinol (NPP, $d_{21} = 85$ pm/V, *Ledoux et al. 1990*), 2-(N-prolinol)-5-nitropyridine (PNP, $d_{21} = 68$ pm/V, *Sutter et al. 1989*), 2-adamantylamino-5-nitropyridine (AANP, $d_{31} = 80$ pm/V, *Tomaru et al. 1991*) and others. When correcting for the different reference materials used, all these materials have optical nonlinearities close to each other.

Table 8.2 clearly shows that there exist many organic crystals with considerably larger nonlinear optical coefficients than can be found in some of the best inorganic materials (e.g. $KNbO_3$ and $LiNbO_3$).

Table 8.2 Crystal data, cut-off wavelength λ_c, and nonlinear optical properties of organic crystals at room temperature.

crystal	point group	λ_c [nm]	powder (x urea)	d_{IJK}[pm/V]	d^2/n^3 [(pm/V)^2]	reference
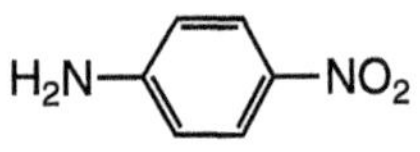 para-nitroaniline (pNA)	2/m	470	0	—	—	*Twieg et al. 1983*
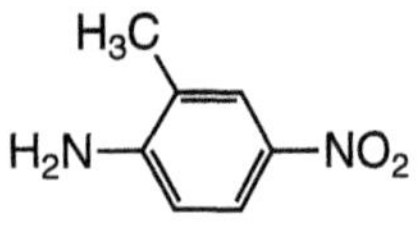 2-methyl-4-nitroaniline (MNA)	m	480	140	$d_{11} = 250$ [a,B,-] $d_{12} = 38$ [a,B,-]	10700[1] 357[1]	*Levine et al. 1979*
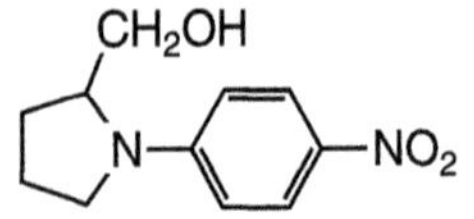 N-(4-nitrophenyl)-(L)-prolinol (NPP)	2	~ 500	150	$d_{21} = 85$ [a,E,+] $d_{22} = 28$ [a,F,-]	834[2] 123[1]	*Ledoux et al. 1990*
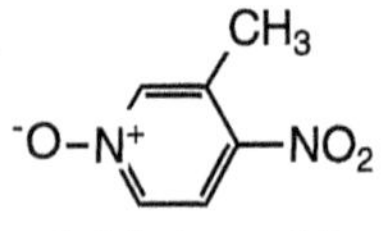 3-methyl-4-nitropyridine-1-oxide (POM)	222	~460	 13	$d_{14} = 9.6$ [a,B,+] $d_{14} = d_{25} = d_{36}$	17[1]	*Zyss et al. 1981* *Twieg et al. 1982*
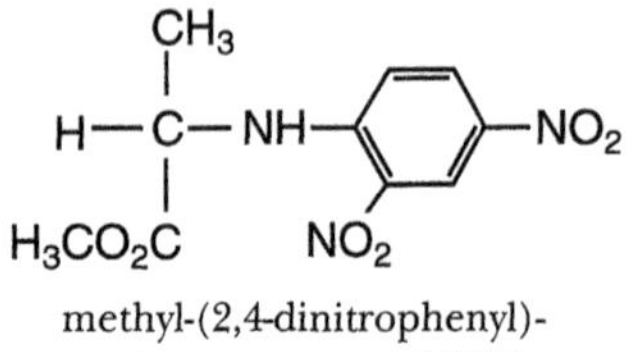 methyl-(2,4-dinitrophenyl)-aminopropanoate (MAP)	2	500	 10	$d_{21} = 17$ [a,B,+] $d_{22} = 18$ [a,B,-]	74[1] 74[2]	*Oudar et al. 1977b* *Twieg et al. 1982*
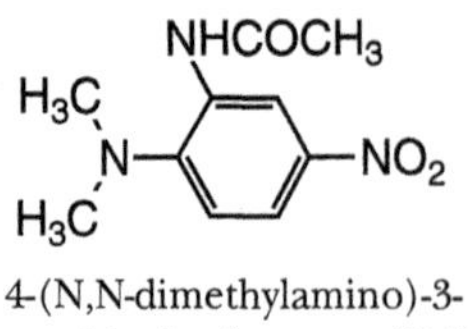 4-(N,N-dimethylamino)-3-acetamidonitrobenzene (DAN)	2	485	115 to 200	$d_{23} = 50$ [a,E,+] $d_{22} = 5.2$ [a,A,-]	425[1] 125[2] 6[1]	*Kerkoc et al. 1989*
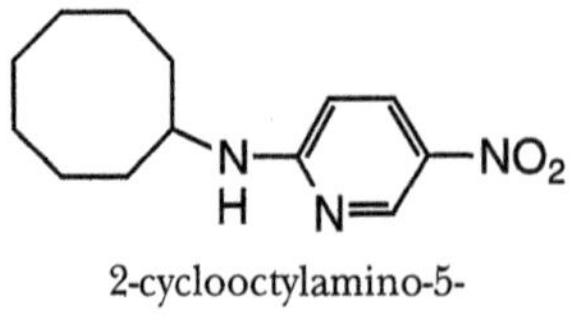 2-cyclooctylamino-5-nitropyridine (COANP)	mm2	490	50	$d_{32} = 43$ [a,A,+] $d_{33} = 19$ [a,A,-]	371[1] 119[2] 82[1]	*Bosshard 1991a*

Table 8.2 Continued.

crystal	point group	λ_c [nm]	powder (x urea)	d_{IJK}[pm/V]	d^2/n^3 [(pm/V)^2]	reference
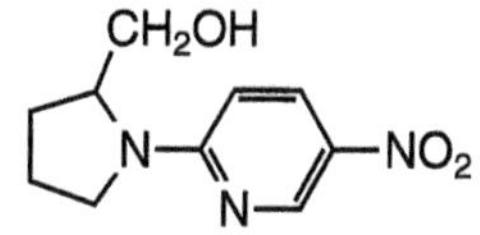 2-(N-prolinol)-5-nitropyridine (PNP)	2	490	140	$d_{21} = 68$ [a,A,+] $d_{23} = 22$ [a,A,–]	722 [1] 83[2] 123 [1]	*Sutter et al. 1988b* *Sutter et al. 1989*
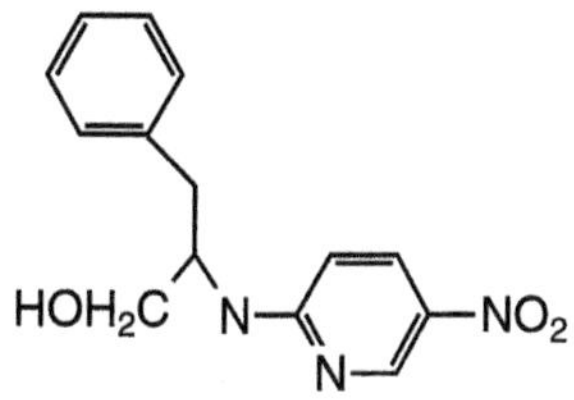 N-(4-nitro-2-pyridinyl)-phenylaninol (NPPA)	2	500	130	$d_{23} = 31$ [a,A,+]	186[1]	*Sutter et al. 1991a*
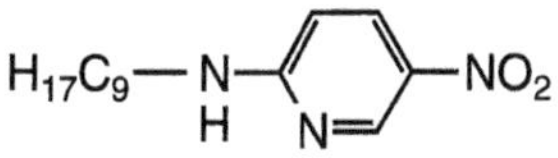 2-adamantylamino-5-nitropyridine (AANP)	mm2	460	300	$d_{31} = 80$ [a,B,+] $d_{33} = 60$ [a,B,–]	1230[1] 667[2] 662[1]	*Tomaru et al. 1991*
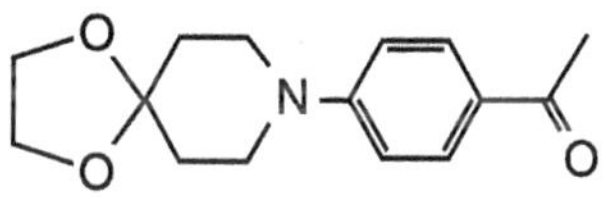 8-(4′-acetylphenyl)-1,4-dioxa-8-azaspiro[4,5] decane (APDA)	mm2	384	—	$d_{31} = 7$ [a,F,+] $d_{33} = 50$ [a,F,–]	54[1] 540[1]	*Sagawa et et al. 1993*
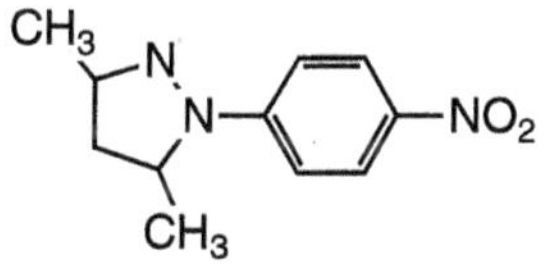 3,5-dimethyl-1-(4-nitrophenyl) pyrazole (DMNP)	mm2	450	16	$d_{32} = 90$ [b,F,+] $d_{33} = 29$ [b,F,–]	1390[1] 172[2]	*Harada et al. 1991*
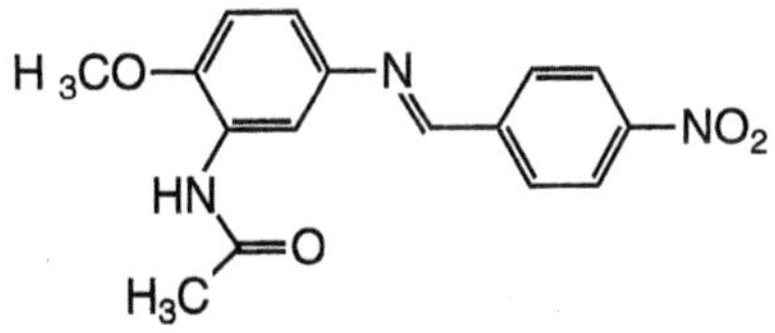 4′-nitrobenzylidene-3-acetamino-4-methoxyaniline (MNBA)	m	~ 520	230	$d_{11} = 175$ [a,A,–]	4400[1]	*Knöpfle et al. 1994*

Table 8.2 Continued.

crystal	point group	λ_c [nm]	powder (x urea)	d_{IJK}[pm/V]	d^2/n^3 [(pm/V)^2]	reference
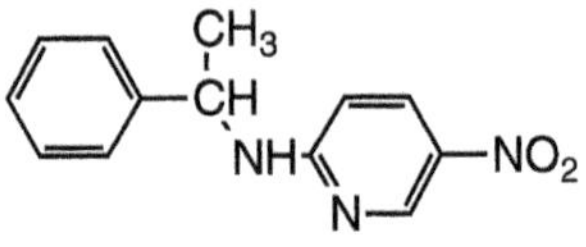 (-)2-(α-methylbenzylamino)-5-nitropyridine (MBANP)	2	450	25	$d_{22} = 63$ [a,B,–] (d_{14}^{+})	600[1]	*Kondo et al. 1989b* *Bailey et al. 1988*
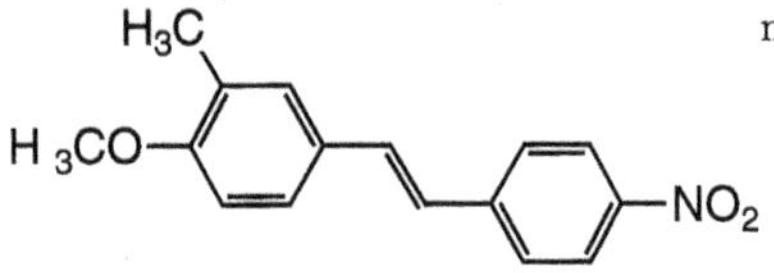 3-Methyl-4-methoxy-4′-nitrostilbene (MMONS)	mm2	515	1250	$d_{33} = 184$ [a,C,–] $d_{24} = 71$ [a,C,+]	3828[1] 850[2]	*Bierlein et al. 1990*
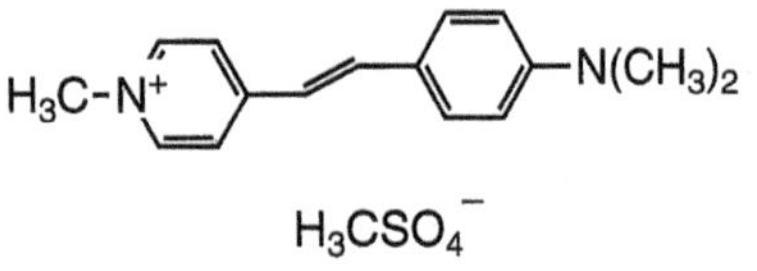 4-dimethylamino-N-methyl-4-stilbazolium methylsulfate (DMSM)	—	620	—	$d = 3800$ [F]	—	*Yoshimura et al. 1989* *Meredith et al. 1986*
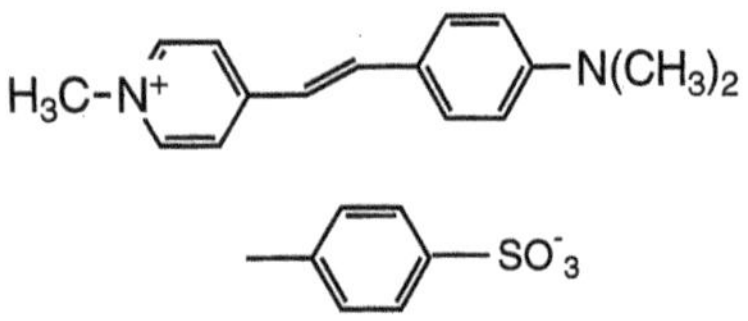 4′-dimethylamino-N-methyl-4-stilbazolium tosylate (DAST)	m	700	1000 (1.9 μm)	$d_{11} = 600$ [c,D,–]	33100[1]	*Perry et al. 1991* *Lawrence 1992*
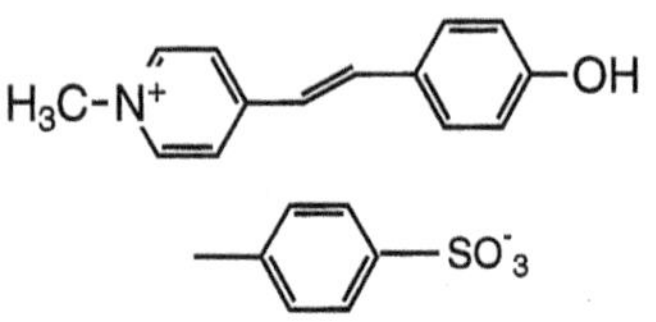 4′-hydroxy-N-methyl-4-stilbazolium tosylate	1	510	—	$d_{11} = 500$ [a,F,–]	49100[1]	*Okada et al. 1990*
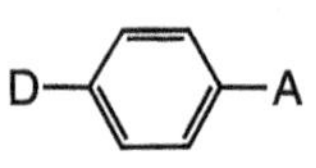 optimized benzene (λ_{eg} (solid state) = 401 nm)	—	501	—	$d_{IJ} = 182^{+}$ $d_{II} = 474^{-}$ (at 1002 nm)	3100[1] 21100[2]	*Bosshard et al. 1993*

Table 8.2 Concluded.

crystal	point group	λ_c [nm]	powder (x urea)	d_{IJK}[pm/V]	d^2/n^3 [(pm/V)^2]	reference
D–(stilbene)–A optimized stilbene (λ_{eg} (solid state) = 479 nm)	—	579	—	$d_{IJ} = 1330^{+}$ $d_{II} = 3450^{-}$ (at 1158 nm)	128000[1] 861000[2]	*Bosshard et al. 1993*
$LiNbO_3$	3m	400	13[&]	$d_{33} = 34$ [a,–] $d_{31} = 5.9$ [a,+]	110[1]	*Choy et al. 1976*
$KNbO_3$	mm2	400	50[&]	$d_{33} = 27.4$ [a,A,–] $d_{31} = 15.8$ [a,A,+] $d_{32} = 18.3$ [a,A,+]	80[1] 21[2]	*Baumert 1985*

[a] measured at λ = 1,064 nm [b] measured at around λ = 950 nm [c] measured at λ = 1,907 nm
[+] phase-matchable [–] not phase-matchable
reference: [A] d_{11} = 0.4 pm/V (quartz) [B] d_{11} = 0.5 pm/V (quartz) [C] d_{33} = 13.7 pm/V (KTP)
[D] d_{33} = 30 pm/V ($LiNbO_3$) [E] phase-matching [F] not specified
[1] $d_{IJ}^2/(n_I n_J^2)$ [2] d_{eff}^2/n^3 (phase-matching)
[&] powder efficiency urea (urea assumed to be 40 × quartz (literature: 20–70 × quartz))

Recently intracavity frequency-doubling has been obtained using molecular crystals (COANP, (*Looser 1990*), DAN, (*Ducharme et al. 1990*)).

Our simple expression for the estimation of the upper limits for electro-optic and nonlinear optical effects shows that the materials known up to now are not optimized (see Chapter 3). There is one optimized benzene derivative with a cut-off at λ = 450 nm, DMNP (d_{32} = 90 pm/V at around λ = 900 nm, *Harada et al. 1991*), which is in good agreement with the theoretical value of 82 pm/V. Other materials with a cut-off at λ = 500 nm such as e.g. NPP (d_{21} = 85 pm/V at λ = 1,064 nm (*Ledoux et al. 1990*)) could have phase-matchable nonlinear optical coefficients up to 150 pm/V. It should be noted, however, that especially for benzene and pyridine derivatives the number of molecules per unit volume may vary considerably. In the case of NPP we have $N = 3.7 \cdot 10^{27}$ m^{-3} in contrast to $N = 6 \cdot 10^{27}$ m^{-3} for MNA. Correcting for this difference leads to d_{max} = 93 pm/V (NPP) which is in good agreement with the measured value. Taking therefore into account the number of molecules per unit volume, we see that NPP is already optimized for its nonlinear optical response. Similar remarks also apply for COANP and PNP. Looking at stilbene derivatives there is only one example, MMONS (*Bierlein et al. 1990*), with a cut-off wavelength as low as λ = 515 nm. Its largest phase-matchable coefficient d_{24} = 71 pm/V (λ = 1,064 nm) is small compared to an optimized value of d_{24} = 250 pm/V (λ = 1,030 nm, see Chapter 3).

The size of the electro-optic coefficients of crystals based on the molecules shown in Table 3.1 and Figure 3.3 can be discussed further:

The largest electro-optic coefficients published so far are also found in DAST (*Perry et al. 1991*). As mentioned above, the molecular arrangement (θ_p = 20°) is almost ideal for electro-optics since the electro-optic coefficient is only decreased by a factor $1/\cos^3\theta_p \approx 1.21$ compared to the ideal configuration with $\theta_p = 0$ (in this estimation local field effects are neglected). Perry measured an electro-optic coefficient r_{11} = 400 pm/V at 820 nm leading to $n_1^3 r_{11}$ = 4,200 pm/V which is a value comparable to the most efficient inorganic electro-optic and photorefractive materials ($BaTiO_3$ (*Landolt-Börnstein 1979a*) and $KNbO_3$ (*Günter 1976*)).

Large electro-optic coefficients have also been found in SPCD (styrilpyridinium cyanine dye) also abbreviated as DMSM (4′-dimethylamino-N-methyl-4-stilbazolium methylsulfate, *Yoshimura 1987, Yoshimura et al. 1989*). Crystals of SPCD have point group symmetry mm2 and the charge transfer axis of the cations is inclined by θ_p = 34° towards the symmetry axis. This leads to a decrease by a factor $1/\cos^3\theta_p \approx 1.75$ compared to the ideal configuration for the electro-optic coefficient. Yoshimura (*1987*) measured an electro-optic figure of merit $n_3^3 r_{33}$ = 1,200 pm/V at 632.8 nm.

DCNP (3-(1,1-dicyanoethenyl)-1-phenyl-4,5-dihydro-1H-pyrazole) is another material with very strong electro-optic response (*Allen et al. 1988, Allen 1989*). Its electro-optic figure of merit $n^3 r$ reaches values around 1,720 pm/V at 632.8 nm. However, this is a wavelength very close to the absorption edge, and measurements at longer wavelength show a drastic decrease by nearly a factor of 5 over a wavelength range of 300 nm (*Allen 1989*). The molecular charge transfer axis of DCNP lies nearly parallel to the polar axis of the crystal which is an ideal orientation for electro-optic applications.

MMONS (3-methyl-4-methoxy-4′-nitrostilbene) (*Bierlein et al. 1990*) has a molecular arrangement (θ_p = 34°) which is not ideal for electro-optics but comparable to DMSM. However, its electro-optic figure of merit is less than half of that of DMSM.

MNBA (4′-nitrobenzylidene-3-acetamino-4-methoxyaniline) and NMBA (4-nitro-4′-methylbenzylidene aniline) are very similar two ring systems. Both have an aza substitution in the 1-position that should lower the second-order polarizability by about a factor of two as compared to the corresponding stilbene derivatives (*Cheng et al. 1989*). Both have a very similar angle between the molecular charge transfer axis and the polar crystal axis (θ_p = 18.7°, MNBA, *Knöpfle et al. 1994*, $\theta_p \sim 18°$, NMBA, *Bailey et al. 1992*). Therefore the values for r_{11} of 29 pm/V (MNBA, *Knöpfle et al. 1994*) and 25.2 pm/V (NMBA, *Bailey et al. 1992*) are reasonable when comparing them with microscopic data.

MNA (2-methyl-4-nitroaniline) (*Lipscomb et al. 1981*) is the best electro-optic material among the group of compounds with benzene and pyridine rings. Its molecular alignment in the crystal lattice is nearly perfect for electro-optics (θ_p = 21°).

All other compounds (PNP, DAN, mNA, COANP, POM, DBNMNA) shown in Figure 3.3 are single π electron ring systems with molecular hyperpolarizabilities comparable to MNA. However, the molecular arrangement is far from ideal for electro-optics. Both COANP and PNP are materials optimized for nonlinear optical applications. As an example the angle between the molecular charge transfer axis and the polar crystal axis of PNP (point group 2) is 59.6° and this reduces its electro-optic coefficient r_{22} by $(1/\cos 59.6)^3 \approx 7.7$ which, with r_{22} (632.8 nm) = 12.8 pm/V, would give a hypothetical value of r around 100 pm/V for the optimized configuration.

The two dotted lines in Figure 3.3 represent the maximum figure of merit to be expected from a stilbene (with λ_{max} (solid state) = 479 nm) and a benzene (λ_{max} (solid state) = 401 nm) derivative. Corresponding molecular second-order polarizabilities of the correct magnitude have been measured (*Cheng et al. 1989*).

Also for electro-optic coefficients our simple expression for the estimation of the upper limits shows that the materials known up to now are not optimized. Optimized benzene derivatives with a cut-off at λ = 500 nm such as e.g. MNA (r_{11} = 67 pm/V at λ = 632.8 nm) could have electro-optic coefficients around 210 pm/V (λ = 500 nm) or 83 pm/V (λ = 632.8 nm) for $\theta_p = 0°$. Correcting the value of 83 pm/V by multiplying with $\cos^3(21°) \approx 0.81$ (projection factor for MNA) we get 67 pm/V. Taking therefore θ_p into account, we see that the difference between the experimental and the optimized value of MNA is due to the projection factor $\cos^3\theta_p$. As was mentioned above COANP and PNP can be discussed in the same way. In a similar manner materials (stilbene derivatives) with a cut-off at λ = 580 nm such as e.g. DMSM (r_{33} = 430 pm/V at λ = 632.8 nm, DMSM is a salt which means that strong coulombic interactions are present!) could show electro-optic coefficients around 1,150 pm/V.

As can be seen from sections 8.1 and 8.2 the major problem in the tailoring of organic materials for electro-optic and nonlinear optical applications is not the molecular but the crystal engineering. Many molecules with exceptionally large nonlinearities are already known. The growth of optimized noncentrosymmetric crystals was, however, not yet successful. The incorporation in LB films or polymers where structural problems can be overcome more easily may be more promising.

8.3 OVERVIEW ON PRESENTLY KNOWN NONLINEAR OPTICAL AND ELECTRO-OPTIC LB FILMS

8.3.1 Molecules and polar LB films

So far more than a hundred molecules have been synthesized in order to prepare LB films with second-order nonlinear optical and electro-optic properties. Examples are represented in Table 8.3. The x_3 axis of the sample coordinate system is defined to be normal to the film surface for all examples except of 2-docosylamino-5-nitropyridine (DCANP). As mentioned in Chapter 5.2 the Y-type films of DCANP have an overall in-plane polarization induced during the dipping process. Hence the x_3 axis is defined to be parallel to the dipping direction. The charge-transfer axes of the DCANP molecules are oriented herringbone-like in the substrate plane (x_1/x_3 plane) with a mean tilt angle $\Phi_{ip} = \pm 27°$. A similar angle Φ_{oop} is defined for the x_2/x_3 plane. One tilt angle Φ of the chromophores with respect to the surface normal is used for the description of the other LB films. There is no tilt azimuth reported to be preferred in the films with normal overall polarization listed in Table 8.3.

Some of the nonlinear optical and electro-optic coefficients reported in the table are very large. However, such data have to be interpreted carefully. The data of the table have not been corrected for resonance enhancement. Coefficients measured at various wavelengths would be needed for the calculation of the long wavelength limits. So far such measurements have been carried out for only a small number of surfactants and LB films. In order to account for effects of resonance enhancement the longest wavelength absorption maxima λ_{eg} are listed in Table 8.3 if reported in the original literature. The wavelengths of the light sources used are also given if they differ from 1,064 nm (nonlinear optical experiment) or from 632.8 nm (electro-optic experiment).

Moreover it should be noticed that some of the coefficients are reported only for monolayers. It is not guaranteed that molecules which form LB monolayers also form well ordered multilayers. The monofilm data, however, are listed in the table since they are valuable for comparison between monolayers of different molecules and for the development of possible guidelines for further research.

The dependence of the generated second-harmonic intensity on the thickness of an LB film is also mentioned in the table if reported in the literature. A quadratic dependence on the number of layers is a reliable argument for a uniform film order maintained over the whole thickness. The second-harmonic efficiency depends much more strongly on the symmetry and the order of the sample as e.g. optical absorption data.

The nonlinear optical and electro-optic characterization of a material requires linear optical data. Therefore experiments which have been carried out for the determination of refractive indices or dielectric constants are listed in the table. Moreover the calibration standards for the second-harmonic intensities and susceptibilities are given in Table 8.3 if reported quantitatively in the original literature.

Additionally some important preparation parameters are mentioned. One such parameter is the pH index of the subphase used for the LB preparation. The absorption of dyes often depends on the pH. This effect is routinely used for pH sensoring. The pH influences the nonlinear optical properties of many dyes mediated by various effects such as e.g. the shift of the absorption band (*Steinhoff et al. 1989*) or the change of strengths of donors or acceptors (*Girling et al. 1985*). If no data are given by the authors, the pH of the subphase is assumed to be about 5.5 due to a saturation of the water with carbon dioxide. On the one hand, as mentioned in Chapter 5.3, it is a great advantage of the LB technique that environmental parameters such as the pH index can be changed easily and in a wide range in order to influence the film properties. On the other hand, however, these parameters must be defined and controlled reliably.

The hydrophilic head group of the LB film forming molecules is identical to the nonlinear optical or electro-optic group in most of the examples. Usually the functional group is an internal charge-transfer system consisting of two aromatic or heteroaromatic rings, connected via conjugated double bonds, and substituted by a donor and an acceptor group in 4,4′ position as e.g.

a) stilbenes (bridge between the aromatic rings: -C=C-)
b) stilbazolium salts (positively charged pyridinium ring, bridge: -C=C-)
c) diazo dyes (bridge: -N=N-)
d) azomethines (bridge: -N=C-)

In some cases the conjugation between the two ring-systems is broken. An example are phenylhydrazones (bridge between the aromatic rings: -CH=N-NH-). Such molecules may show a relatively low maximum absorption wavelength connected with medium hyperpolarizabilities (see the three phenylhydrazones in Table 8.3 described by Bubeck et al. and Laschewsky et al.). The most outstanding nonlinear optical or electro-optic properties have been detected in films of stilbazolium salts. The pyridinium group acts as a very good acceptor. Within the series of the three merocyanines published by Bubeck et al. and Laschewsky et al. the different donor properties of the amino, the thioester and the ester group can be compared. The strongest donor is the dialkylamino group. The hyperpolarizability of N-methyl-4-(4′-N,N-dihexadecylamino) — styrylpyridinium iodine ($\beta = 1300 \cdot 10^{-40} m^4/V$) is resonance enhanced. After changing the amino function by a weaker donor (alkylether or thioether) medium hyperpolarizabilities can still be achieved, but are now combined with a remarkable blue shift of the maximum absorption wavelength.

The advantages of thioester donor groups were pointed out previously (*Morley et al. 1987, Hutchings et al. 1989, Li et al. 1988*). The alkylthioester is a stronger donor than the alkylester. This can also be seen comparing the two 4,4′ substituted phenylhydrazones in Table 8.3. The hyperpolarizability enhancement of the hydroxy derivative within the series of phenylhydrazones can be attributed to an intramolecular hydrogen bond increasing the planarity of the conjugated system. Similar effects have been reviewed elsewhere (*Broussoux et al. 1989*).

Some remarkable results have been obtained by Ashwell and co-workers through a combination of molecular modelling and chemical synthesis. High quality noncentrosymmetric Y-type LB films with more than 150 active layers of the stilbazolium salt N-octadecyl-4-{2-(4-dimethylamino-phenyl)ethenyl}pyridinium iodide could be deposited. The active films were alternated with a specially designed amphiphile (4,4′-dioctadecyl-3,5,3′,5′-tetramethyldipyrrylmethene hydrobromide, DPM). The two hydrocarbon chains of DPM are separated from each other so that the space between them can be occupied by the tail of the active molecule. Adjacent layers were designed to fasten to each other like a 'molecular zip'. So far no optical waveguiding experiments and not much structural data have been published which could confirm the proposed order. However, the quadratic dependence of the second-harmonic intensity on the number of deposited layers up to large numbers of films is quite promising.

An increase of the coefficient d_{33} by an order of magnitude was achieved for monolayers of the same stilbazolium compound by mixing it with octadecylsulfate in a 1:1 molar ratio. The cosurfactant was designed to act as a counterion and a

Table 8.3 Second-order nonlinear optical and electro-optic LB film molecules

If not specified, nonlinear optical experiments were performed at 1,064 nm (fundamental wavelength), electro-optic experiments at 632.8 nm. The x_3 axis is normal to the films, except for the DCANP samples where x_3 corresponds to the dipping direction of the LB preparation process. Susceptibilities which are mentioned without values have been detected but not determined quantitatively.

Φ tilt angle of chromophore with respect to x_3, for DCANP an in-plane and an out-of-plane tilt (Φ_{ip}, Φ_{oop}) has to be considered

λ_{eg} longest wavelength absorption maximum * assuming $d = 1/2\chi$

molecule	λ_{eg} [nm]	film type/ maximum number of deposited layers	d_{IJK} [pm V^{-1}] β [10^{-40} $m^4 V^{-1}$] Φ [deg] r_{IJK} [pm V^{-1}]	remarks SH intensity ∝ (number of layers)^2?	references
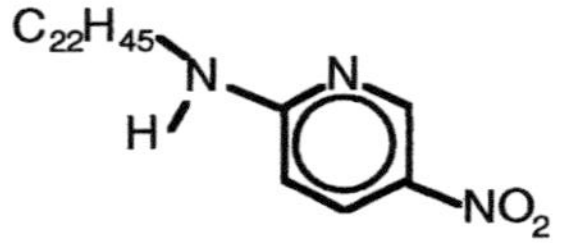 $C_{22}H_{45}$, N, H, N, NO_2 DCANP	375	Y/540 (pyrex glass)	$d_{33} = 7.8 \pm 1$ $d_{32} = 2.6 \pm 0.6$ $d_{31} = 2.0 \pm 0.5$ $d_{23} = 1.0 \pm 0.3$ d_{22}, d_{21}	SH quadratic $d_{quartz} = 0.4$ pm V^{-1}	*Decher et al.* *1988, 1989a* *Bosshard et al.* *1990, 1991a* *1992d*
			$d_{33} = 5.6 \pm 1$ (1318 nm) $d_{33} = 3.7 \pm 0.5$ $d_{31} = 1.1 \pm 0.3$ (1907 nm) $\Phi_{ip} = \pm 27$ $\Phi_{oop} = \pm 27$	n by mode spectroscopy and ellipsometry phase-matched waveguiding —Cerenkov-type — mode-conversion	*Bosshard et al.* *1991b, 1992c,e* *Flörsheimer* *et al. 1992a,b*
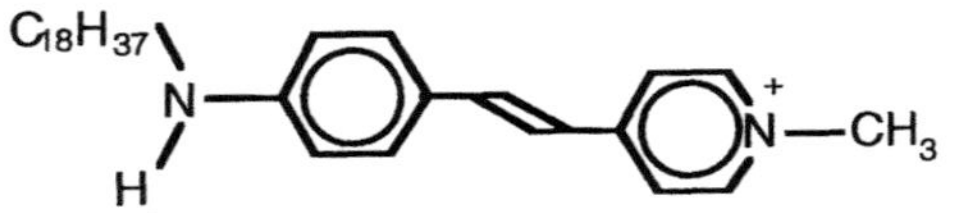 $C_{18}H_{37}$, N, H, N^+—CH_3		1 (silver)	d_{33}, d_{32} $\Phi = 23.7$ $r_{33} = -317.7 - i\ 398.7$	pH = 5.5 $\varepsilon(\omega) = 2$ assumed, $\varepsilon(2\omega)$ by surface plasmon spectroscopy	*Cross et al.* *1988*
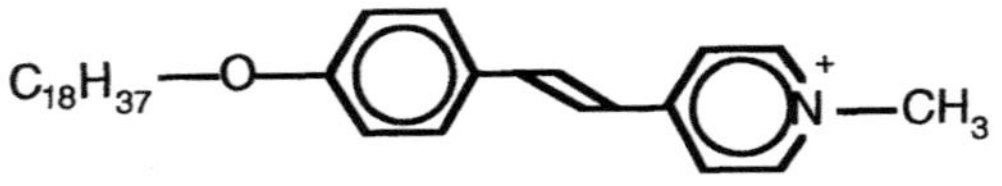 $C_{18}H_{37}$—O, N^+—CH_3		1 (silver)	d_{33}, d_{32} $\Phi = 34.1$ $r_{33} = -37.4 + i\ 32$	pH = 5.5 $\varepsilon(\omega) = 2$ assumed, $\varepsilon(2\omega)$ by surface plasmon spectroscopy	*Cross et al.* *1988*

Table 8.3 Continued.

molecule	λ_{eg} [nm]	film type/ maximum number of deposited layers	d_{IJK} [pm V^{-1}] β [10^{-40} m^4 V^{-1}] Φ [deg] r_{IJK} [pm V^{-1}]	remarks SH intensity ∝ (number of layers)^2?	references
6.0 5.5	360	1 (glass)	$d_{33} = 18*$ $d_{32} = 4.2*$ β = 180 Φ = 35	n(ω) =1.5, n(2ω) = 1.8 assumed	*Bubeck et al. 1991, Laschewsky et al. 1991*
	380	1 (glass)	$d_{33} = 38*$ $d_{32} = 5*$ β = 270 Φ = 26	n(ω) =1.5, n(2ω) = 1.8 assumed	*Bubeck et al. 1991, Laschewsky et al. 1991*
$(C_{10}H_{21})_2N-C_6H_4-CH=CH-C_5H_4N^+-C_2H_4-COO^-$		1 (silver)	$r_{33} = -279 + i\ 13.98$	pH = 6…7 area decrease 42%/h (34 mNm^{-1})	*Cross et al. 1987*
$(C_{14}H_{29})_2N-C_6H_4-CH=CH-C_5H_4N^+-C_2H_4-COO$		1 (silver)	$r_{33} = -169 - i\ 48.6$	pH = 6…7 area decrease 32%/h (34 mNm^{-1})	*Cross et al. 1987*
$(C_{16}H_{33})_2N-C_6H_4-CH=CH-C_5H_4N^+-CH_3$ J^-	475	1 (glass)	$d_{33} = 73*$ $d_{32} = 32*$ β = 1300 Φ = 43	n(ω) =1.5, n(2ω) = 1.8 assumed	*Bubeck et al. 1991, Laschewsky et al. 1991*
$(C_{18}H_{37})_2N-C_6H_4-CH=CH-C_5H_4N^+-C_2H_4-COO^-$		1 (silver)	$r_{33} = -156.1 - i\ 5.82$	pH = 6…7 area decrease 6%/h (34 mNm^{-1})	*Cross et al. 1987*

Table 8.3 Continued.

molecule	λ_{eg} [nm]	film type/ maximum number of deposited layers	d_{IJK} [pm V^{-1}] β [10^{-40} m^4 V^{-1}] Φ [deg] r_{IJK} [pm V^{-1}]	remarks SH intensity ∝ (number of layers)^2?	references
$C_{18}H_{37}$–N^+(pyridinium)–CH=CH–(phenyl)–$N(CH_3)_2$ J^- and DPM (Br^-)	460	1 (glass)	$d_{33} = 500 \pm 250$* $d_{32} = 30 \pm 10$* $\beta = 4000 \pm 2000$ $\Phi = 18$	$d_{quartz} = 0.5$ pm V^{-1} n = 1.7 assumed increase of d33 by an order of magnitude due to co-surfactant	*Ashwell et al. 1992a*
		Y alternating/ 200 active layers	d_{33}, d_{32}		*Ashwell et al. 1991*
				SH quadratic, tested for up to 150 active layers	*Ashwell 1992b*
$C_{22}H_{45}$–N^+(pyridinium)–CH=CH–(phenyl)–OH		Y alternating with 22-tricosenoic acid/ 3 active layers	d_{33}, d_{32} $\beta = 11000 \pm 9000$ $\Phi = 9$	data from more polarizable deprotonated form (NH_3, 12 h) SH subquadratic	*Girling et al. 1985*
			$\beta = 4000$ (EFISH, $-CH_3$ instead of $-C_{22}H_{45}$, 1890 nm)		*Dulcic et al. 1978b*

Table 8.3 Continued.

molecule	λ_{eg} [nm]	film type/ maximum number of deposited layers	d_{IJK} [pm V^{-1}] β [10^{-40} m^4 V^{-1}] Φ [deg] r_{IJK} [pm V^{-1}]	remarks SH intensity $\propto$ (number of layers)^2?	references
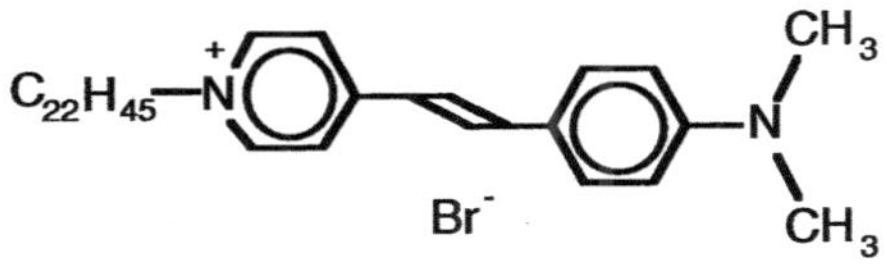	339	1 (glass)	d_{33}	pH = 9.2	
	477			diluting dye in arachidic acid (1:4), d_{33} increases by a factor of 6.6 (diminishes concentration of H-aggregates)	*Schildkraut et al. 1988*
		1 (silver)	d_{33}, d_{32} $\Phi = 22$ $r_{33} = 223.7 + i\,106$	pH = 5.5	*Cross et al. 1988*
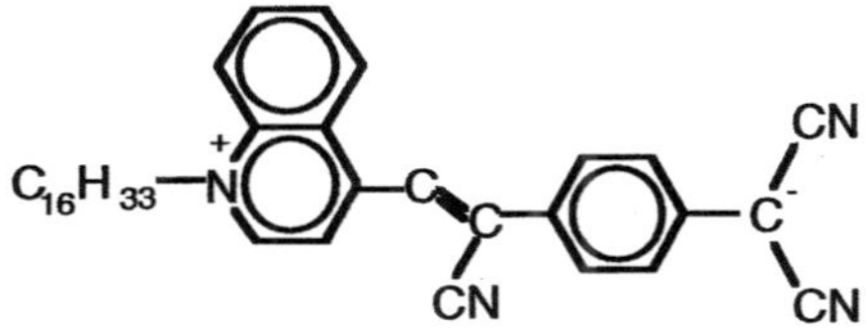 R-Q3CNQ	565	Z/50	d_{33}, d_{32}	SH quadratic pH = 5.5 intramolecular charge transfer in mixed films of R-Q3CNQ derivatives and similar dyes, λ_{eg} tunable by mixing ratio	pH = 5.5 *Ashwell et al. 1990, 1991*
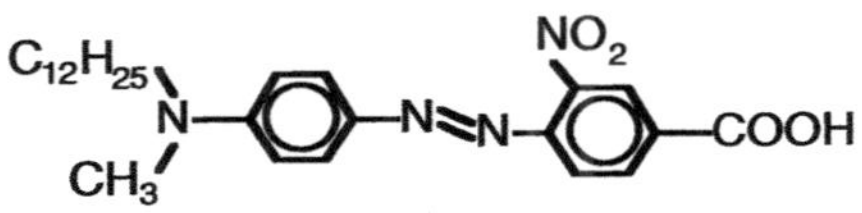 DPNA	460	Y		$n(\omega) = 1.7$, $n(2\omega) = 4$ assumed	*Allen et al. 1986*
		1 (glass)	$d_{33} = 1400$* $\beta = 3700$	on 1 mM $CdCl_2$ or $CaCl_2$, improves film quality (higher collapse pressure, more homog. fringing pattern)	
		Z/20		SH sublinear, especially for first layers	

Table 8.3 Continued.

molecule	λ_{eg} [nm]	film type/ maximum number of deposited layers	d_{IJK} [pm V^{-1}] β [10^{-40} $m^4 V^{-1}$] Φ [deg] r_{IJK} [pm V^{-1}]	remarks SH intensity ∝ (number of layers)^2?	references
	500	Z	$d_{33} = 1300$* $\beta = 5100$	prepared from pure water subphase, unusual isotherm	
	450	1 (glass)	$d_{33} = 340 \pm 30$* $\beta = 1300 \pm 100$ $\Phi = 65 \pm 5$	$d_{quartz} = 0.5$ pm V^{-1} $n(\omega) = 1.60$, $n(2\omega) = 1.90$ by ellipsometry (y/41) pH = 6.5–7.1, 0.75 mM $(CH_3COO)_2Cd$	*Ledoux et al. 1988*
DPNA continued		Y/41		SH signal (not quantified)	
		Z/12		SH subquadratic (≈ const. for 6 layers) spacer film on glass decreases β	
		Y/121	$\lvert r\ (TE)\rvert = 48$ (y/75) $\lvert r\ (TM)\rvert = 250$ (y/115)	guided modes	*Loulergue et al. 1987*
		Z/11		SH const. for 1–11 layers	
$C_{12}H_{25}$–⟨ring with NO_2⟩–N=N–⟨ring⟩–N(C_2H_4COOH)(CH_3)	460	1 (glass)	$d_{33} = 160 \pm 20$* $\beta = 920 \pm 80$* $\Phi = 60 \pm 5$	$d_{quartz} = 0.5$ pm V^{-1} pH = 6.5–7.1, 0.75 mM $(CH_3COO)_2Cd$	*Ledoux et al. 1988*

Table 8.3 Continued.

molecule	λ_{eg} [nm]	film type/ maximum number of deposited layers	d_{IJK} [pm V^{-1}] β [10^{-40} m^4 V^{-1}] Φ [deg] r_{IJK} [pm V^{-1}]	remarks SH intensity ∝ (number of layers)^2?	references
$C_{17}H_{35}$–OC(=O)–C$_6$H$_3$(NO$_2$)–N=N–C$_6$H$_4$–N(C_2H_4COOH)(CH_3)	440	1 (glass) y z	$d_{33} = 160 \pm 20$* $\beta = 1120 \pm 125$ $\Phi = 52 \pm 3$	$d_{quartz} = 0.5$ pm V^{-1} pH = 6.5–7.1, 0.75 mM $(CH_3COO)_2Cd$	*Ledoux et al. 1988*
$C_{17}H_{35}$–OC(=O)–C$_6$H$_3$(NO$_2$)–N=N–C$_6$H$_4$–N(C_2H_4COOH)(CH_3) and DPNA		Y alternating/ 11 active layers	$d_{33} = 170$*	$d_{quartz} = 0.5$ pm V^{-1} n(ω) =1.57, n(2ω) = 1.72 by ellipsometry pH = 6.5–7.1, 0.75 mM $(CH_3COO)_2Cd$ SH subquadratic, especially for first layers	*Ledoux et al. 1988*
–(CH_2–CH(C(=O)–O–$(CH_2)_2$–OH))$_6$–(CH_2–CH(C(=O)–$(O(CH_2)_2)_3$–R))– R = –N(CH_3)–C$_6$H$_4$–N=N–C$_6$H$_4$–S(=O)$_2$–$C_{18}H_{37}$ (random copolymer)		Y alternating with 22-tricosenoic acid/ 30 active layers		SH quadratic	*Penner et al. 1991*
		Y alternating with poly (iso-butyl methacrylate)/ 93 active layers	$d_{33} = 21$* (EFISH, 1907 nm)	n(ω) =1.495, n(2ω) = 1.500 by ellipsometry	

Table 8.3 Continued.

molecule	λ_{eg} [nm]	film type/ maximum number of deposited layers	d_{IJK} [pm V^{-1}] β [10^{-40} m^4 V^{-1}] Φ [deg] r_{IJK} [pm V^{-1}]	remarks SH intensity ∝ (number of layers)^2?	references
$C_{16}H_{33}$–O–C$_6$H$_4$–C(H)=N–N(H)–C$_6$H$_4$–NO_2	420	1 (glass)	$d_{33} = 15^*$ $d_{32} = 11^*$ $\beta = 240$ $\Phi = 56$	n(ω) =1.5, n(2ω) = 1.7 assumed	*Bubeck et al. 1991* *Laschewsky et al. 1991*
$C_{18}H_{37}$–O–C$_6$H$_3$(OH)–C(H)=N–N(H)–C$_6$H$_4$–NO_2	420	1 (glass)	$d_{33} = 22^*$ $d_{32} = 14^*$ $\beta = 280$ $\Phi = 66$	n(ω) =1.5, n(2ω) = 1.7 assumed hydrogen bond increases planarity of conjugated system	*Bubeck et al. 1991* *Laschewsky et al. 1991*
$C_{18}H_{37}$–S–C$_6$H$_4$–C(H)=N–N(H)–C$_6$H$_4$–NO_2	405	1	$d_{33} = 32^*$ $d_{32} = 12^*$ $\beta = 340$ $\Phi = 41$	n(ω) =1.5, n(2ω) = 1.7 assumed	*Bubeck et al. 1991* *Laschewsky et al. 1991*
FA 06		Y alternating with trimethylsilyl-cellulose/ 200 active ayers	$d_{32} = 10^*$	$d_{quartz} = 0.5$ pm V^{-1} SH quadratic	*Hickel et al. 1993*

Table 8.3 Concluded.

molecule	λ_{eg} [nm]	film type/ maximum number of deposited layers	d_{IJK} [pm V^{-1}] β [10^{-40} m^4 V^{-1}] Φ [deg] r_{IJK} [pm V^{-1}]	remarks SH intensity ∝ (number of layers)^2?	references
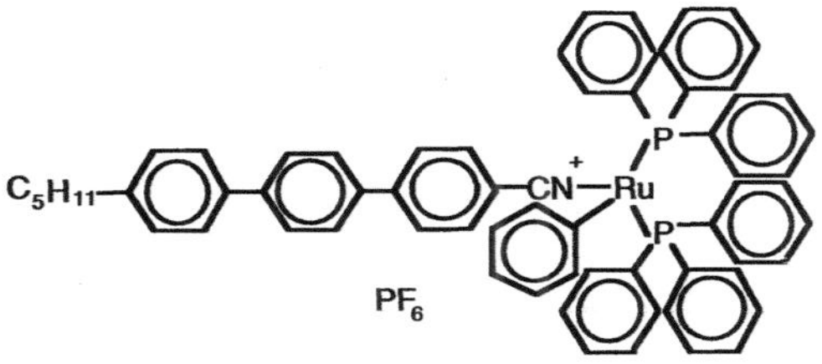 C_5H_{11} … CN–Ru, P, P, PF_6 RuCTP	332	Z/7	d_{33}, d_{32} $\beta = 470$	pH = 5.8 SH superquadratic transfer ratio > 0.95 optical absorption and loaded mass increase linearly with number of layers	*Richardson et al. 1989*
mixture of RuCTP : CTP = 1: 1.7 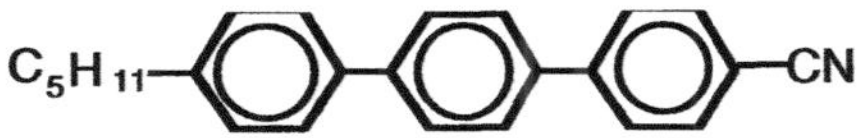 C_5H_{11} … CN CTP		Z/7	d_{33} d_{32} $\beta = 700$	pH = 5.8 higher molecular packing and higher degree of structural order as compared to pure RuCTP	*Richardson et al. 1989*
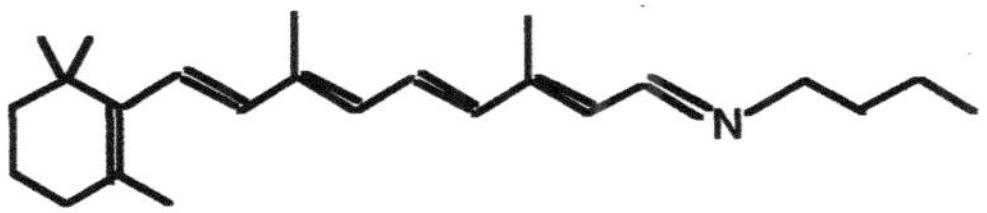 N		1 (water)			*Huang et al. 1988*

spacer molecule. It prevents the active molecule from H-aggregation and allows for a smaller tilt angle of the chromophores which was confirmed by absorption spectroscopic and nonlinear optical orientation measurement. However, it would be interesting to calculate if the increase of d_{33} is mainly due to the decrease of tilt of the dye monomers or if there is an additional effect as e.g. a change of the acceptor strength of the active molecule or a significant change of the monomer/aggregate ratio of the dye. Moreover, as can be seen from Table 8.3, the large nonlinear optical coefficients of these films are partly due to resonance enhancement. Additionally only monolayers could be prepared. However, the considerable increase of the largest nonlinear optical coefficient d_{33}, which is at least partly due to a more favorable molecular alignment, demonstrates well the successful modeling of the molecular packing in a two-dimensional system.

Two other polar materials which are suited for optical waveguiding have been described by Penner et al. and by Hickel et al. Both teams have utilized prepolymerized polymers. These polymers are desired to be transferred from fluid-like states on the water surface in order to avoid grain boundaries. Penner and co-workers used a random copolymer with a diazo dye as chromophore and hydroxyethylacrylate in a 1:6 molar ratio. Y-type multilayers with a normal overall polarization have been prepared by alternating the active polymer films with inactive monolayers. Various polymeric and monomeric species have been used as inactive compounds. LB films with 93 active layers could be prepared with poly (iso-butylmethacrylate) as an inactive layer. A coefficient of d_{33} = 21 pm V^{-1} was calculated for this composite material from electric field induced second-harmonic generation data (EFISH, see Chapter 7.2.2) of the chromophore at a fundamental wavelength of 1,907 nm. Attenuation losses as low as 1 dB cm^{-1} were measured for similar LB films (iso (tert-butylmethacrylate) as interlayer).

Hickel et al. used a phenylhydrazone (FA 06) as a monomeric active compound. Alternating Y-type multilayers with a normal overall polarization were prepared with interlayers of a series of inactive compounds. The largest second-order susceptibilities (d_{32} = 10 pm V^{-1}) were obtained utilizing the polymer trimethylsilylcellulose. LB films with 200 active layers could be prepared with a quadratic dependence of the generated second-harmonic signal on the number of layers.

Attenuation losses of 3.5 dB cm^{-1} were measured in an LB film of this material. Waveguide properties of LB films will be discussed in more detail in Section 8.3.2.

A different type of interesting compound are the metal organics. An example is given in Table 8.3: (η^5-cyclopentadienyl)-(bistriphenylphosphine)-4-cyano-4″-n-pentyl-p-terphenyl-hexafluorphosphate (RUCTP). The high hyperpolarizability, combined with a very low maximum absorption wavelength, is striking. The thermal stability of similar complexes was shown to be quite good (*Richardson et al. 1989*).

Most of the monomeric species used so far for the preparation of polar LB films have a single hydrocarbon chain. Some examples of molecules with two tails are listed in Table 8.3. Usually all the chains of a molecule are fixed on one site of the head group, e.g. as substituents of an amino- or methyl-group. The reason is that most scientists have tried to orient the charge-transfer axes as steeply as possible, i.e. LB films with a normal overall polarization have been preferred. However, as will be pointed out in Chapter 9.1, an in-plane overall polarization is advantageous for efficient frequency-doubling in a waveguide. It was demonstrated that an in-plane orientation of the molecular charge-transfer axes can be obtained during the LB transfer process of DCANP and other films, mentioned in Chapter 5.2. DCANP is the first material where phase-matched frequency-doubling was achieved (Cerenkov-type and mode-conversion, see Chapter 9.1). Unfortunately the magnitude of the susceptibilities of DCANP LB films are not sufficient for technical applications. Moreover the optical attenuation losses are too high. Better materials are desired. A possibility for future development of in-plane polar materials would be the substitution of a larger chromophore by two or more chains connected with the head group at different sites, such that molecular packing considerations (Chapter 5.3) would allow for a large tilt angle of the charge-transfer axis.

8.3.2 Waveguiding properties of LB films

Presently, only a few molecules are known to form LB multilayers thick enough for optical waveguiding. Examples of linear and nonlinear optical LB films are listed in Table 8.4. Attenuation losses of guided modes smaller than 1 dB cm^{-1} are desired for technical applications. The first successful waveguide experiments with LB films were reported by Pitt and Walpita. Using films of N-(n-octadecyl)-acylamide they measured attenuations down to 1 dB cm^{-1} for a TM_0-mode at a wavelength of 632.8 nm. However, it should be considered that the effective refractive index N_{eff} in these experiments was very close to the substrate index n_S (see Table 8.4). Hence the modes were very weakly guided. Consequently the attenuation measured in these experiments was predominantly due to the substrate and not to the intrinsic losses of the LB films, which are much higher.

In order to compare the attenuation reported in Table 8.4, the refractive index data and the thicknesses of the LB films are given if mentioned in the cited publications. Additionally the wavelengths of the probing light, the types of the modes and the methods of coupling light into the waveguides are listed. Different types of surfactants are presented: conventional amphiphiles with a hydrocarbon chain and a hydrophilic head group ('lollipop'-type monomers), 'hairy rod' molecules and preformed polymers.

Most monomeric LB films are transferred from liquid crystalline or crystalline states of monolayers with large numbers of domains on the water surface. This leads to high densities of grain or twin boundaries and disclinations in the LB films (see Chapter 5.3). One approach to overcome this problem is to transfer monolayers from fluid-like states. This can be done, utilizing performed polymers which usually have side-chains of different length. Tredgold et al. were the first who reported on waveguiding in such LB films.

Two polyglutamates (GD 10, GD 14) are mentioned in Table 8.4 as examples of 'hairy rod' molecules. The polypeptide backbone of these molecules forms a stiff α-helical structure. The conformationally mobile side-chains are wrapping each

Table 8.4 Waveguiding properties of LB films

molecule	λ [nm]	thickness [nm]	mode	attenuation [dB cm^{-1}]	substrate index n_S	film index n_F	effective index N_{eff}	coupling method	references
$C_{18}H_{37}$-N(H)-C(=O)-CH=CH_2	632.8	440	TE_0 TM_0	< 1 ($N_{eff} \sim n_S$)	1.4573 1.4573	1.5127 1.5266	1.47978 1.45748	prism coupling	*Pitt et al. 1980*
DCANP ($C_{22}H_{45}$(H)N–pyridine–NO_2)	632.8 632.8	265 265	TM_0 TE_0	12–20 25–35	1.470 1.470	x 1.598	1.472 1.492	grating and prism coupling	*Bosshard et al. 1990*
	926.5	199	TE_0	5.5	1.451	1.587	1.604		*Flörsheimer et al. 1992b*
GD 14: (N(H)-CH((CH$_2$)$_2$-C(=O)-OR$_1$)-C(=O)-)$_x$-(-N(H)-CH((CH$_2$)$_2$-C(=O)-OR)-C(=O)-)$_{1-x}$; R_1 = $-C_{20}H_{41}$, R_2 = CH_3	632.8	600	TE_0	2.5–5.5	1.470	1.498	x	grating coupling	*Hickel et al. 1990*

Table 8.4 Continued.

molecule	λ [nm]	thickness [nm]	mode	attenuation [dB cm^{-1}]	substrate index n_S	film index n_F	effective index N_{eff}	coupling method	references
$(NH-CH(CH_2)_2COOR_1-CO-)_x-(-NH-CH(CH_2)_2COOR-CO-)_{1-x}$ $R_1 = -C_{18}H_{37}$ $R_2 = CH_3$ $x \sim 0.3$ GD 10	530.9	542.5	TM_0 TE_0 TE_0	 2.0 ± 0.5 intrinsic: 6 ± 1	1.4611	1.4944 $n_{F3} = 1.5042$ $n_{F1} = 1.4937$ (in-plane anisotropy, optical main axes parallel and normal to dipping direction)	x	prism coupling	*Mathy et al. 1992*
$-(CH_2-CHR-CH(COOH)-CH(COOCH_3))_n-$ R= $-(CH_2)_2-C(=O)-C_6H_4-C_6H_4-C_5H_{11}$ (random copolymer)	632.8	544	x	11	x	x	x	prism coupling	*Tredgold et al.* 1987
$-(CH_2-C(CH_3)(COOR))_n-$ R= $-(CH_2)_2-O-C(=O)-(CH_2)_2-C(=O)-N(C_{17}H_{35})(C_2H_5)$	632.8	800	TE_0	2	1.46 (74% of intensity in LB film)	1.503	1.480	prism coupling	*Falk et al. 1992*

Table 8.4 Continued.

molecule	λ [nm]	thickness [nm]	mode	attenuation [dB cm^{-1}]	substrate index n_S	film index n_F	effective index N_{eff}	coupling method	references
$-(CH_2-CH(-C(=O)-O-(CH_2)_2-OH))_6-(CH_2-CH(-C(=O)-(O(CH_2)_2)_3-R))-$ R = $-N(CH_3)-C_6H_4-N{=}N-C_6H_4-SO_2-C_{18}H_{37}$ (random copolymer) alternating with poly (*tert*-butylmethacrylate)	632.8	560	x	1	x	x	x	prism coupling	*Penner et al. 1991*
$C_{18}H_{37}-S-C_6H_4-CH{=}N-NH-C_6H_4-NO_2$ alternating with trimethylsilylcellulose	632.8	720	TE_0	3.5	x (80% of intensity in LB film)	x	x	prism coupling	*Hickel et al. 1993*

Table 8.4 Concluded.

molecule	λ [nm]	thickness [nm]	mode	attenuation [dB cm^{-1}]	substrate index n_S	film index n_F	effective index N_{eff}	coupling method	references
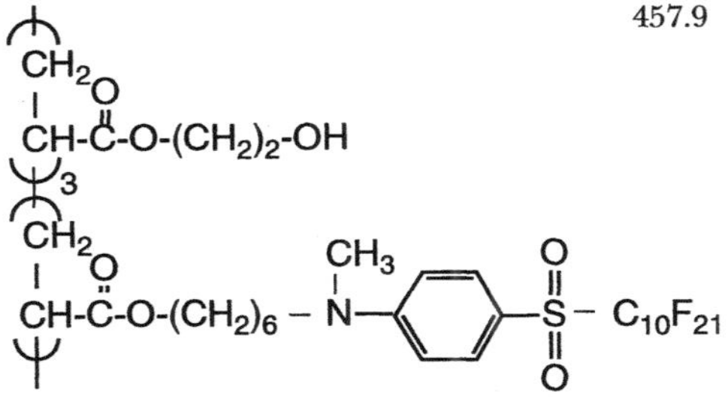 (a) 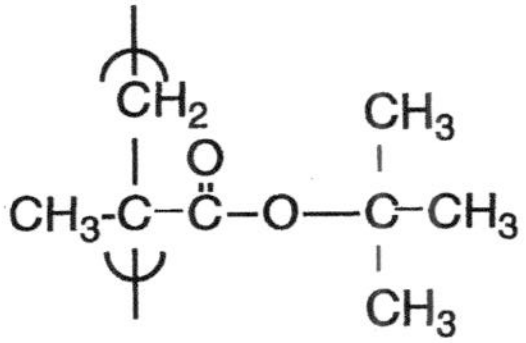 (b) NLO-active polymer (a) with poly (*tert*-butyl methacrylate (b)	457.9	440	TM_0	2.6	1.481	n_{ord} =1.5075	—	prism coupling	*Clays et al. 1993a*
			TE_0	5.8	1.481	n_{ext} =1.506	1.483		

x values not mentioned in the publications

backbone in a fluid-like skin. There are two conceptions combined with the design of such molecules: one aim is to prepare multilayers without grain boundaries for low loss waveguides. The other aim was reported in more detail in Chapter 5.3: the rod-like molecules are examples of elongated objects which can be oriented in the film plane during the LB transfer procedure.

The attenuation losses measured in LB film waveguides of the polyglutamates (Table 8.4) are lower than the losses in most examples of monomeric LB film waveguides published so far. In the publication of Mathy et al., refractive index data for various wavelengths can be found. Moreover a strong in-plane anisotropy was measured. (n_{F3} in Table 8.4 is the refractive index of the LB film for the electric field oriented parallel to dipping direction, n_{F1} is the in-plane index for the electric field normal to the dipping direction.) It is astonishing that the GD 14 derivative did not show an index anisotropy.

So far it has not been shown that the in-plane anisotropy of 'hairy rod' compounds can be utilized for second-order nonlinear optical and electro-optic applications. For this hyperpolarizable backbones have to be used. However, it could be a problem to prevent the polar rods from a random antiparallel alignment.

Considering the nonlinear optical and the waveguide properties, the best LB film materials reported so far are clearly the polymeric alternating Y-type materials developed by Penner and co-workers and by Hickel et al. This is a promising result since these films are the first nonlinear optically active materials based on the preformed polymer concept and further progress can be expected. A drawback of alternatingY-type LB films with passive interlayers is the dilution of the chromophores. Hence the development of active/active alternating Y-type or Z- and X-type films is desirable if one looks for materials with a normal overall polarization. An additional drawback of preformed polymers with chromophores in a part of the side-chains is the dilution with disordered inactive parts of the polymer.

However, the concept of preformed polymers is not the only promising concept for further progress. Future materials could also be prepared from liquid crystalline or solid states of monomeric Langmuir monolayers. Possibilities for the increase of the range of orientational order and for the reduction of the densities of grain boundaries and disclinations have been discussed comprehensively in Chapter 5.3. So far lipid films are the best characterized films. A systematic combination of molecular modelling of new compounds, chemical synthesis and physical investigation is desirable. Especially the *in situ* characterization of Langmuir monolayers in dependence of the environmental parameters seems promising for the discovery of favorable ordered films with low defect densities.

8.4 OVERVIEW ON PRESENTLY KNOWN ELECTRO-OPTIC AND NONLINEAR OPTICAL POLYMER FILMS

Table 8.5 summarizes electro-optic and nonlinear optical coefficients for some important polymer systems. According to Eqs. (6.14, 6.15) the nonlinear response should be proportional to the poling field strength and the number density of the incorporated active groups.

The fractional amount of poling can be estimated in the following way. If all the chromophores are perfectly aligned, $\langle cos^3\vartheta \rangle$ in Eq. (6.1) equals 1. Knowing the number density of the nonlinear optical molecules and their microscopic hyperpolarizability β, the second-harmonic coefficient is given by

$$d_{33} = N \cdot f^{2\omega} \cdot f^{\omega} \cdot f^{\omega} \cdot \beta_{zzz} \tag{8.3}$$

with local field correction factors as in Eq. (3.27). For the guest-host system PMMA/DR1 (*Singer 1987, 1988*) one has $N = 2.74 \cdot 10^{26} m^{-3}$. The hyperpolarizability β_{zzz} equals $525 \cdot 10^{-40} m^4/V$ at $\lambda = 1.36\ \mu m$ (see Table 8.1). Assuming refractive indices of about 1.52 at both frequencies the maximum value for d_{33} is 43 pm/V. From the measured value $d_{33} = 2.5$ pm/V at $\lambda = 1.58\ \mu m$ one gets a fractional amount of poling, $\langle cos^3\vartheta \rangle$, of approximately 6% (assuming a δ-distribution).

At higher field strengths saturation effects have to be taken into account. The exact solution for the isotropic model in Eq. (6.10) is

$$\left\langle cos^3\vartheta \right\rangle_{isotropic} = \frac{e^a(a^3 - 3a^2 + 6a - 6) + e^{-a}(a^3 + 3a^2 + 6a + 6)}{(e^a - e^{-a})a^3} \tag{8.4}$$

with $a = \dfrac{\mu^*_{g,z} E_p}{k_B T}$

Figure 8.2 is a plot of this function. The dashed line corresponds to the linear regime, Eq. (6.10). Saturation effects start at $a \approx 1$. This means for a corrected dipole moment $\mu^*_{g,z} = 1.63 \times \mu_{g,z} = 4.7 \times 10^{-29}$ Cm (= 14.2D) of DR1 ($\mu_g = 2.9 \times 10^{-29}$ Cm (=8.7 D), $\varepsilon = 3.6$, n = 1.52 (*Sohn 1989*)) that at elevated temperatures (125°C $\simeq T_g$)

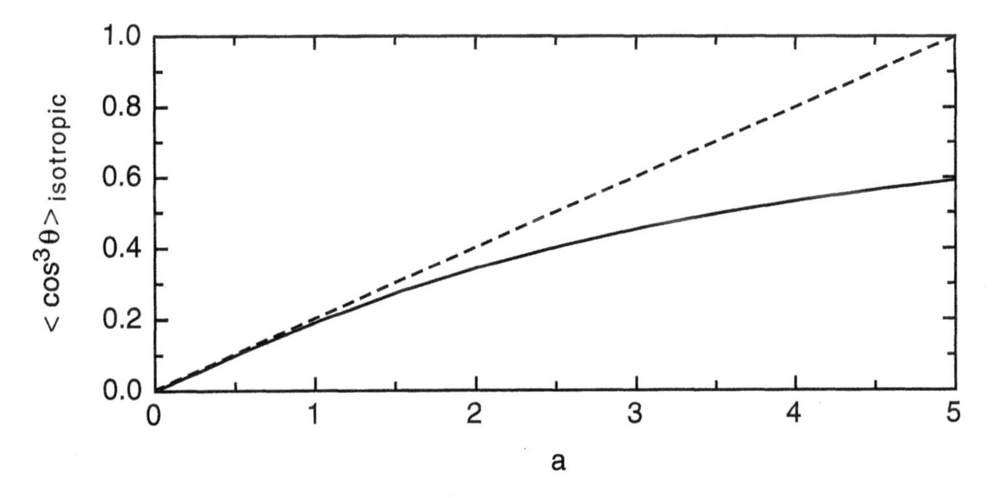

Figure 8.2 Fractional amount of poling of a polymer film for the isotropic model.

Table 8.5 Electro-optic and nonlinear optical properties of selected polymer systems.

Typ[a)]	Matrix and nonlinear optical moiety	E_p [V/μm]	d[pm/V] @ λ [nm]	r[pm/V] @ λ [nm]	T_g [°C]	references
GH	poly(methyl methacrylate) PMMA 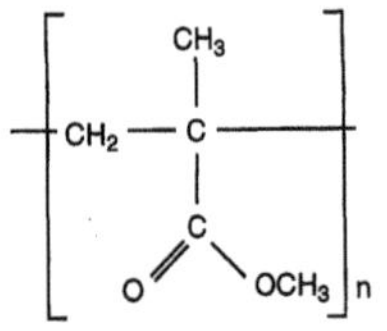 4[N-ethyl-N-(2 hydroxyethyl)] amino-4′-nitrozobenzene (Disperse Red 1) DRI 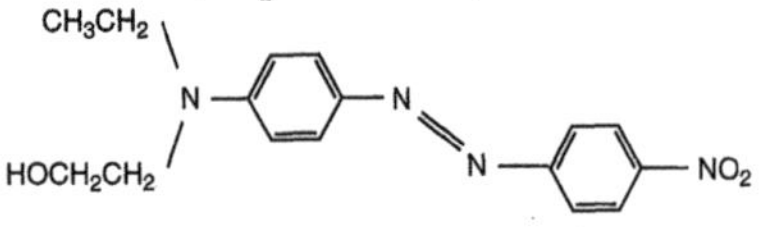	37	$d_{33} = 2.5$ (1580)	$r_{33} = 2.5$ (633)	—	*Singer et al. 1987*
GH	PMMA + 4-tricyanovinyl-4′-(diethylamino) azobenzene TCV 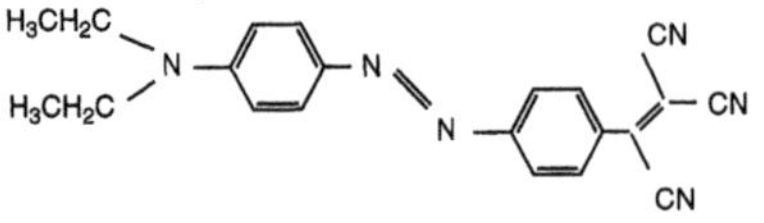	—	$d_{33} = 84$ (1356)	—	—	*Katz et al. 1988*
GH	Pyralin 2611D (polyimide) 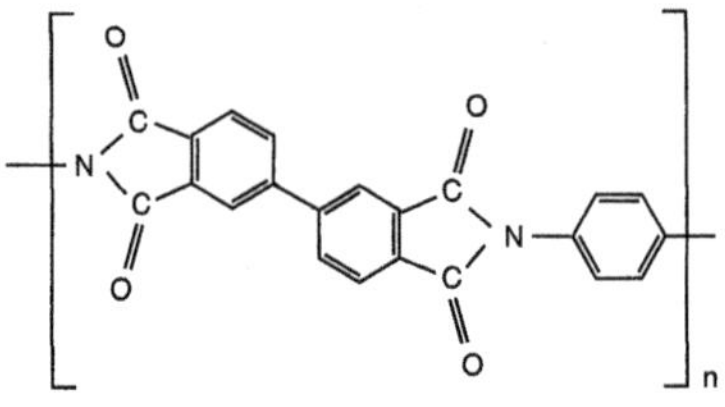 + Eriochrome Black T 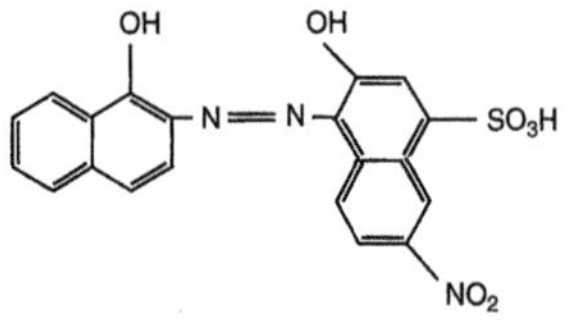	50	—	—	≈ 320	*Wu et al. 1991*

poling field strengths higher than 1.1×10^8 V/m are necessary to leave the linear range.

Table 8.5 also gives a nice overview of the developments made in the field of nonlinear optical polymers. At the beginning people started with relatively simple guest-host systems in order to demonstrate the advantages of this new class of

Table 8.5 Continued.

Typ[a)]	Matrix and nonlinear optical moiety	E_p [V/μm]	d[pm/V] @ λ [nm]	r[pm/V] @ λ [nm]	T_g [°C]	references
CL	N,N-diglycidyl-4 nitroaniline NNDN + N-(2-aminophenyl)-4-nitroaniline NAN 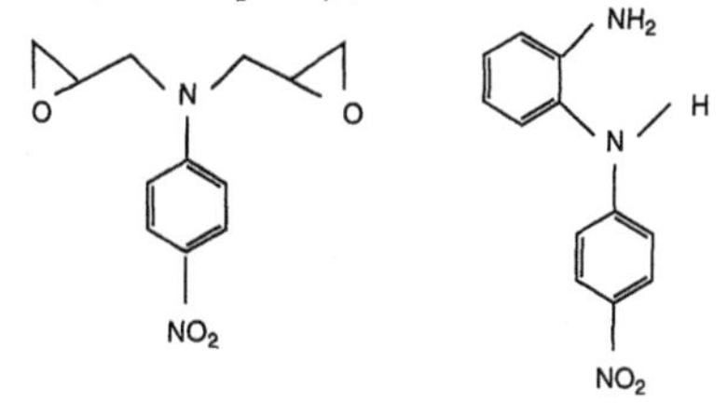	—	$d_{31} = 14$ (1064)	$r_{13} = 6.5$ (530.9)	≈ 110	*Jungbauer et al. 1990*
SC	PMMA + 4-dicyanovinyl-4′-[N-ethyl-N-(2 hydroxyethyl)] azobenzene DCV 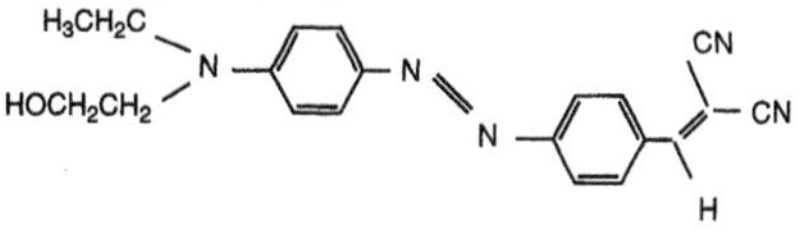	—	$d_{33} = 21$ (1580)	$r_{33} = 18$ (799)	127	*Singer et al. 1988*
SC	b) + 4-dimethylamino-4′-nitrostylbene DANS 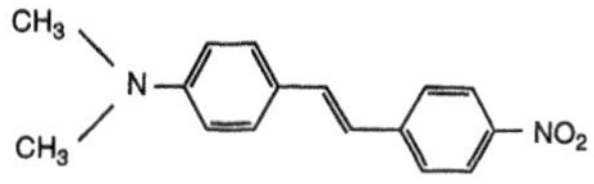	170	—	$r_{33} = 34$ (1300)	140	*Möhlmann et al. 1990*
SC	b) + DANS based	≈ 200		$r_{33} = 38$ (1300) $r_{33} = 98$ (633)		*Haas et al.* 1989
SC	PMMA + 4-dicyanovinyl-4′-(diethylamino) diazobenzene 3RDCVXY 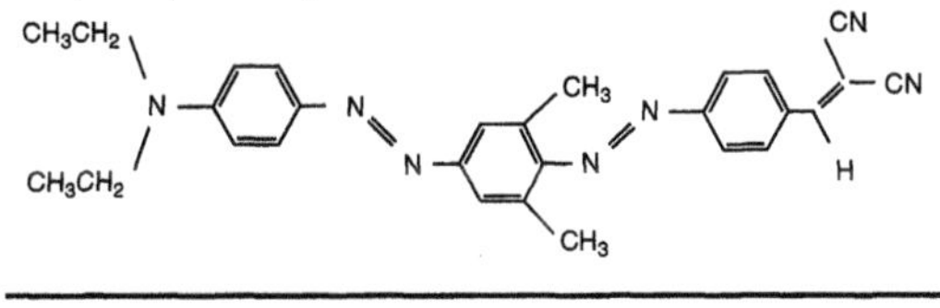	≈ 200	$d_{33} = 420$ (1064)	$r_{33} = 40$ (633)	< 140	*Shuto et al. 1991*

a) GH: guest-host system, CL: cross-linked polymer, SC: side-chain polymer
b) backbone polymer not known

materials. The next step was the search for better nonlinear optical dyes on one side and the amelioration of the long time stability through high T_g or cross linked systems on the other hand. At the moment much effort is made toward integrated electro-optic devices with regard to a fast introduction of such products to the market.

8.5 APPLICATION OF PHYSICAL MODELS

8.5.1 Linear electronic polarizabilities and refractive indices

Similarly to the nonlinear optical hyperpolarizabilities and coefficients (see Chapters 3 and 8.5.2) the linear molecular polarizabilities α_i (see Eq. (3.1)) can also be related to macroscopic data, in this case to the refractive indices. We will illustrate this by calculating the linear polarizability of COANP. The following relations are used for the calculations (see e.g. *Zyss et al. 1987*): first the Lorenz-Lorentz relation between a_{II} and n_I

$$\frac{n_I^2 - 1}{n_I^2 + 2} = \frac{N a_{II}}{3} \tag{8.5}$$

where $N = 3.00 \cdot 10^{27}$ molecules/m^3 (for COANP) and n_I are the refractive indices. And second the relation between α_{ij} and a_{II} (all intermolecular contributions are neglected):

$$a_{II} = \frac{1}{n(g)} \sum_{s}^{n(g)} \sum_{i,j} cos(\theta_{Ii}^{S})\, cos(\theta_{Ij}^{S})\, \alpha_{ij} \tag{8.6}$$

$n(g)$ is the number of equivalent positions in the unit cell ($n(g)$ = 4 for COANP), s denotes a site in the unit cell and θ_{Ii}^{s} and θ_{Ij}^{s} are the angles between the crystallographic X, Y, Z and the molecular x, y, z axes. We make the following definition for the molecular coordinate system in COANP. The z-axis is along the charge-transfer axis, y is perpendicular to z, but in the plane of the pyridine ring and x is perpendicular to both y and z (see Chapter 8.5.2). The corresponding angles are listed in Table 8.6.

It should be noted that, in contrast to the nonlinear optical effects, contributions from σ-electrons are important for the linear optical polarizability α. In order to solve Eqs. (8.5) and (8.6), we assume that $\alpha_{xy} \sim \alpha_{xz} \sim \alpha_{yz} \sim 0$. With this assumption we have three equations with three unknowns.

The solutions can be fitted with a Sellmeier one-oscillator model (Figure 8.3)

$$\alpha(\omega) = \frac{\omega_{pm}^2 \gamma_{om}}{\omega_{eg,m}^2 - \omega^2} + A_m = \frac{q_m}{\lambda_{eg,m}^{-2} - \lambda^{-2}} + A_m \tag{8.7}$$

Table 8.6 Angles between the crystallographic X, Y, Z and the molecular x, y, z axes. (for COANP).

projection on x	projection on y	projection on z
$\theta_{Xx} = 25.0°$	$\theta_{Xy} = 97.8°$	$\theta_{Xz} = 66.4°$
$\theta_{Yx} = 114.4°$	$\theta_{Yy} = 118.1°$	$\theta_{Yz} = 38.8°$
$\theta_{Zx} = 85.3°$	$\theta_{Zy} = 150.7°$	$\theta_{Zz} = 118.8°$

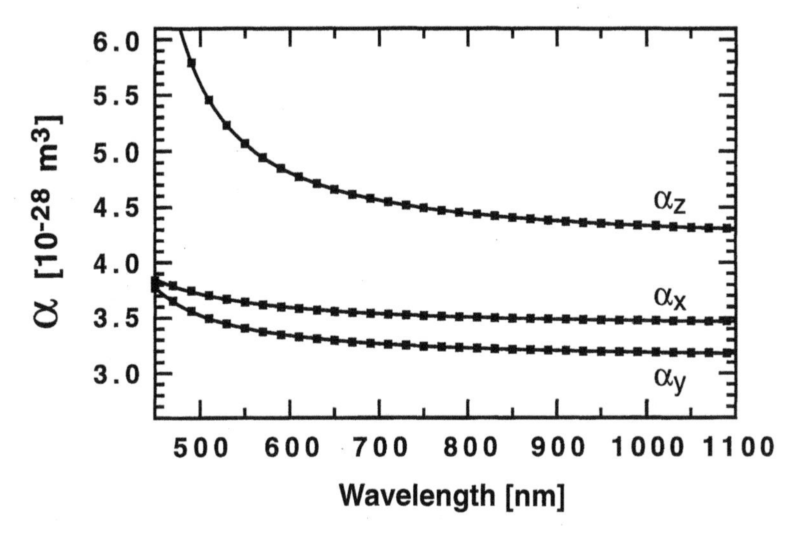

Figure 8.3 Dispersion of the linear polarizabilities α_x (= α_{xx}), α_y (= α_{yy}), and α_z (= α_{zz}) of COANP. The points are calculated from the refractive indices (Eqs. (8.5) and (8.6)). The continuous curve is a fit with the Sellmeier one-oscillator model (Eq. (8.5)).

where ω_{pm} is the plasma frequency, γ_{om} is a local field parameter, $q_m = \omega_{pm}^2\gamma_{om}/(2\pi c)^2$, $\omega_{eg,\,\mathrm{m}}$ is the eigen frequency of the dominant oscillator, $\lambda_{eg,\,\mathrm{m}} = 2\pi c/\omega_{eg,\,\mathrm{m}}$, and A_m is a constant containing the sum of all the other oscillators (see e.g. *Born 1980b*).

The Sellmeier parameters and the diagonal elements of the linear polarizability tensor at $\lambda = 632.8$ nm are listed in Table 8.7.

It is obvious that also for the refractive indices the charge-transfer transition is dominant. However, the other tensor components cannot be neglected for the linear polarizabilities. A crossing of α_x and α_y is present for the polarizabilities as it is for the refractive indices (Figure 7.7a, *Bosshard et al. 1989*). The only major difference between the macroscopic and microscopic values lies in the lower absorption wavelengths λ_{eg}. This red shift of the refractive index n with respect to the microscopic polarizability α can be explained (see e.g. *Born 1980b*): α is a property of the molecule. If we bring the molecule into a polarizable environment (e.g. into a crystal), this environment induces a shift of the wavelength of maximum absorption. The particularly low value $\lambda_{eg,m}$ for α_x is not very surprising:

Table 8.7 Linear polarizabilities of COANP at $\lambda = 632.8$ nm and dispersion parameters of a one-oscillator model.

	α_x [10^{-28} m^3]	α_y [10^{-28} m^3]	α_z [10^{-28} m^3]
$\lambda = 632.8$ nm	3.572	3.310	4.708
q_m [10^{-16} m]	4.2602	2.7913	4.3931
$\lambda_{eg,m}$ [nm]	317.01	359.80	406.47
A_m [10^{-28} m^3]	3.0001	2.7761	3.4674

since x is the axis perpendicular to the pyridine ring, α_x contains only contributions from σ-electrons.

It should be noted that the calculation of the refractive indices n from the linear polarizabilities is quite difficult since, unlike in the case of second-order nonlinearities, there is generally not a single dominant tensor element. The determination of the refractive indices, however, only gives an average over several tensor elements.

8.5.2 Molecular second-order polarizabilities and nonlinear optical susceptibilities

The oriented gas model discussed in Chapter 3.4 allows the comparison of microscopic and macroscopic nonlinear optical properties. In this section we will apply it to several examples in order to demonstrate its validity. As was discussed in Chapter 3.7 the resonance frequency of the transition ω_{eg} of the nonlinear optical molecules is red shifted in a dielectric medium due to local field effects. This implies that if e.g. a nonlinear optical coefficient d_{IJK} at λ = 1,064 nm should be calculated from molecular data, a corresponding second-order polarizability β_{zzz} (or β_z) at λ = (1,064 – 80) nm (or λ = (1,064 – 120) nm for stilbene derivatives) should be used (see Chapter 3.7). The measured values of β_{zzz} in this chapter are therefore converted to a value at a wavelength 80 nm (or 120 nm) below the corresponding nonlinear optical susceptibility using the two-level model (Eq. (3.2)) and the data of β_0 in Table 8.1. Note, however, that given the experimental uncertainties in the determination of β and d the shift of λ_{eg} is not very important.

8.5.2.1 COANP. Taking the one-dimensional charge-transfer axis along the axis defined by the nitrogen atom of the nitro group and the nitrogen atom of the amine group (Figure 8.4), the following angles between the charge-transfer axis and the crystal axes a, b and c are obtained (crystallographic data can be found in *Bosshard et al. 1989*)

$$\theta_{X,z} = 66.4°$$

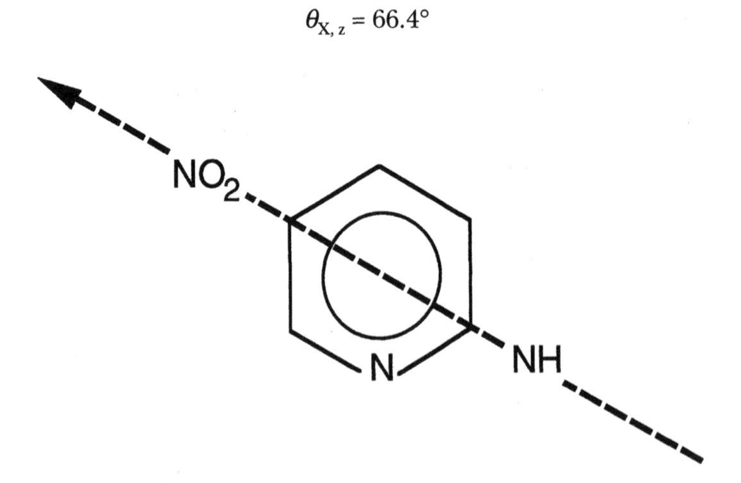

Figure 8.4 Charge-transfer axis between donor (NH) and acceptor (NO_2) of COANP.

$$\theta_{Y,z} = 38.8° \qquad (8.8)$$

$$\theta_{Z,z} = 118.8°$$

The angle $\theta_{Z,z} = 118.8°$ is close to the optimum angle $\theta_p = 54.7°$ (or 125.3°) for phase-matchable nonlinear optical coefficients as determined by Zyss (*1982*). The structural dependence between d_{IJK} and β_{zzz} is (see Chapter 3.4)

$$d_{32} = Nf_3^{2\omega}(f_2^{\omega})^2 cos^2(\theta_{Y,z})\, cos(\theta_{Z,z})\beta_{zzz}$$

$$d_{33} = Nf_3^{2\omega}(f_3^{\omega})^2 cos^3(\theta_{Z,z})\beta_{zzz} \qquad (8.9)$$

$$d_{24} = Nf_2^{2\omega} f_2^{\omega} f_3^{\omega} cos(\theta_{Z,z})\, cos^2(\theta_{Y,z})\beta_{zzz}$$

$$f_i^{\omega,2\omega} = \frac{(n^{\omega,2\omega})^2 + 2}{3} \qquad (8.10)$$

$f^{\omega,2\omega}$ are the local field corrections in the Lorentz approximation

$$N = 3.00 \cdot 10^{27}\ \mathrm{m}^{-3}$$

$$n_y^{\omega} = 1.699,\ n_y^{2\omega} = 1.720,\ n_z^{\omega} = 1.599,\ n_z^{2\omega} = 1.612\ (\lambda = 1907\ \mathrm{nm})$$

All intermolecular contributions to the second-order nonlinearity are neglected and the nonlinear optical hyperpolarizability tensor is described by one single element β_{zzz}. According to Eq. (8.9) the nonlinear optical coefficients d_{IJK} can be calculated from β_{zzz}. If we take the values at e.g. λ = 1,907 nm (Table 8.8), we see that the results agree surprisingly well. Only d_{33} does not fit. Therefore the assumption that the second-order polarizability tensor β_{ijk} can be described by one single element probably does not apply to COANP, and a second component within the pyridine ring should be considered. Such a model is reasonable because the nitrogen atom in the ring pulls electron density out of the ring. If we have e.g. an additional tensor element β_{yyy}, it will influence d_{33} of COANP about 3.4 times more strongly than it will influence d_{32} and d_{24}. Representative examples of other such compounds are e.g. o-nitroaniline (o-NA), m-nitroaniline (m-NA) and 2,4-dinitroaniline (*Zyss et al. 1987*).

Table 8.8 Nonlinear optical coefficients d_{IJK} of COANP calculated from the molecular second-order polarizabilities β_{zzz} (= β_z from EFISH) at λ = (1,907 – 80) nm (charge transfer along donor-acceptor axis).

d_{IJK}	$\beta_{zzz}[10^{-40}\ \mathrm{m^4/V}]$	d_{IJK}^{calc} [pm/V]	d_{IJK}^{meas}[pm/V]
d_{33}	52 ± 7	6.2 ± 0.9	8.5 ± 2
d_{32}	52 ± 7	19 ± 3	16 ± 2
d_{24}	52 ± 7	19 ± 4	16 ± 2

8.5.2.2 PNP. PNP crystals belong to point group 2. Therefore the dielectric 2 axis coincides with the crystallographic b axis (*Sutter et al. 1988b*). The molecule also contains a nitropyridine ring. If we again assume a one-dimensional charge-transfer axis from the nitrogen atom of the nitro group and the nitrogen atom of the amino group, the following angles between the charge-transfer axis and the crystal axes a and b are obtained

$$\theta_{X,z} = 149.6°$$
$$\theta_{Y,z} = 59.6° \tag{8.11}$$

The angle $\theta_{Y,z} = 59.6°$ is again close to the optimum angle $\theta_p = 54.7°$ (or 125.3°) for phase-matchable nonlinear optical coefficients as determined by Zyss (*1982*). The structural dependence between d_{IJK} and β_{zzz} is

$$d_{22} = Nf_2^{2\omega}(f_2^{\omega})^2 cos^3(\theta_{Y,z})\beta_{zzz}$$
$$d_{21} = Nf_2^{2\omega}(f_1^{\omega})^2 cos^2(\theta_{X,z})\, cos(\theta_{Y,z})\beta_{zzz} \tag{8.12}$$
$$d_{16} = Nf_1^{2\omega} f_1^{\omega} f_2^{\omega} cos^2(\theta_{X,z})\, cos(\theta_{Y,z})\beta_{zzz}$$

where $N = 3.83 \cdot 10^{27}\ m^{-3}$

$$n_x^{\omega} = 1.849,\ n_x^{2\omega} = 1.890,\ n_y^{\omega} = 1.720,\ n_y^{2\omega} = 1.738\ (\lambda = 1{,}907\ nm)$$

Table 8.9 shows the comparison between experiment and theory.

The agreement is excellent for d_{22} but not for d_{21} and d_{16}. For the same reason as for COANP the use of two tensor components may yield better results.

Table 8.9 Nonlinear optical coefficients d_{IJK} of PNP calculated from the molecular hyperpolarizabilities β_{zzz} (= β_z from EFISH) at λ = (1,907 – 80) nm.

d_{IJK}	$\beta_{zzz}[10^{-40}\ m^4/V]$	d_{IJK}^{calc} [pm/V]	d_{IJK}^{meas}[pm/V]
d_{22}	47 ± 7	11 ± 2	9 ± 2
d_{21}	47 ± 7	37 ± 6	29 ± 5
d_{16}	47 ± 7	38 ± 6	24 ± 5

8.5.2.3 MNA. In analogy we can obtain d_{11} and d_{12} of MNA (point group m). According to Levine (*Levine et al. 1979*) the structural relations are given by

$$d_{11} = Nf_1^{2\omega}(f_1^{\omega})^2 cos^3(\psi)\beta_{zzz}$$
$$d_{12} = Nf_1^{2\omega}(f_2^{\omega})^2 cos(\psi)\, sin^2(\psi)\beta_{zzz} \tag{8.13}$$

with the following values needed:

$$\psi \approx 21°,\ N = 5.94 \times 10^{27}\ m^{-3}$$

$$n_x^{\omega} = 1.8,\ n_x^{2\omega} = 2.2,\ n_y^{\omega} = 1.5,\ n_y^{2\omega} = 1.85\ (\lambda = 1{,}064\text{ nm})$$

This leads to the following results:

Table 8.10 Nonlinear optical coefficients d_{IJK} of MNA calculated from the molecular hyperpolarizabilities β_{zzz} (= β_z from EFISH) at λ = 1,907 – 80) nm.

d_{IJK}	$\beta_{zzz}[10^{-40}\text{ m}^4/\text{V}]$	d_{IJK}^{calc} [pm/V]	d_{IJK}^{meas}[pm/V]
d_{11}	83[b]	279 ± 60	250 ± 50[b]
d_{12}	83[b]	29 ± 8	38 ± 10[b]

[a] measured by *Teng et al. 1983*
[b] d_{11} = 0.5 pm/V of quartz as reference

The agreement between theory and experiment is remarkably good. This indicates that a single tensor element β_{zzz} is indeed sufficient to describe the nonlinearity of the MNA molecules.

8.5.2.4 MMONS For MMONS (point group mm2) the structural relations are given by

$$\begin{aligned} d_{32} &= Nf_3^{2\omega}(f_2^{\omega})^2 cos^2(\theta_{Y,z})\, cos(\theta_{Z,z})\,\beta_{zzz} \\ d_{33} &= Nf_3^{2\omega}(f_3^{\omega})^2 cos^3(\theta_{Z,z})\,\beta_{zzz} \\ d_{24} &= Nf_2^{2\omega} f_2^{\omega} f_3^{\omega} cos(\theta_{Z,z})\, cos^2(\theta_{Y,z})\,\beta_{zzz} \end{aligned} \tag{8.14}$$

with the following values needed (*Bierlein et al. 1990*):

$$\theta_{Y,z} = 61°,\ \theta_{Z,z} = 34°,\ N = 2.87 \cdot 10^{27}\text{ m}^{-3}$$

$$n_y^{\omega} = 1.630,\ n_y^{2\omega} = 1.770,\ n_z^{\omega} = 1.961,\ n_z^{2\omega} = 2.352\ (\lambda = 1{,}064\text{ nm})$$

This leads to the following results:

Table 8.11 Nonlinear optical coefficients d_{IJK} of MMONS calculated from the molecular hyperpolarizabilities β_{zzz} (= β_z from EFISH) at λ = (1,064 – 120) nm.

d_{IJK}	β_{zzz}[a]$[10^{-40}\text{ m}^4/\text{V}]$	d_{IJK}^{calc} [pm/V]	$d_{IJK}^{meas,b}$[pm/V]
d_{33}	266	415	184 ± 137
d_{32}	266	90	41 ± 8
d_{24}	266	77	55 ± 28

[a] *Cheng et al. 1989*
[b] d_{33} = 13.6 pm/V of KTP as reference

Whereas the ratio between the coefficients e.g. d_{33}/d_{32} is in excellent agreement, the absolute values of d_{IJK} calculated from β are considerably higher than the actual measured ones. It is not clear if the discrepancy arises from a failure of the oriented gas model or if there exist other reasons.

8.5.3 Molecular second-order polarizabilities and electro-optic coefficients

Using Eqs. (3.2) and (3.26) electronic contributions to the electro-optic coefficients can also be calculated from molecular second-order polarizabilities. The parameters needed for this evaluation and comparison come from Tables 3.1 and 8.1 and from Section 8.5.2. Table 8.12 lists the β values together with the resulting electro-optic coefficients r for COANP, PNP, MNA, and MMONS. Again the microscopic quantities are used at $\lambda_1 = (\lambda - \Delta\lambda)$ ($\Delta\lambda = 40$ nm for benzenes and $\Delta\lambda = 60$ nm for stilbenes).

The correspondence between calculated and measured electro-optic coefficients is good for r_{23} of COANP (as for d_{32}, see section 8.5.2.1) and r_{11} of MNA (as for d_{11}, see section 8.5.2.3). For MMONS the same observation as for the nonlinear optical coefficients can be made: the values disagree considerably. The reason for this is not known. The disagreement for r_{33} of COANP (as for d_{33}, see section 8.5.2.1) and the values of PNP may arise from two effects. On one hand two tensor components of β may have to be used, since the nitrogen atom in the ring pulls electron density out of the ring. On the other hand we believe that the discrepancies are at least partially due to non-negligible optic phonon

Table 8.12 Molecular second-order polarizabilities $\beta_{zzz}(-\omega, \omega, 0)$ and calculated and measured electro-optic coefficients of COANP, PNP, MNA, and MMONS.

r_{IJK}	n_I ($\omega \approx 0$)	wavelength of β_{zzz}[nm]	β_{zzz} [10^{-40} m^4/V]	wavelength of r_{IJK} [nm]	r^{calc}_{IJ} [pm/V]	r^{meas}_{IJ} [pm/V]
			COANP			
r_{33}	1.595	474.5	191 ± 30	514.5	12 ± 2	28 ± 5
r_{23}	1.595	474.5	191 ± 30	514.5	27 ± 4	26 ± 4
r_{33}	1.595	592.8	92 ± 14	632.8	6 ± 1	15 ± 2
r_{23}	1.595	592.8	92 ± 14	632.8	14 ± 2	13 ± 2
			PNP			
r_{22}	1.715	474.5	211 ± 32	514.5	19 ± 3	28.3 ± 3.4
r_{12}	1.715	474.5	211 ± 32	514.5	45 ± 7	20.2 ± 2.4
r_{22}	1.715	592.8	90 ± 14	632.8	9 ± 2	12.8 ± 1.3
r_{12}	1.715	592.8	90 ± 14	632.8	22 ± 4	13.1 ± 1.3
			MNA			
r_{11}	1.682	592.8	73	632.8	57	67 ± 25
			MMONS			
r_{33}	1.907	572.8	222	632.8	63	39.9 ± 8
r_{23}	1.907	572.8	222	632.8	30	19.3 ± 4

contributions to the electro-optic coefficients. In crystals with almost parallel alignment of the molecules, and therefore large electronic value of r, such effects are not as important (see e.g. MNA) because the electronic values are dominant. Contributions from acoustic phonons to the static electro-optic coefficients can be ruled out as can be seen e.g. from Figure 7.20 (similar values of $n^3 r$ for 1 kHz and 1 MHz are observed).

8.5.4 Comparison of electro-optic and nonlinear optical coefficients

Using Eqs. (3.28) and (3.29) electro-optic and nonlinear optical coefficients can be compared. Table 8.13 shows the results for COANP, PNP, MNA and MMONS.

In this case the correspondence is good for MMONS (rules out optic phonon contributions) and MNA. For COANP and PNP it is not satisfactory. We therefore believe that indeed optic phonon contributions to the electro-optic coefficients play an important role in the case of COANP and PNP. Such effects may be especially important if θ_p deviates considerably from zero as in these cases.

Table 8.13 Comparison of electro-optic and nonlinear optical coefficients of COANP, PNP, MNA and MMONS at different wavelengths. r_{IJ} is the measured and r^e_{IJ} the calculated value from Eqs. (3.28) and (3.29). The nonlinear coefficients (at λ = 1,064 nm) used for this calculation are also listed.

wavelength [nm]	r_{IJ}[pm/V]	r^e_{IJ}[pm/V]	d_{IJ} [pm/V]	r_{II} [pm/V]	r^e_{II} [pm/V]	d_{II} [pm/V]
			COANP			
514.5	+ 26 ± 4	31 ± 6	43 ± 9	+ 28 ± 5	19 ± 3	19.3 ± 3
632.8	+ 13 ± 2	16 ± 4	43 ± 9	+ 15 ± 2	9.3 ± 1.5	19.3 ± 3
			PNP			
514.5	+ 20.2 ± 2.4	32 ± 6	68 ± 8	+ 28.3 ± 3.4	15 ± 3	21.6 ± 4
632.8	+ 13.1 ± 1.3	16 ± 2.2	68 ± 8	+ 12.8 ± 1.3	7.4 ± 1.4	21.6 ± 4
			MNA			
632.8	—	—	—	+ 67 ± 25	70 ± 14	250 ± 50[a]
			MMONS			
632.8	+ 19.3 ± 4	19.3 ± 3.9	41 ± 8	+ 39.9 ± 8	35.9 ± 7	184 ± 37

MNA: d_{11} = 0.5 pm/V of quartz as reference
MMONS: d_{33} = 13.6 pm/V of KTP as reference
COANP, PNP: d_{11} = 0.4 pm/V of quartz as reference

9. ELECTRO-OPTICS AND NONLINEAR OPTICS IN WAVEGUIDES

9.1 GUIDED WAVE NONLINEAR OPTICS

Optical waveguides are ideal for nonlinear optical applications because they provide strong beam confinement over long propagation distances (*Stegeman et al. 1985*). Even for low power laser beams large efficiencies in e. g. nonlinear wave mixing are possible. Possible materials for nonlinear optical organic thin films are: electric-field oriented molecules in a polymeric host (*Singer et al. 1986*), organic thin films on a glass substrate (*Hewig et al. 1983*), organic crystal-cored fibers (*Kerkoc et al. 1989*) and Langmuir-Blodgett (LB) films (*Aktsipetrov et al. 1985*). In all these material types, guided wave nonlinear optical effects have been successfully demonstrated (see below and Table 9.1). Note that LB films have a specific advantage for low attenuation guided wave propagation arising from the fact that the thickness variation can be controlled very accurately (*Pitt et al. 1976*) to within a single bilayer thickness of about 4 nm. Specifically the stability for phase-matching requires the waveguide dimension to be controlled to better than ±5 nm (*Stegeman et al. 1989*).

There are several good review articles on the subject of nonlinear effects in optical waveguides by Stegeman, giving a general overview on this matter (*1986, 1989*), or focusing on planar (*1985*) or organic waveguides (*1987*). Many of the references not cited explicitly in this text can be found therein. Recent progress in nonlinear optical waveguide devices using organic materials has been summarized by Stegeman (*1993*). Higher efficiencies $\eta = P^{2\omega}/P^{\omega}$ are obtained by using the film instead of the substrate as the nonlinear medium, even though in many of the cases the film consisted of crystallites which were not all oriented in the same direction. The largest conversion efficiencies have been obtained in cases where the overlap integral (defined in Eq. (9.2)) was optimized.

There are three important possibilities for phase-matched frequency-doubling in waveguides:

a) phase-matching by conversion of a guided fundamental mode into a guided second-harmonic mode using modal dispersion and/or birefringence,
b) conversion by periodic modulation of either the refractive index of the waveguide structure or the nonlinear optical susceptibility,
c) conversion of a guided fundamental mode into a second-harmonic radiation mode in the form of Cerenkov radiation.

The first method allows the use of the nonlinear optical coefficients, d_{11}, d_{22} and d_{33} which cannot be utilized in bulk materials. On the other hand the thickness has to be perfectly adjusted for phase-matching and the overlap integral is sometimes difficult to optimize (see below). The modulation procedure has the advantage that many phase-matching conditions can be fulfilled by adjusting the modulation period. Partial modulation in an organic material has been achieved in polymers of oxy-nitrostilbene (*Khanarian et al. 1990*). Second-harmonic generation of the Cerenkov-type has less stringent requirements on the design of the waveguides. The phase-matching does not depend on the film thickness and there is no overlap integral.

The approach of the coupled-mode formalism is useful for treating problems which involve energy exchange between modes. The general theory and applications to electro-optic modulation, photoelastic modulation and nonlinear optical applications, such as second-harmonic generation for the case of planar waveguides are treated by e.g. *Yariv et al. (1973)*. The basic configuration of a slab dielectric waveguide is shown in Figure 9.1.

In the case of second-harmonic generation the coupled-wave theory yields

$$\frac{P^{2\omega}}{P^{\omega}} = \frac{\omega^2 \varepsilon_o^2}{4} \frac{P^{\omega}}{bt} |S|^2 L^2 sinc^2\left(\frac{\Delta\beta L}{2}\right) \tag{9.1}$$

$$S = \sqrt{t} \int_0^t d_{zzz} \left(E_z^{n,\omega}\right)^2 E_z^{m,2\omega} dy \tag{9.2}$$

for the case $TE^{\omega}_{input} \rightarrow TE^{2\omega}_{output}$. $\beta = (2\pi/\lambda) N_{eff}$, $\Delta\beta = \beta^{2\omega} - \beta^{\omega}$, S is the overlap integral. m, n are the mode numbers, b is the illuminated width of the waveguide, t is the thickness and L the length of the waveguide. In a similar way expressions for other conversions (TE $\rightarrow$ TM, TM $\rightarrow$ TM, TM $\rightarrow$ TE) can be obtained.

The magnitude of the overlap integral is important for efficient frequency-doubling in waveguides. If the product of the field distributions changes sign across the waveguide the value of S is reduced. The overlap integral is usually small unless all the interacting modes have the same mode number.

The phase-matching condition is given by $\Delta\beta\ L/2 = 0$. This corresponds to the condition $N^{\omega}_{eff} = N^{2\omega}_{eff}$. The effective index of waveguide modes depends critically on the waveguide thickness t. This requires very good control of the waveguide thickness to ensure that the phase-matching condition is met along the full

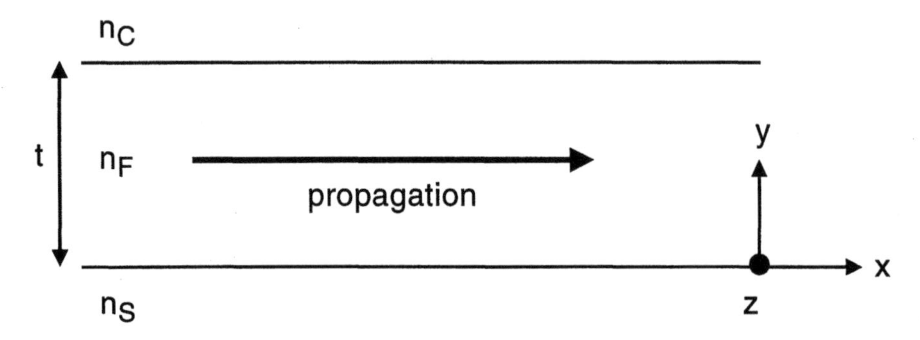

Figure 9.1 Basic configuration of a slab dielectric waveguide.

waveguide length. In the absence of birefringence and assuming normal dispersion of the material refractive index, we have $\beta(TE_m, 2\omega) > \beta(TE_{m'}, \omega)$, that is, phase-matched second-harmonic conversion between two identical modes ($m = m'$) of the same polarization is not possible. On the other hand, with phase-matching between different mode numbers the overlap integral is considerably reduced. In birefringent materials this problem can be overcome.

Looking at Eq. (9.1) we can see that for channel waveguides where S is optimized and where $b \sim \lambda$ the ratio between phase-matched second-harmonic generation efficiency in waveguides and bulk crystals is approximately

$$\frac{\eta_{wg}}{\eta_{bulk}} \sim \frac{L}{\lambda} \tag{9.3}$$

For typical values of $L = 1$ cm and $\lambda = 1$ μm this yields a ratio of 10^4 (*Zyss, 1985*). This ratio is too optimistic considering the many simplifications made such as e.g. neglection of pump depletion. A ratio of 100 is more realistic.

For Cerenkov-type second-harmonic generation the phase-matching condition is different (see e.g. *Chikuma et al. 1990*). Figure 9.2 shows a simple physical picture of the process for a conversion using the nonlinear coefficient d_{33}. The fundamental mode is generating light at 2ω which is not guided. At point A the frequency-doubled light enters the substrate under an angle θ and propagates to C. The fundamental wave also generates frequency-doubled light at point B. Constructive interference at the front surface of the second-harmonic wave (BC) and thus Cerenkov radiation takes place if

$$v_F^{\omega} \cos\theta = v_S^{2\omega} \quad \text{or} \quad \cos\theta = \frac{N_{eff}^{\omega}}{n_S^{2\omega}} \tag{9.4}$$

v_F^{ω} and $v_S^{2\omega}$ are the velocity of the fundamental mode and of the second-harmonic wave in film and substrate, respectively. This condition can only be met for$v_F^{\omega} > v_S^{2\omega}$. Generally Cerenkov radiation is possible for

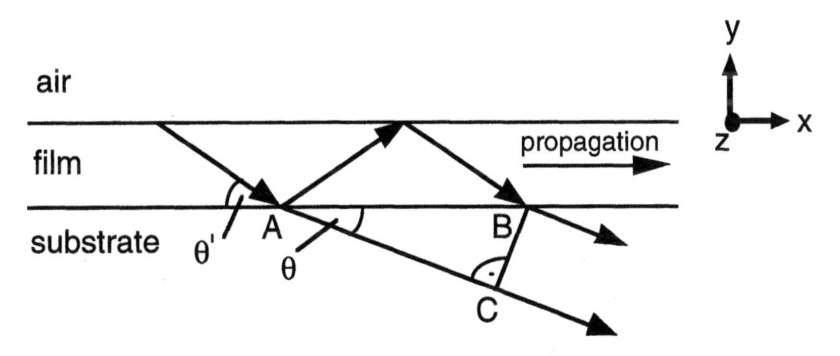

Figure 9.2 Cerenkov-type phase-matching due to constructive interference at points B and C.

$$n_S^{2\omega} > N_{eff}^{\omega} > n_S^{\omega} \tag{9.5}$$

The second-harmonic intensity for Cerenkov-type phase-matching is only proportional to the interaction length L since the generated light is spatially non-overlapping. The main advantage of a Cerenkov-type configuration is the possibility of operating within the absorption band of the material, thus utilizing the strong enhancement of the nonlinear optical coefficients.

Different approaches to obtain the correct thickness for phase-matched nonlinear optical mixing (for conversion between guided modes) have been attempted. One possibility is to taper the film slightly in a direction perpendicular to the propagation direction and to translate the film to get the optimum phase-matching condition. Another approach is to use a liquid as cladding material and to tune for phase-matching by varying the refractive index of the liquid. In addition gratings have been used to obtain the extra wave vector component needed for phase-matching. Waveguides with the optimum thickness for phase-matching can be prepared utilizing Langmuir-Blodgett films (see below).

The first report on organic thin films deposited on a glass substrate (by vacuum evaporation) was by Hewig and Jain (*1983*) who fabricated thin polycrystalline film of parachloro phenylurea. They obtained phase-matched second-harmonic generation ($TM_0 \rightarrow TM_2$) in the wavelength range from 0.84 to 0.92 μm (no specification of the input power levels). Due to the low value of the overlap integral and the high losses the conversion efficiency was small. Sasaki and co-workers used 2-methyl-4-nitroaniline (MNA, also on a glass substrate) to obtain $\eta = 10^{-2}$ with input powers of 10 mW ($TE_0 \rightarrow TE_1$) (*Sasaki et al. 1984*). In this experiment a tapered waveguide was used to obtain the correct phase-matching thickness.

Phase-matched second-harmonic generation by mode conversion from LB films was first demonstrated using 2-docosylamino-5-nitropyridine (DCANP, *Flörsheimer et al. 1992b*). A four-layer waveguide configuration (*Ito et al. 1978*) was utilized in order to increase the overlap integral of a fundamental TE_0 mode and a second harmonic TE_1 mode. Figure 9.3 shows the calculated transverse electric field distribution of these modes for a three-layer configuration (Figure 9.3b, substrate — LB film — cover) in comparison with the field distribution for the four-layer geometry (Figure 9.3a, substrate — linear optical waveguide — LB film — cover). The thicknesses of the LB films are calculated to meet the phase-matching conditions for a fundamental wavelength of $\lambda = 926$ nm. The modes were normalized for equal power/unit width, i. e. for 1 W/m.

In the example of the four-layer geometry, the fraction p of the fundamental power within the nonlinear optical material (p = 20%) is smaller than in the case of the three-layer configuration (p = 32%). However, the overlap integral for the former geometry ($S = 16 \cdot 10^{-9}\ V^{1/2}A^{-3/2}m$) is distinctly larger than for the latter configuration ($S = 1.7 \cdot 10^{-9}\ V^{1/2}A^{-3/2}m$) so that the theoretical frequency doubling efficiency (Eqs. (9.1), (9.2)) for the four-layer geometry exceeds the efficiency for the three-layer configuration by a factor of more than 80! The small overlap integral for the three-layer geometry is due to the change of sign of the integrand of S near the nodal line of the TE_1 mode (dotted line in Figure 9.3b). In contrast to

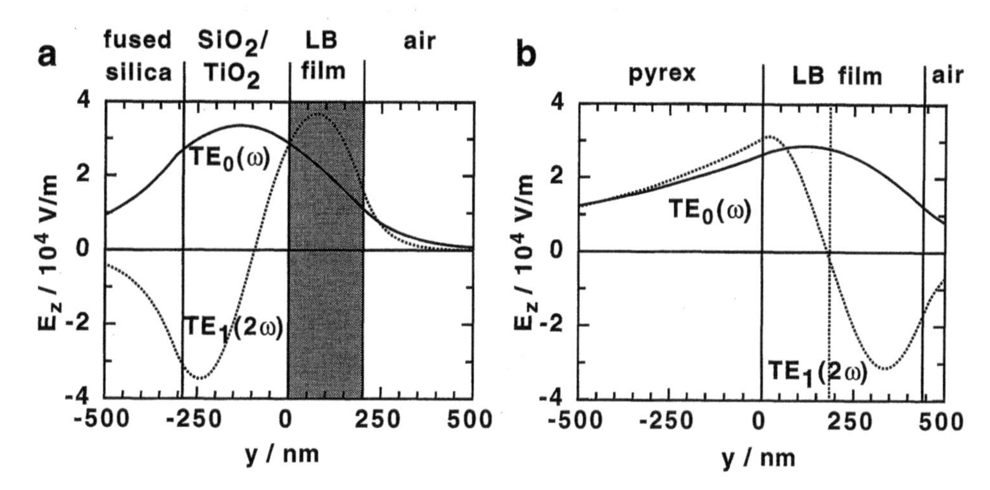

Figure 9.3 Electric field distribution of a fundamental TE_0 mode and a phase-matched second-harmonic TE_1 mode a) for the four-layer configuration (substrate-SiO2/TiO2-DCANP-air). In b) the overlap integral S is small since the integrand of S changes its sign at the nodal line of the TE_1 mode (dotted line). In a) the cancellation of positive and negative parts of the integrand of S is avoided, since the only contribution to S originates from the nonlinear optical film (shaded area). As can be seen additionally from a) a thickness reduction of the SiO_2/TiO_2 layer in favour of the LB film would further increase the overlap integral (*Flörsheimer et al. 1992b*).

this the cancellation of positive and negative parts of the integrand of S is avoided using the four-layer geometry since there the only contribution to S originates from the nonlinear optical region (shaded in Figure 9.3a, in the linear optical region the second order susceptibility vanishes). Phase-matching was observed for the calculated thickness of the LB film in the four-layer geometry (t_{LB} = 199 nm, 90 monolayers). The conversion efficiency was η = 0.6% (P^{ω} = 30 W inside of the film).

A special advantage of DCANP is the preferential alignment of the molecular change-transfer axes in a plane parallel to the dipping direction of the LB transfer process (*Decher et al. 1988,1989a*). Thus the largest nonlinear optical coefficient d_{33} can be utilized optimally with TE modes. The mechanism of the organizing process is still under investigation. A detailed analysis of the molecular orientation of the transferred films is given in the literature (*Bosshard et al. 1992d*). A simplified and idealized model is shown in Figure 9.4a together with the coordinate system used. The z axis is the dipping direction. In Figure 9.4b additionally a TE mode of a guided wave is sketched. This illustration does not give information on the cross-sectional field distribution of a guided mode, but it represents the orientation of the electric field. The electric field of the TE mode is parallel to the in-plane dipoles of the molecules so that they can be optimally excited to oscillate. Orthogonally oriented dipoles cannot be utilized for second-harmonic generation with fundamental TE modes. For comparison a TM mode is sketched in Figure 9.4c. This mode can be used with a material showing an overall polarization normal to the waveguide or an in-plane polarization along the x axis. Most LB films which have been prepared so far show a normal overall polarization (see Chapters 5 and 8.3).

Second-harmonic generation in the form of Cerenkov radiation was reported in 1970 (*Tien et al. 1970*). The method has been successfully applied to such inorganic materials as e.g. $LiNbO_3$ (*Taniuchi et al. 1986, Tohman et al. 1990*). Planar organic waveguides have been realized e. g. with (-)2-(α-methylbenzylamino)-5-nitropyridine (MBANP, *Kondo et al. 1989a*), with copolymers of polymethylmethacrylate (PMMA) and 4-(N-ethyl-N-(2-hydroxyethyl))-amino-4′-nitroazobenzene (Disperse Red 1, *Kinoshita et al. 1992*) and with DCANP (*Bosshard et al. 1991b, 1992c, Asai et al. 1992*).

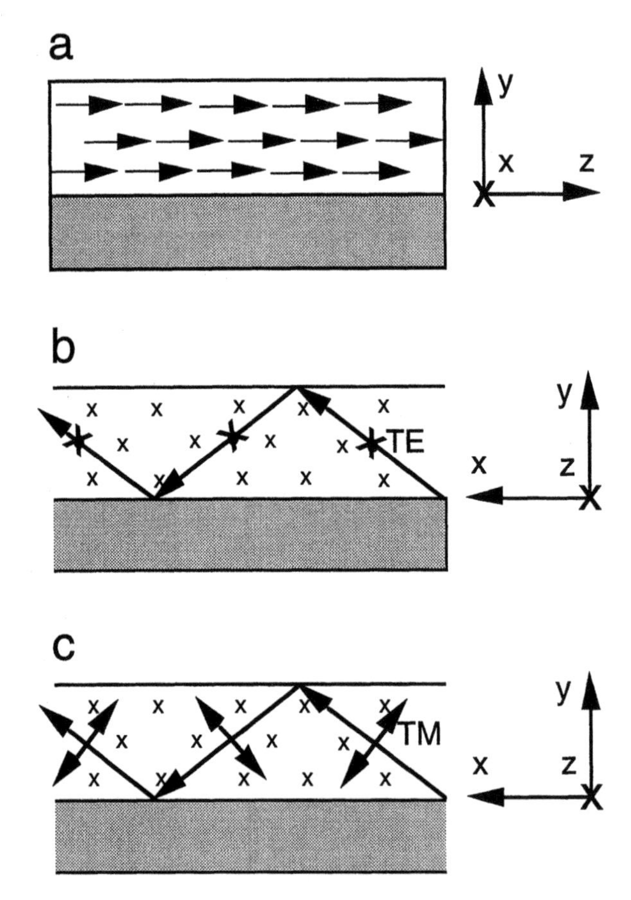

Figure 9.4 Side view of a DCANP LB film with orientation of the molecules (simplified and idealized) and coordinate system. The molecules are symbolized with the ground state dipoles of their charge transfer groups. The substrate is shaded. a) The dipoles are aligned preferentially along the dipping direction z. b) The electric field of a TE mode propagating in x direction is parallel to the dipoles (small crosses) and can be fully utilized. c) The electric field of a TM mode propagating in x direction is normal to the dipoles and cannot be utilized. Materials with an overall polarization in y direction are not suitable for TE modes but they exploit TM modes.

Fibers cored with organic crystals have been prepared e. g. from benzil (*Nayer et al. 1983*).

With DCANP planar LB film waveguides Cerenkov efficiencies of $\eta = 0.2$ % have been obtained at a fundamental wavelength of $\lambda = 820$ nm. The harmonic wavelength was well within the absorption region of the material ($d_{33} = 27$ pm/V, *Bosshard et al. 1992c*). For a smaller wavelength ($\lambda = 795$ nm, $d_{33} = 44$ pm/V), an optimized substrate material and a better beam confinement (strip waveguide, width $W_{strip} = 2$ μm, optimized thickness $t = 380$ nm) a theoretical efficiency of $\eta = 10\%$ W^{-1} (normalized with the fundamental power) was estimated (*Flörsheimer et al. 1992a*).

The best results with organic fibers in glass capillaries have so far been obtained with 4-(N,N-dimethylamino)-3-acetamidonitrobenzene (DAN). Conversion efficiencies of up to $\eta = 25\%$ in pulsed experiments have been measured in multimode fibers (*Kerkoc, 1991*). The main problem there arises from the difficulty to get good phase-matching in monomode fibers since it is difficult to make glass capillaries with core diameters of around 1 μm and appropriate refractive index. Generation of continuous wave blue coherent light from a semiconductor laser using crystal-cored fibers of 3,5-dimethyl-1-(4-nitrophenyl) pyrazole (DMNP) has been reported by Harada (*1991*). He obtained 64.5 μW (at $\lambda = 442$ nm) for a 16.6 mW input for cw second-harmonic generation using Cerenkov-type phase-matching. This material is interesting due to its low absorption cut-off at about $\lambda = 450$ nm.

Recently quasi-phase-matching has also been observed in organic materials. Khanarian (*Khanarian et al. 1990, Norwood et al. 1990*) reported on periodically poled polymer waveguides. They used a copolymer of methylmethacrylate and a monomer containing 4-oxy-4′-nitrostilbene. Quasi-phase-matching in poled polymer waveguides has also been demonstrated by other groups (*Rikken et al. 1991, 1993*).

Noncollinear phase-matching was achieved in a stilbene-dye attached polymer waveguide (*Shuto et al. 1989*). Sugihara (*1991b*) demonstrated phase matching by mode conversion in a MNA doped methacrylate polymer.

What can be expected from an organic waveguide? Choosing 2-(N-prolinol)-5-nitropyridine (PNP) with its highest nonlinear optical coefficient $d_{23} = 68$ pm/V at $\lambda = 1{,}064$ nm (*Sutter et al. 1989*) in a 0.5 μm × 0.5 μm channel waveguide (propagation length $L = 1$ mm) with optimized overlap integral, values of η of tens of percent should be expected for a 1 W fundamental power. Therefore these materials are very interesting for low power integrated optics. Table 9.1 lists some of the nonlinear optical waveguides so far known.

It should be noted that the best results on SHG and related $\chi^{(2)}$ nonlinear interactions have been obtained with inorganic waveguides. In terms of low input power/high efficiency Ti-indiffused $LiNbO_3$ channel waveguides still yield some of the best results (*Regener et al. 1988*). Some data:

input power [mW]	interaction length [mm]	efficiency [%]	used modes
1	24	0.1	$TE_{oo} \rightarrow TM_{oo}$
0.1	40	0.1	$TE_{oo} \rightarrow TM_{oo}$

In addition optical parametric oscillation with a conversion efficiency of 15% was obtained for 15 W pump power at $\lambda = 0.65\ \mu m$. Difference-frequency experiments carried out illustrated the problem of guided waves at wavelengths differing by a factor of 2 or 3. It is not possible to have single-mode waveguides over such a long wavelength range.

Advantages of $LiNbO_3$-waveguides are the low losses on the order of 0.1 dB/cm allowing multicentimeter interaction lengths. Temperature tuning and electro-optic effects facilitate phase-matching. The main problem in $LiNbO_3$-waveguides is the photorefractive damage due to exposure to light at mW power levels (visible spectrum). $LiTaO_3$ shows less susceptibility to photorefractive damage. 8 mW of stable blue light output have been generated in a proton exchanged, periodically poled waveguide with 72 mW of coupled cw diode laser light (*Yamamoto et al. 1993a*). Research continues to get better materials that are cheap, stable, and easily fabricated.

To characterize the nonlinear optical efficiency of the waveguides, we introduce the following figures of merit η':

$$\eta' = \frac{P^{2\omega}}{(P^{\omega})^2 \cdot L^2} \tag{9.6}$$

$$\eta' = \frac{P^{2\omega}}{(P^{\omega})^2 \cdot L} \tag{9.7}$$

depending on whether the nonlinear optical interaction is proportional to L (as in Cerenkov second-harmonic generation) or L^2 (as in mode to mode conversion) (L:interaction length).

For applications, the absorption coefficients must be small to avoid thermal detuning of phase-matching, light damage, and to insure low losses. The uniformity of the refractive indices must be better than ±0.0004. The waveguide dimension tolerances and stability must be better than ±5 nm. In addition good mechanical and chemical properties are necessary.

The major drawback of nonlinear optical waveguides with organic materials lies in the optical quality leading to strong light attenuation. Thus far these materials cannot compare to e.g. $LiNbO_3$ waveguides with low losses on the order of 0.1 dB/cm.

9.2 Guided wave electro-optics

There already exist various electro-optic waveguide applications (using $LiNbO_3$). Typical devices are the Mach-Zehnder interferometer and the directional coupler (Figure 9.5, see e.g. *Voges 1987*). Figure 9.6 shows an example of constructive and destructive interference at the output of a Mach-Zehnder interferometer.

One of the major problems arising in the context of electro-optic amplitude modulation in waveguides is the bandwidth. The wavelength of the modulating

Table 9.1 Figures of merit for some nonlinear optical waveguides.

material	phase-matching technique	λ[nm]	$\eta = P(2\omega)/P(\omega)^2$ [$\%W^{-1}$]	η'	reference
organic crystals and crystal cored fibers					
DAN	birefringence	1064	3.6 [a]	14 [a,1]	*Kerkoc 1991*
			3.2 [b]	13 [b,1]	
	Cerenkov	1064	4.4 [b]	9.4 [b,2]	*Uemiya et al. 1992*
DMNP	Cerenkov	884	58 [b]	39 [b,2]	*Harada et al. 1991*
MNA	birefringence	1064	$1.24 \cdot 10^{-4}$ [a]	$4.9 \cdot 10^{-2}$ [a,1]	*Sugihara 1991a*
poled polymers					
poled polymer*	noncollinear PM	1064	$3.8 \cdot 10^{-7}$ [a]	$6.1 \cdot 10^{-6}$ [a,1]	*Shuto et al. 1989*
poled polymer*	quasi-PM	1340	$1.2 \cdot 10^{-4}$ [a]	$1.2 \cdot 10^{-8}$ [a,1]	*Khanarian et al. 1990*
poled polymer*	Cerenkov	812	$1.5 \cdot 10^{-4}$ [a]	—	*Kinoshita et al. 1992*
poled polymer*	mode conversion	1064	$1.7 \cdot 10^{-4}$ [a]	$1.7 \cdot 10^{-2}$ [a,1]	*Sugihara et al. 1991b*
LB films					
DCANP	Cerenkov	1064	$1.2 \cdot 10^{-7}$ [a]	$1.4 \cdot 10^{-7}$ [a,2]	*Bosshard et al. 1991b, 1992c*
		860	$2 \cdot 10^{-4}$ [a]	$4 \cdot 10^{-4}$ [a,2]	
		820	$2 \cdot 10^{-4}$ [a]	$4 \cdot 10^{-4}$ [a,2]	
		795	10 [a, c]	10 [a, c,2]	*Flörsheimer et al. 1992a,b*
	mode conversion	926	$2 \cdot 10^{-2}$ [a]	$5 \cdot 10^{-1}$ [a,1]	
all polymeric LB film*	Cerenkov	1064 860	—	—	*Clays et al. 1993b*
inorganic crystals					
$LiNbO_3$	birefringence	1064	1000 [b]	63 [b,1]	*Regener et al. 1988*
	quasi-PM	851.7	0.54 [b]	600 [b,1]	*Yamada et al. 1993*
$LiTaO_3$	quasi-PM	870	220b	220 [b,1]	*Yamamoto et al. 1993b*
KTP	quasi-PM	800	30 [b]	120 [b,1]	*van der Poel et al. 1990*
	quasi-PM	850	43 [b]	214 [b,1]	*Risk et al. 1993*
$KNbO_3$	mode conversion	866	8.8 [b]	16 [b,2]	*Fluck et al. 1993*

[a] pulsed   [1] $\eta' = \dfrac{P^{2\omega}}{(P^{\omega})^2 \cdot L^2} [\% W^{-1} cm^{-2}]$ [2] $\eta' = \dfrac{P^{2\omega}}{(P^{\omega})^2 \cdot L} [\% W^{-1} cm^{-1}]$

[b] continuous wave   [c] estimated for optimized substrate

* see references for material description

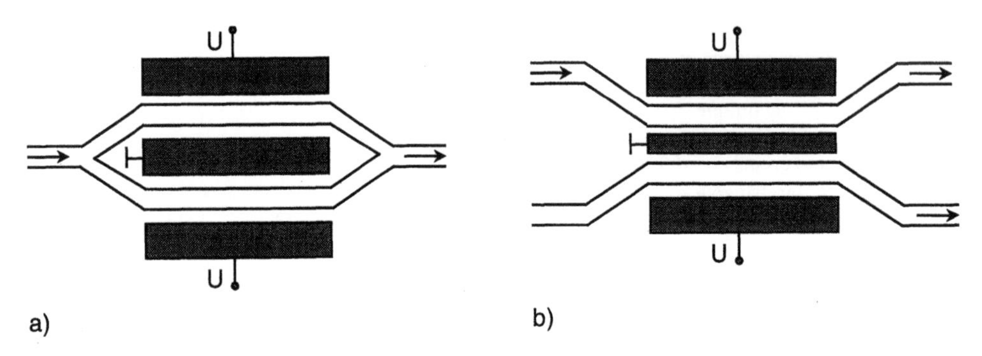

Figure 9.5 Configuration for (a) a Mach-Zender interferometer and (b)a directional coupler (after *Voges 1987*).

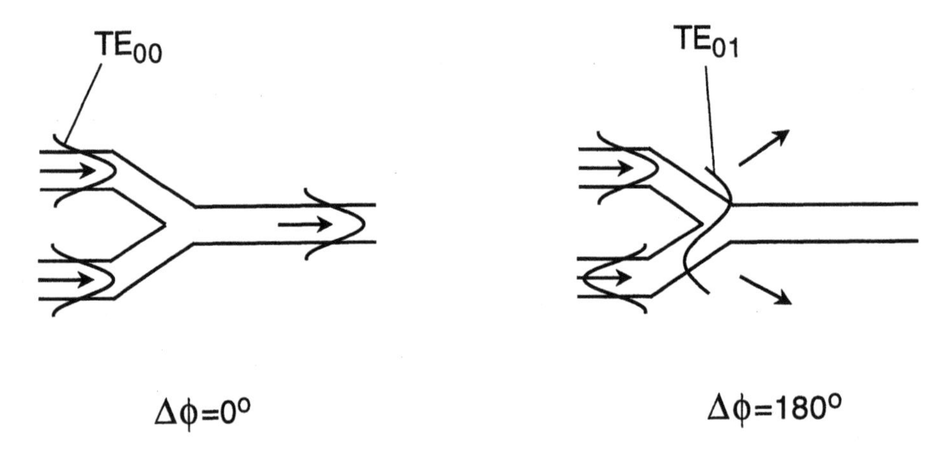

Figure 9.6 Amplitude modulation through electro-optically induced interference in a Mach-Zender interferometer.

electric field becomes shorter than the modulator length at high frequencies. In this case the modulation of the optical beam is achieved with travelling microwaves. There are limits to the modulation bandwidth due to a refractive index mismatch between microwaves and optical waves. The 3dB bandwidth of the modulator (frequency at which the power in the optical side bands is reduced by one-half) is given by (see e.g. *Voges 1987*)

$$\Delta f_{3dB} = \frac{1.4c}{\pi |n_O - n_M| L} \tag{9.8}$$

where n_O and n_M are the refractive indices at optical and microwave frequencies, respectively, c is the speed of light and L is the waveguide length. Given here are some examples.

$LiNbO_3$: $n_O \approx 2.2$, $n_M \approx 4.2 \rightarrow$ bandwidth of 6.6 GHz cm$/L$

GaAs: $n_O \approx 3.3$ (1.3 μm), $n_M \approx 2.4 \rightarrow$ bandwidth of 14.9 GHz cm$/L$

COANP: $n_O \approx 1.65$, $n_M \approx 1.61 \rightarrow$ bandwidth of 334 GHz cm$/L$

MNBA: $n_O \approx 2.02$, $n_M \approx 2.0 \rightarrow$ bandwidth of 669 GHz cm$/L$

m-NA: $n_O \approx 1.7$, $n_M \approx 1.8 \rightarrow$ bandwidth of 134 GHz cm$/L$

Here the advantage of semiconductor (e.g. GaAs) and organic materials come into play. Both have a small dispersion in ε and therefore a small phase-mismatch in waveguide modulators (Table 9.2). In ionic crystals, lattice vibrations contribute significantly to the electro-optic coefficients and the dielectric constant leading to a strong frequency dependence of both. These ionic contributions increase the electro-optic effects considerably, but, due to the increase of the dielectric constants they also limit the bandwidth (see Eq. (9.8)). By using special electrode-waveguide geometries, the microwave speed can be increased to a certain degree. A 40 GHz Ti:$LiNbO_3$ modulator has been demonstrated (*Noguchi et al. 1993*). On the other hand in organic substances, for example, the contributions are mainly of electronic origin yielding smaller electro-optic effects but also show a small frequency dependence.

The research for electro-optic applications using organic substances has not advanced very far. The best results have been achieved with polymers doped with

Table 9.2 Electro-optic and optical properties of inorganic and organic materials (r^T, r^s and r^e are the electro-optic coefficients at constant stress, constant strain and at optical frequencies (electronic contribution, calculated from nonlinear optical susceptibilities)). Data at 632.8 nm (for electro-optics) and 1,064 nm (nonlinear optics) except where noted. A more extensive list of organic electro-optic materials can be found in Table 3.1.

material	point group symmetry	refractive index	r^T[pm/V]	r^s[pm/V]	r^e[pm/V]	d [pm/V]	ε^T	$n^2 = \varepsilon_{optical}$
	inorganic							
$BaTiO_3$	4 mm	$n_3 = 2.480$[1]	$r_{33} = 103 \pm 3$[1]	28	+0.85	$d_{33} = -6.7$	$\varepsilon_{11} = 4300$	6.2
			$r_{13} = 14.5 \pm 0.5$[1]	8	+2.3	$d_{31} = -18$	$\varepsilon_{33} = 168$	
			$r_{42} = 1700$	820	+2.3	$d_{24} = -18$		
$KNbO_3$	mm2	$n_1 = 2.279$	$r_{33} = 63$	34	+5.0	$d_{33} = -27.4$	$\varepsilon_{11} = 154$	5.2
		$n_2 = 2.329$	$r_{13} = 34$	20	+2.7	$d_{31} = -18.3$	$\varepsilon_{22} = 985$	5.4
		$n_3 = 2.167$	$r_{42} = 450$	360	2.7	$d_{24} = -17.1$	$\varepsilon_{33} = 44$	4.7
$LiNbO_3$	3m	$n_1 = 2.272$	$r_{33} = 30.8$	30.8	+5.8	$d_{33} = -36$	$\varepsilon_{33} = 29$	5.2
		$n_3 = 2.187$	$r_{13} = 9.6$	8.6	+0.85	$d_{31} = -5.3$		4.8
KD_2PO_4	$\bar{4}2$ m	$n_1 = 1.51$	$r_{63} = 26.4$	24	0.39	$d_{36} = 0.53$	50	2.3
		$n_3 = 1.47$	$r_{41} = 8.8$	—	0.39	$d_{14} = 0.53$		2.2
GaAs	$\bar{4}3$ m	$n = 3.60$	$r_{41} = -1.5$	-1.5[3]		$d_{14} = +140$[3]	$\varepsilon_{11} = 13.2$	13

Table 9.2 Continued

material	point group symmetry	refractive index	r^T[pm/V]	r^s[pm/V]	r^e[pm/V]	d [pm/V]	ε^T	$n^2 = \varepsilon_{optical}$
	organic							
MNA	m	$n_1 = 2.0$	$r_{11} = 67 \pm 25$	—	70 ± 14	$d_{11} = 250 \pm 50$	3.5	4
MMONS	mm2	$n_1 = 1.569$	$r_{33} = +19.3 \pm 4$	—	19.3 ± 3.9	$d_{33} = 184 \pm 37$	—	2.5
		$n_2 = 1.693$	$r_{31} = +39.9 \pm 8$		35.9 ± 7	$d_{32} = 41 \pm 8$		2.9
		$n_3 = 2.129$				$d_{24} = 55 \pm 28$		4.9
m-NA	mm2	$n_1 = 1.805$	$r_{33} = 16.7 \pm 0.2$	—	—	$d_{33} = 21$	$\varepsilon_{11} = 3.2$	3.3
		$n_2 = 1.715$	$r_{23} = 0.1 \pm 0.6$			$d_{32} = 1.6$	$\varepsilon_{22} = 3.3$	2.9
		$n_3 = 1.675$	$r_{13} = 7.4 \pm 0.7$			$d_{31} = 20$	$\varepsilon_{33} = 3.2$	2.8
POM	222	$n_1 = 1.712$	$r_{41} = 3.6 \pm 0.6$	—	4.0	$d_{14} = 10$	$\varepsilon_{11} = 3.77$	2.9
		$n_2 = 1.919$	$r_{52} = 5.1 \pm 0.4$		5.1	$d_{25} = 10$	$\varepsilon_{22} = 5.41$	3.7
		$n_3 = 1.638$	$r_{63} = 2.6 \pm 0.3$		3.7	$d_{36} = 10$	$\varepsilon_{33} = 3.77$	2.7
urea	$\bar{4}2$ m	$n_1 = 1.48$	$r_{63} = 0.83$	—	—	$d_{36} = 2.3^2$	—	2.2
		$n_3 = 1.59$	$r_{41} = 1.9$			$d_{14} = 2.3^2$		2.5
COANP	mm2	$n_1 = 1.672$	$r_{33} = 15 \pm 2$	—	9.3 ± 1.5	$d_{33} = 19.3 \pm 3$	$\varepsilon_{11} = 2.8$	2.65
		$n_2 = 1.781$	$r_{23} = 13 \pm 2$		15.4 ± 3.2	$d_{32} = 43 \pm 9$	$\varepsilon_{22} = 2.7$	3.87
		$n_3 = 1.647$	$r_{13} = 3.4 \pm 0.5$			$d_{24} = 40 \pm 6$	$\varepsilon_{33} = 2.6$	2.54
PNP	2	$n_1 = 1.990$	$r_{22} = 12.8 \pm 1.3$	—	8.0 ± 1.4	$d_{22} = 70 \pm 8$	—	4.0
		$n_2 = 1.788$	$r_{12} = 13.1 \pm 1.3$		13.0 ± 2.2	$d_{23} = 21.6 \pm 4$		3.2
		$n_3 = 1.467$				$d_{16} = 53.4 \pm 6$		2.2
MNBA	m	$n_1 = 2.024$	$r_{11} = 30 \pm 3$	—	—	$d_{11} = 175$	$\varepsilon_{11} \sim 4$	4.1
		$n_2 = 1.585$						
		$n_3 = 1.652$						

[1] at $\lambda = 514.5$ nm [3] at $\lambda = 10600$ nm
[2] at $\lambda = 600$ nm

For further details of this table see text, Landolt-Börnstein (*1979a*), or Zgonik *et al.* 1993.

nonlinear active molecules (see Chapter 6). Half-wave voltages as low as 7 V have been demonstrated in such organic waveguides (*Lytel et al. 1989b*). A 20 GHz electro-optic Mach-Zehnder modulator, based on a DANS-containing side-chain polymer (*Girton et al. 1991*), showed the potential of organic materials for high speed applications. Recently, a travelling-wave polymeric optical intensity modulator was reported with more than 40 GHz of 3-dB electrical bandwidth (*Teng 1992*).

10. PHOTOREFRACTIVE EFFECTS IN ORGANIC MATERIALS

10.1. INTRODUCTION

Many materials change their index of refraction when they are irradiated with light. This effect is generally known as photorefraction. Several mechanisms can lead to photorefraction, most of which are not reversible. However, there also exist some reversible mechanisms, which are of utmost importance for a number of proposed applications.

Reversible photorefraction in solids can be due to several microscopic mechanisms. Besides the space charge induced photorefractive effect, which will be explained in detail below, other possible processes are e.g. photodimerization (*Tomlinson et al. 1972*), photoisomerization (*Ottelenghi et al. 1987*), thermo-optic effects (*Eichler et al. 1986*), or photoinduced inter- or intramolecular structural changes (*Tomlinson et al. 1980, Cohen et al. 1957, Todorov et al. 1984*).

This chapter will focus on the space charge induced photorefractive effect. This effect is often simply called *the photorefractive effect*, a convention that will also be used here. The photorefractive effect is a consequence of the interaction of different physical processes (photoinduced charge carrier generation, charge transport and trapping, and the electro-optic effect). A more detailed discussion is given in section 10.3.

This type of photorefractive effect has been found in a large number of inorganic materials, mostly oxides and semiconductors. Typical examples are $LiNbO_3$, $KNbO_3$, $BaTiO_3$, $Bi_{12}SiO_{20}$ (BSO), $Bi_{12}GeO_{20}$ (BGO), GaAs, or InP:Fe.

The key characteristics of the photorefractive effect are a high diffraction efficiency for low power writing beams, a wide range of response times, and complete reversibility. The effect has been proposed for a large number of applications, such as dynamic holography (*Bartolini et al. 1976*), spatial light modulation (*Shi et al. 1983, Voit et al. 1987*), image amplification, and optical phase conjugation (*Kukhtarev et al. 1979a, Hellwarth 1977, Günter 1982*).

While the photorefractive effect in inorganic oxides was found more than twenty years ago and in semiconductors in the early eighties, a closer examination of organic materials was only started in this decade. This new interest was mainly stimulated by the very powerful nonlinear optical and electro-optic organic materials that emerged in the last ten years, as described in the previous chapters of this work. A first observation of the photorefractive effect in an organic material was reported in 1990 (*Sutter et al. 1990a, b*). In 1991 first photorefractive effects were also found in polymers (*Ducharme et al. 1991, Schildkraut 1991*).

This chapter presents a discussion of the potential of organic photorefractive materials, a comparison with the performance of inorganic compounds, and a summary of some experimental data on the presently known photorefractive single crystals.

10.2 THE PHOTOREFRACTIVE EFFECT

The photorefractive effect occurs only in materials that are *electro-optic* (i.e. that change their refractive index under an applied electric field) and show *light induced charge transport* (due to diffusion, photovoltaic effect, or photoconductivity). If such a material is illuminated by an inhomogeneous light pattern, the following processes will lead to a refractive index change:

— In the illuminated parts of the material charge carriers will be generated. The carriers can be electrons, holes, or ions.
— These charge carriers will drift or diffuse into dark zones leaving behind a stationary ionic site of opposite charge. In the dark areas the carriers will be trapped, e.g. by defects, impurities, empty generation sites, or self trapping.
— The field that is generated by this charge redistribution leads to a change of the refractive index via the electro-optic effect.

For a detailed discussion of these processes see e.g. *Hall et al. (1985)*.

10.2.1 Basic theory: the transport equations

The experimental situation for a standard photorefractive experiment is shown in Figure 10.1. A photorefractive crystal is illuminated by two interfering beams. The direction of the grating vector $\boldsymbol{K}$ is denoted by **x**, the surface normal by **z**. **y** is perpendicular to **x** and **z**. The (internal) angle of incidence of the writing beams is given by θ (and is equal for both beams). The axes of the dielectric reference system of the crystal are assumed to be parallel to **x**, **y** and **z**.

The photorefractive process in inorganic crystals is usually described in terms of the *band transport model.* This model assumes that charges move between their trapping sites in well defined conduction (or valence) bands, where their

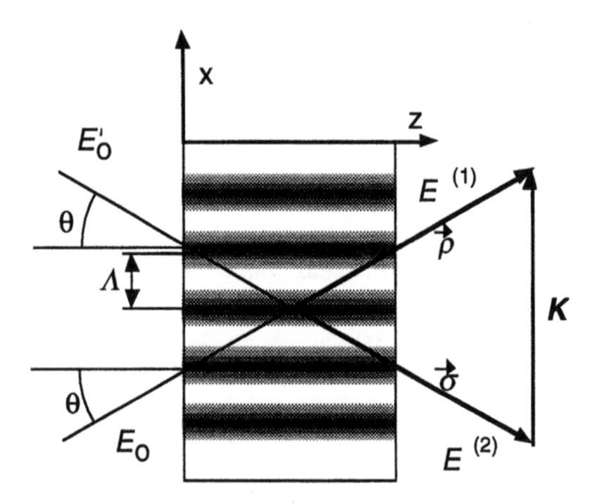

Figure 10.1 Experimental set-up for the formation of a photorefractive grating.

movement can be fully described by the carrier mobility. As will be explained in section 10.3, the band transport model often fails to give a satisfactory description of charge transport processes in organic materials, and it must be replaced by the *hopping model*. The hopping model assumes that the movement of a charge carrier takes place in instantaneous hops between trapping centers. The macroscopic concept of mobility is replaced by a microscopic transition probability between trapping centers.

It is presently not clear which model is more appropriate for the description of the photorefractive effect in organic compounds, especially in single crystals, where charge transport seems to fall into an intermediate region between hopping and band transport (see section 10.3). However, it can be shown that both models work equally well in the absence of applied electric fields and their differences are generally only manifested in temperature-dependence of the macroscopic parameters or in measurements with applied ac-fields (*Mullen 1988*).

In its simplest form, the band-transport model starts from a semiconductor or insulator with two in-gap electronic states, a donor state and an acceptor state. In the case of electrons as dominant carriers, there are more electron donors than acceptors and all acceptors are filled, independently of illumination and temperature. The remaining electrons are either in the conduction band or on a donor site. Thermal or light induced transitions between a donor state and the conduction band are allowed. Note that the only role of the acceptors in this model is to reduce the number of available charge carriers. They are not involved in the charge transport dynamics.

The following equations (generally referred to as 'Kukhtarev's equations') fully describe the photorefractive process for one type of charge carrier. A complete discussion of these formulas is given in the standard literature (see e.g. *Kukhtarev et al. 1979b, Solymar 1987, Hall et al 1985*). For the derivation of these equations the following important assumptions were made:

(1) Charge transport is dominated by only one carrier species.
(2) Only one trap level exists, no shallow traps.
(3) The efficiency of carrier generation does not depend on the illumination intensity I.
(4) The efficiency of carrier generation does not depend on the space charge field.
(5) No photovoltaic currents.

$$\frac{\partial N_D^+}{\partial t} = (\beta + sI)(N_D - N_D^+) - \gamma N_D^+ n, \qquad (10.1a)$$

$$q\frac{\partial}{\partial t}(n - N_D^+) + \frac{\partial J}{\partial x} = 0, \qquad (10.1b)$$

$$J = e\mu nE - qD\frac{\partial n}{\partial x}, \qquad (10.1c)$$

$$\frac{\partial}{\partial x}(\varepsilon\varepsilon_o E) = q(n + N_A - N_D^+) \qquad (10.1d)$$

with

n: free carrier density
J: current density
E: electric field
N_A: acceptor density
N_D: donor density
N_D^+: density of ionized donors
D: diffusion constant ($D = \mu\, k_B T/e$, k_B: Boltzmann's constant)
μ: carrier mobility
e: elementary charge
q: carrier charge (for electrons: $q = -e$)
β: thermal generation rate
s: photoionization constant
γ: recombination constant
ε: dielectric constant of the photorefractive material
ε_o: vacuum permittivity
I: illumination intensity

Note that point (4) of the assumptions made above might break for many organic compounds, especially for polymers, where the quantum efficiency for the carrier generation is increased by orders of magnitude under fields larger than some kV/cm. This is due to a partial suppression of geminate recombination by carrier acceleration. More details on this effect are given in section 10.4. (For an analytical treatment of the photorefractive effect under the influence of geminate recombination see e.g. *Twarowski 1989*). In the case of COANP:TCNQ and MNBA, the organic photorefractive single crystals that will be discussed in more detail in sections 10.6 and 10.7, it was found that the quantum efficiency is independent of the applied electric field for fields smaller than 5 kV/cm.

10.2.2 Grating amplitude

For the discussion of grating formation, we assume the illumination to be of the form

$$I = I_o(1 + m\cos(Kx)) = I_o\mathrm{Re}(1 + m\,e^{iKx}) \qquad (10.2)$$

where $m = 2\sqrt{I_1 I_2}/I_o$ is the modulation factor and K is the grating vector ($K = 2\pi/\Lambda$, Λ: grating spacing).

To solve Eqs. (10.1), the following approximations are generally made:

(a) Small modulation: $m \ll 1$

(b) Small dark conductivity: $\beta \ll sI$
(c) Low free carrier density (low optical power): $n \ll N_D^+$, $n \ll N_D$.

Using these approximations, the solution in the steady state ($\partial/\partial t = 0$) of Eqs. (10.1) can be linearized and the resulting electric field takes the form

$$E = \mathrm{Re}(E_o + E_1\, e^{iKx}) \tag{10.3}$$

where E_o is the applied electric field and E_1 is the modulation of the field, that gives rise to the photorefractive grating.

The calculation of E_1 from Kukhtarev's equations using the approximations (a)–(c) is described in the literature (see e.g. *Solymar 1987, Fluck et al. 1991*) and will not be repeated here. The result is the well known formula (10.4), which describes the first Fourier component E_1 of the space charge field as a function of the applied field E_o, the trap density limited field E_q, and the diffusion field E_d:

$$E_1 = m\frac{E_q\,(iE_d - E_o)}{(E_q + E_d) + iE_o} = mE_q\frac{i\,(E_qE_d + E_d^2 + E_o^2) - E_qE_o}{(E_q + E_d)^2 + E_o^2} \tag{10.4}$$

where

$$E_q = \frac{q}{K\varepsilon\varepsilon_o}\frac{N_A(N_D - N_A)}{N_D} \quad \text{trap density limited field and}$$

$$E_d = \frac{q}{e}\frac{k_BT}{e}K \quad \text{diffusion field}$$

Note that E_1 can be a complex number. Equations (10.2, 10.3) show that its complex argument corresponds to the phase shift between illumination and field grating.

10.2.3 Response time

Besides the grating amplitude, the response time is a second relevant quantity for photorefractive applications. Its dependence on the material parameters is also fully described by Kukhtarev's equations. The basic parameter for the response time in the regime of approximations (a)–(c) and in absence of an applied electric field is the dielectric response time τ_{di}

$$\tau_{di} = \frac{\varepsilon\varepsilon_o}{e\mu n_o} = \frac{\varepsilon\varepsilon_o}{\sigma(I_o)} \tag{10.5}$$

where $\sigma(I_0)$ is the photoconductivity at the average intensity I_0.

For times t larger than the recombination time τ_{Re}, but smaller than the dielectric relaxation time $\tau_{Re} \ll t \ll \tau_{di}$, the time dependence of E_1 is linear, i.e.

$$E_1(t) = E_1(t = \infty)\frac{t}{\tau} \tag{10.6}$$

where the response time τ is given by (*Valley et al. 1988*)

$$\tau = \frac{1 + K^2 L_D^2}{1 + K^2 L_S^2} \times \tau_{di} \tag{10.7}$$

with $$L_S = \sqrt{\frac{\varepsilon\varepsilon_o k_B T}{N^* e^2}} \quad \text{Debye screening length}$$

and $$L_D = \sqrt{\frac{\mu k_B T}{e}\tau_{Re}} \quad \text{diffusion length,}$$

where $N^* = N_A\,(N_D - N_A)/N_D$.

Equation (10.7) shows that the response time depends on the grating spacing. At a large grating spacing Λ the response time will be the dielectric response time. For short grating spacings τ depends on the relative magnitude of diffusion and screening lengths, L_D and L_S. If $L_D \gg L_S$ the time constant will increase for small values of Λ, if $L_D \ll L_S$ it will decrease.

10.2.4 Beam coupling

Beam coupling by two-wave mixing can be observed in a set-up as shown in Figure 10.2. A strong pump beam and a weak signal beam are used to create an interference grating in a photorefractive material. If the refractive index grating has non-zero phase shift with respect to the intensity grating or if one of the beams

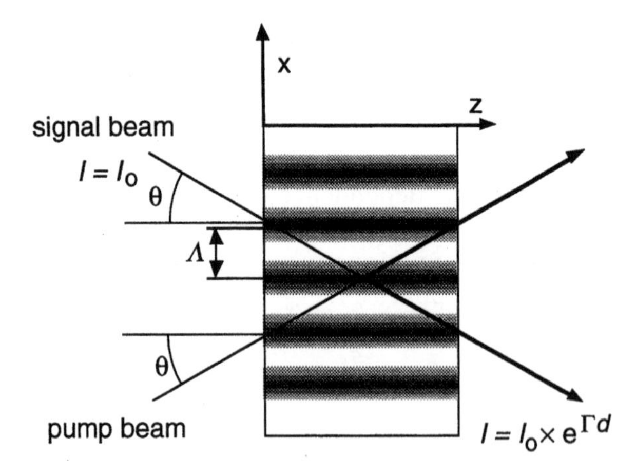

Figure 10.2 The arrangement for beam coupling. The intensity of the pump beam is assumed to be much larger than the intensity of the signal beam.

is much stronger than the other, energy of the pump beam can be transferred to the weak signal beam. This allows an efficient amplification of the signal beam.

The exponential gain factor Γ (beam coupling gain) for photorefractive isotropic beam coupling neglecting absorption or reflection losses is given by (*Hall et al. 1985*)

$$\Gamma = p \times \frac{2\pi}{\lambda} \times n^3 r \times \mathrm{Im}\,(E_1)\,/\,m \tag{10.8}$$

where λ is the wavelength of the writing beams and p a projection factor. p depends on the geometry of the experimental arrangement. When using a diagonal element of the electro-optic tensor (i.e. r_{IIJ}), the geometry can generally be chosen such that p is close to 1. For a non-diagonal electro-optic coefficient (r_{IJK} with $I \neq J$) p can be considerably smaller when the beams must be polarized in their plane of incidence (see e.g. *Fainman et al. 1986*). Since the largest electro-optic coefficient of optimized organic compounds is usually a fully diagonal element r_{ZZZ} (see Chapter 3), p will always be close to 1 for these materials.

The electro-optic coefficient that should be used in Eq. (10.8) is some intermediate value between the clamped and the unclamped coefficient, depending on crystal symmetry and direction of the grating vector (*Günter et al. 1991*). For organic materials this distinction is of little importance since the electro-optic coefficients are dominated by the electronic contributions and clamped and unclamped coefficients do not differ greatly. For inorganic crystals, however, there can be a difference of more than a factor of two between clamped and unclamped electro-optic effects.

For large grating spacings in the absence of applied electric fields Eq. (10.8) becomes (with $E_d = q/e \cdot Kk_B T/e$)

$$\Gamma = p \times \frac{2\pi}{\lambda} \times n^3 r \times \frac{Kk_B T}{e} \tag{10.9}$$

10.2.5 Photorefractive sensitivity

We will use two figures of merit for characterizing the photorefractive sensitivity of an organic material. The first one, S_n, describes how much optical energy absorbed per unit volume is needed to produce a given refractive index change, the second one, S_η, gives the energy needed for a diffraction efficiency of 1% for a 1 mm thick storage material. The first one is given by (*Günter 1982*).

$$S_n = \frac{n^3 r_{\text{eff}}}{2\varepsilon\varepsilon_o} \times \frac{\phi}{h\nu} \times eL_{\text{eff}}\, m \tag{10.10}$$

where L_{eff} is the effective charge transport length and ϕ the quantum efficiency of carrier generation. For no applied fields and no photovoltaic effect L_{eff} can be found from the diffusion length L_D and the grating vector K:

$$L_{\text{eff}} = \frac{KL_{\text{D}}}{1 + (KL_{\text{D}})^2} L_{\text{D}} \tag{10.11}$$

where L_{D} is defined in Eq. (10.7). L_{eff} has the maximum $L_{\text{eff}} = 1/2\ L_{\text{D}}$ at $L_{\text{eff}}K = 1$.

Note that if the dielectric constant is large ($\varepsilon \gg 1$), Eq. (10.10) can be expressed in terms of the polarization-optic coefficient f, defined by $f = r/(\varepsilon_o(\varepsilon - 1))$, and we obtain

$$S_{\text{n}} \cong \frac{n^3 f}{2} \times \frac{\phi}{h\nu} \times eL_{\text{eff}}\, m \tag{10.12}$$

The polarization-optic coefficient f is found to be similar in different inorganic electro-optic materials (*Wemple et al. 1972*). An upper limit of f in most inorganic oxides was estimated to be 0.25 m^2/C (*Günter 1982*). Organic compounds, however, can have considerably larger polarization-optic coefficients. Values in the order of 1 m^2/C are no exception (*Bosshard et al. 1993*).

The second figure of merit, S_η, is given by (*Günter 1982*)

$$S_\eta = \frac{\pi}{\lambda \cos\theta} S_{\text{n}} \tag{10.13}$$

The material parameters going into Eqs. (10.10, 10.13) will be discussed in the next chapters.

10.3 CHARGE TRANSPORT IN ORGANIC SOLIDS

Charge transport is one of the basic mechanisms that define the photorefractive performance of a material. The nature of charge transport varies strongly in different classes of solids. In organic single crystals charge transport profits from the high degree of order, which gives rise to high mobilities, while in unordered systems (such as non-crystalline polymers) charge transport is severely disturbed by disorder and carrier mobilities are very low. In the following, the basic properties of charge transport in organic single crystals and in non-crystalline polymers will be discussed.

10.3.1 Charge transport in organic single crystals

The charge transport properties of molecular single crystals differ strongly from those of the conventional inorganic photorefractive crystals. Inorganic single crystalline semiconductors or oxides have strong covalent or ionic bonds, leading to wide conduction and valence bands. Also, the atoms and ions are rigidly held in place and the lattice is not strongly deformed by charged sites within the crystal. In contrast to this, the molecules of organic crystals are usually held together by Van der Waals forces or hydrogen bonds. Since these are weaker interactions, the electronic states of a molecule in a solid do not differ strongly from the vacuum states of an isolated molecule. This results in very narrow conduction bands with widths comparable to the thermal energy k_BT at room temperature. Also, the Coulomb forces of free carriers can induce a local distortion of the crystal lattice, which strongly perturbs the carriers' Bloch state. For these reasons the classical band model in its simple form is inadequate for a complete description of the charge transport in organic single crystals (*Silinsh 1980*).

The exact nature of the states of free carriers in a molecular crystal is not yet completely understood (*Kao et al. 1981*). Most authors agree that the dielectric response of the environment to an excess charge leads to a considerable stabilization of a free carrier state, resulting in a decrease of electron energies and an increase of hole energies. Figure 10.3 shows schematically the electronic energy states of an isolated molecule and the energy levels of an excess carrier in a molecular crystal.

Holes will have a higher energy (by P_h) in the crystal than in the molecule since the polarization of the environment makes the creation of a positive ion easier, while electron states will have lower energy (by A_c) since energy is won by bringing an excess electron into a dielectric (*Pope et al. 1982, Silinsh 1980*). I_g is the ionization energy of a free molecule and E_g is the gap energy of the solid. Note that E_g does not denote the spectral position of an absorption edge as in semiconductors, because the hole and electron levels are nearly unoccupied in equilibrium state.

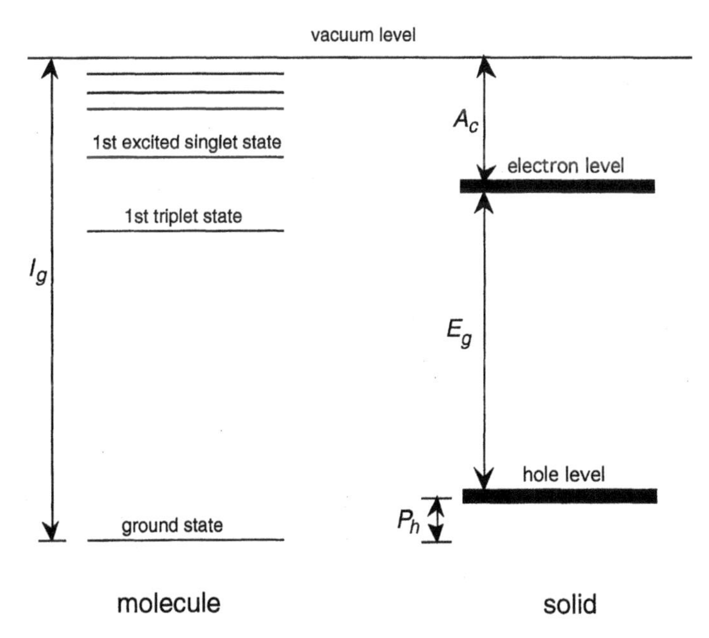

Figure 10.3 A schematic comparison of the energy levels of an isolated molecule and of an excess carrier in a molecular solid.

The nature of charge transport determines the carrier mobility μ. In organic single crystals at room temperature carrier transport can either be band-like or hopping (*Pope et al. 1982*). Typical carrier mobilities are around 1 cm^2/Vs (*Schein et al. 1982*).

At low temperatures (<100K) band like transport often dominates the conduction properties and very pure molecular single crystals can show mobilities in the order of several 100 cm^2/Vs (*Schein et al. 1982*).

The carrier mobility is also found to be strongly anisotropic in many organic single crystals. A large electron overlap of neighboring molecules in one direction increases the mobility considerably. Best mobilities are found in compounds with flat, π-electron rich molecules that are arranged in stacks. Such compounds can have mobilities that are one order of magnitude higher along the stack direction than perpendicular to it.

Strong doping or poor crystalline quality can reduce the carrier mobility considerably. For example doping of pure anthracene crystals with 40 ppm of tetracene was found to decrease the hole drift mobility by one order of magnitude (*Leblanc 1967*) and doping with 10 ppm of naphtacene decreased it by two orders of magnitude (*Hoesterey et al. 1963*). Defects and dopant sites are believed to introduce shallow traps, which lead to this decrease of the mobility (*Kao et al. 1981*).

10.3.2 Charge transport in non-crystalline polymers

Charge transport in unordered *saturated* polymers is clearly of the hopping type (*Emin 1986*). The lack of any long-range order makes the formation of extended electron states impossible. The mobilities of the carriers are low (10^{-9} to 10^{-6} cm^2/Vs at low electric fields) and increase with temperature. The charges travel mainly by hopping through side-chain or guest molecules. Note that all known electro-optic and photorefractive polymer materials use saturated polymers as host materials.

The backbone of the polymers only contributes to carrier transport if it contains mobile π-electrons (*unsaturated or conjugated* polymers). Unsaturated backbones can give rise to higher carrier mobilities within one molecule. However, unsaturated polymers are often chemically unstable and strongly absorbing and thus have not been used as hosts for poled electro-optic materials.

Recently, a rather high hole mobility in the order of 10^{-2} cm^2/Vs has been reported in a disordered molecular film of TAPC (1,1-bis(di-4-tolylaminophenyl)-cyclohexane, *Borsenberger et al. 1991*). It should be noted, however, that at least part of the high mobility can be attributed to a lack of molecular reorientation of the unpolar TAPC molecules upon removal or addition of a charge carrier. Since all presently known electro-optic polymers contain highly polar, anisotropic, electro-optic dopant molecules or groups, carrier induced reorientation and polaronic effects will always play an important role in these compounds and it seems unlikely that they can exhibit such high carrier mobilities at room temperature and low electric fields.

10.3.3 Charge trapping

Besides high carrier mobility, trapping of photoexcited carriers is another important factor for photorefractive properties.

Free charge carriers in organic solids can be trapped by the same mechanisms as carriers in inorganic compounds. It is easy to imagine that trapping by defects is a very common mechanism in non-crystalline polymers, but it can also be found in single crystals, where it obviously depends strongly on crystal quality and impurity concentration. For anthracene, for example, it was found that the trap density varied between 4.2×10^{14} cm^{-3} for solution grown crystals and 1.5×10^{19} cm^{-3} for melt grown samples (*Kao et al. 1981*). Doping was also found to increase the trap density and decrease the carrier lifetimes, e.g. doping anthracene with 10 ppm guest molecules decreased the lifetime of carriers by more than two orders of magnitude (*Leblanc 1967*). The extent of self trapping inorganic crystals is still under discussion (*Kao et al. 1981*).

10.4 PHOTOCONDUCTIVITY OF ORGANIC MATERIALS

Photoconductivity in organic compounds has been widely investigated. Due to their potential in electrophotography, most data available are for polymeric systems. Experiments on single crystals are mostly restricted to some few standard materials such as anthracene or naphthalene. We will discuss photoconductivity in polymers and single crystals separately.

10.4.1 Photoconducting polymers

We will restrict our discussion to saturated polymers, as they are used as host systems for electro-optic polymers.

Saturated backbone polymers are mostly transparent in the visible and near infrared and intrinsic photoconductivity is only found in spectral regions of strong absorption. Photoconductivity in the visible can be induced by adding appropriate sensitizer molecules or atoms (*Stolka et al. 1978, Mort et al. 1982*). Two groups of guest molecule can be distinguished:

— *Dyes*: polymers are often doped with dyes to induce visible photoconductivity. The mechanism of carrier generation is not fully understood (*Fox 1976*), but it is assumed that the energy absorbed by a dye molecule travels as an exciton until it can excite a carrier residing at a trap or defect site, or that electron transfer occurs between the photoexcited dye molecule and a nearby donor (or acceptor) molecule (*Mort et al. 1982*).

— *Charge transfer compounds*: in this case the polymers are doped with electron-accepting or donating molecules, which form a charge transfer state with the host material. Doping with electron acceptors is more common since polymers can easily be synthesized with donor side-chains. The resulting charge transfer states are generally characterized by an absorption peak in the visible or near infrared, which does not appear in the pure dopant or polymer. Illumination of the doped polymer at this wavelength leads to a charge transfer between the acceptor and the donor. The excited charge transfer state can dissociate into two free carriers or recombine (geminate recombination).

Note that a sensitizer molecule can also act as a potential trap for mobile charge carriers. On the other hand, the charge transport properties of polymers can often be improved by adding molecules that act as transport moiety. An example for the latter case is bisphenol A polycarbonate (Lexan) doped with triphenylamine (TPA) (Figure 10.4), where the charge carriers hop among localized states associated with the TPA molecules (*Mort et al. 1982*).

The most commonly used types of photoconductive polymers are compounds with aromatic amine groups incorporated in the polymer chain or in the side chain (*Stolka et al. 1978*). These polymers (which have an electron-donating character) are doped with moderately strong electron-acceptors. A typical representative of this family is polyvinylcarbazole (PVK) doped with trinitrofluorenone (TNF) (Figure 10.4).

In PVK:TNF the side chain of the polymers (which is a weak donor) acts simultaneously as transport medium for the holes and the acceptor TNF acts as transport medium for the electrons. In this case the sensitizer molecule is also a transport medium for electrons but a trap for holes: an increase of the TNF concentration increases the electron-mobility while decreasing the hole-mobility.

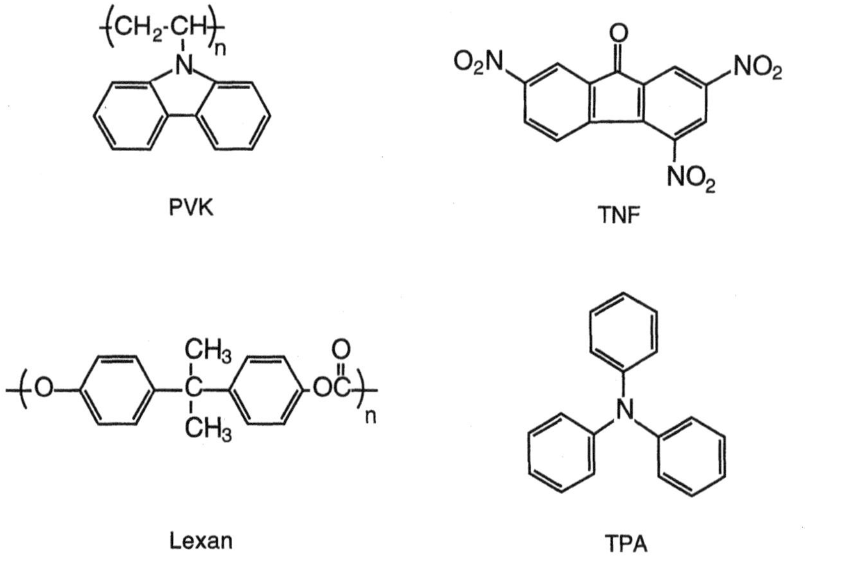

PVK TNF

Lexan TPA

Figure 10.4 The photoconductive polymer systems polyvinylcarbazole (PVK) doped with trinitrofluorenone (TNF) and bisphenol A polycarbonate (Lexan) doped with triphenylamine (TPA).

Since the charge generation mechanism is not fully understood and since sensitizer molecules can also change the charge transport properties, the criteria for selection of efficient sensitizing molecules remain unclear (*Mort 1980*). Dye-sensitizers are generally added in lower concentrations than charge transfer molecules, which are sometimes used at molecular ratios of more than 1:1 (*Fox 1976*).

The efficiency of carrier generation in polymers depends strongly on the applied electric fields (*Mort 1980*). At low fields geminate recombination of a carrier pair is very likely to occur, while at higher fields the probability that one of the carriers can escape increases. The strong influence of geminate recombination in non-crystalline polymers can be attributed to two factors: the low dielectric constants in the order of 2.5 to 6 and the absence of clear conduction bands as they are found in classic semiconductors.

The low dielectric constants lead to a larger capture radius r_c. r_c is defined by the distance at which the kinetic energy of a thermalized carrier is equal to the attractive potential energy of a stationary ion of opposite charge. r_c can be calculated easily from the relation

$$r_c = \frac{e^2}{4\pi\varepsilon\varepsilon_0 kT} \tag{10.14}$$

The capture radius is thus inversely proportional to the dielectric constant of the enclosing medium, which is generally smaller in molecular compounds ($\varepsilon \cong 2$ to 4 corresponding to $r_c = 135$ to 270 Å) than in semiconductors or oxides ($\varepsilon \cong 10$ to 1,000, $r_c = 0.5$ to 50 Å).

The absence of any long-range conduction bands in non-crystalline polymers leads to a small diffusion length that does not allow the carriers to gain a sufficient initial separation distance before thermalization, unless high electric fields are applied, which decrease the capture radius, as shown in Figure 10.5.

An example for the influence of geminate recombination on the charge generation efficiency (number of carriers generated by an absorbed photon) is found in polyvinylcarbazole (PVK) doped with trinitrofluorenone (TNF). In this photoconductor the carrier generation efficiency increases from 10^{-5} at low fields (< 10 kV/cm) to more than 10^{-2} at high fields (1,000 kV/cm) (*Perlstein et al. 1982*).

10.4.2 Photoconducting organic single crystals

As mentioned above, photoconductivity in molecular single crystals has been carefully investigated for some model materials such as anthracene. Most of the research is concentrated on photoinduced carrier generation in very pure crystals.

For most organic molecules the energy of the lowest lying singlet state (which dominates the absorption edge and the nonlinearities in most nonlinear optical molecules) is generally lower than the energy required to create free carriers (*Mort 1980*), so the mechanisms of charge carrier generation near the absorption edge involves mostly biphotonic excitations or carrier generation from trap-sites,

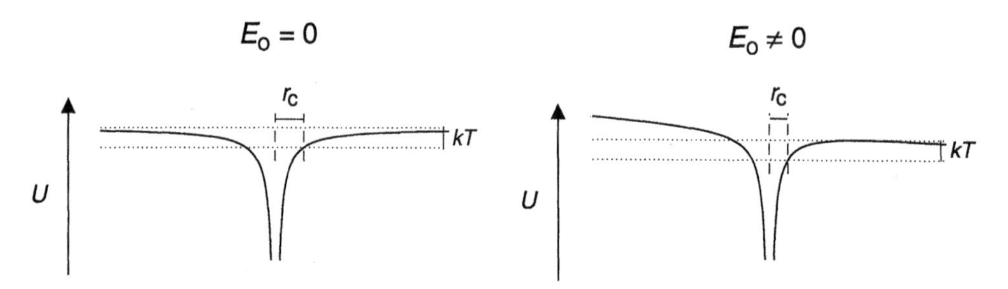

Figure 10.5 The potential U created by an ionic site after carrier generation with and without applied electric field. The capture radius r_c is given by the distance at which a thermalized carrier has sufficient energy to escape the potential. The applied field leads to a decrease of r_c.

defects, or surface-states. At shorter wavelengths (in the order of $h\nu = 4$ eV) direct electron-hole production becomes possible. The efficiency of carrier generation is rather low (10^{-4} to 10^{-2}) because of geminate recombination (*Pope 1989*) and strong competition from the relaxation processes to triplet and singlet ground states (*Fox 1976*). However, single crystals show generally better photogeneration efficiency than unordered systems, thanks to better mobilities and a smaller influence of geminate recombination (*Perlstein et al. 1982*).

For photorefractive applications, photoconductivity is desired in spectral regions of weak absorption, where the low photon energy makes an efficient carrier generation difficult. Thus intrinsic photoconduction is often weak. As in polymeric systems, the photoconductivity can be increased by the addition of appropriate dopant molecules.

10.5 ORGANIC PHOTOREFRACTIVE MATERIALS

In the following we will discuss the potential of organic materials for photorefractive applications using known data on photoconductivity in these compounds as well as the results of Chapters 3 and 8 on the electro-optic effect.

In order to calculate the *dielectric response time* τ_{di} from Eq. (10.5), an estimate of the photoconductivity must be made. The photoconductivity $\sigma(I)$ is given by

$$\sigma(I) = \phi\mu\, \tau_R\, e\, \alpha\, I/h\nu \tag{10.15}$$

Using Eqs. (10.5) and (10.15), the dielectric response time τ_{di} can be calculated as

$$\tau_{di} = \frac{\varepsilon\varepsilon_o h\nu}{e\phi\mu\tau_R e\alpha I} \tag{10.16}$$

Another relevant parameter for photorefractive applications is the beam coupling gain Γ. For an estimate of the upper limit of Γ (without applied electric field E_o), we calculate the diffusion limited maximum beam coupling gain Γ (see Eq. (10.9)) from

$$\Gamma = p \times \frac{2\pi}{\lambda} \times n^3 r \times \frac{K k_B T}{e} \tag{10.17}$$

The projection factor p is taken as $p = 1$ for diagonal electro-optic coefficients and $p = 0.76$ for the coefficients r_{42} or r_{51} in point groups mm2 and 4mm. It must be noted that the gain factors calculated from this equation are only approximate upper limits. In an experiment lower gains are generally measured. Reasons for lower values of Γ can e.g. come from the fact that the actual space charge field is smaller than the diffusion limited field E_d (because of large dielectric constants or a low trap density), that not the full unclamped electro-optic coefficients can be used for the photorefractive process, that contributions from the second carrier species cancel at least part of the space charge field or that the investigated crystals are not fully monodomain. The distinction between unclamped and clamped electro-optic coefficients will have only weak influence on the gain of organic crystals where the electro-optic coefficients are mainly of electronic origin, but will reduce the results of the inorganic compounds considerably.

Beam coupling gain Γ and response time τ are not independent quantities. If the electro-optic figure of merit $n^3 r$ of a material is known, we can calculate the required electric field or the corresponding electric charge density for a given value of Γ (*Yeh 1987*). Depending on the quantum efficiency ϕ and the absorption constant α, a certain number of incident photons per area is required to generate this charge. This allows the calculation of a minimum illumination time τ as a function of illumination intensity I. It can be shown that the following relation (*Yeh 1987*) must be fulfilled:

$$\Gamma / \tau \leq \frac{\alpha e \Lambda \pi \phi I n^3 r}{2 \lambda \varepsilon \varepsilon_o h \nu} \tag{10.18}$$

We will try to estimate typical and limiting values for τ_{di}, Γ, and Γ/τ for single crystals and polymers at an average writing intensity $I = 1\ \mathrm{W/cm^2}$ (corresponding to a photon flux of $I/h\nu = 2.60 \cdot 10^{22}\ \mathrm{m^{-2}s^{-1}}$ at 515 nm). We assume the absorption coefficient α to be 1 $\mathrm{cm^{-1}}$, which is a reasonable value for most photorefractive applications.

Typical values for the electro-optic figure of merit $n^3 r$ and the dielectric constant ε were given in Chapters 3 and 8. The material parameters ϕ, μ,and τ_R will be discussed in the following.

10.5.1 Photorefractive polymers

As mentioned earlier, many polymers can be made photoconductive by doping with appropriate sensitizers. Electro-optic effects can be induced by mixing with nonlinear optical chromophores (or adding the chromophores to the polymer's side chain) and alignment of thin films under high electric fields. A combination of these two methods can be used for the synthesis of a photorefractive polymer.

A first report of an electro-optic polymer sensitized for photoconduction was published in 1991 (*Schildkraut 1991*). The polymer had an acrylic backbone with an electro-optic chromophore in the side chain. Photoconductivity was induced by adding two additional compounds — a sensitizer and a hole transport molecule.

In another work (*Ducharme et al. 1991*) a first photorefractive polymer was presented. It consisted of an epoxy based nonlinear optical polymer (BisA-NPDA (*Eich et al. 1989b*) or NA-APNA (*Jungbauer et al. 1990*) mixed with a hole transport agent. Photorefractive diffraction efficiencies up to 0.1% were observed under very high applied fields (~ 100 kV/cm). The structures of the used sensitizer (DEH), the chromophore (NPDA), and the polymerizer (Bis-A) are shown in Figure 10.6.

In the following years, a large number of publications on photorefractive properties of polymeric systems appeared (*Tamura et al. 1992, Li et al. 1991, Walsh et al. 1992, Silence et al. 1992a, b, Scott et al. 1992, Ducharme et al. 1993, Orczyk et al. 1993, Kawakami et al. 1993*). Most of this work relates to electro-optic polymeric systems that were sensitized for photoconductivity.

In the following the basic parameters ($\phi\mu\tau_R$ and n^3r) for photorefractive polymeric systems are discussed.

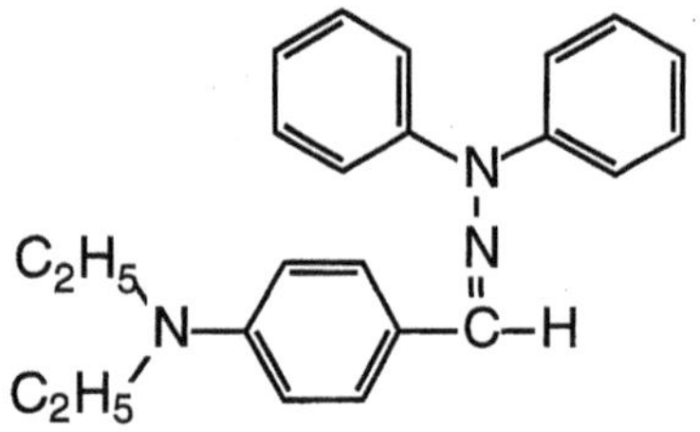

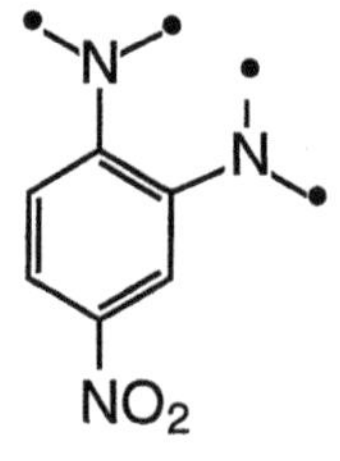

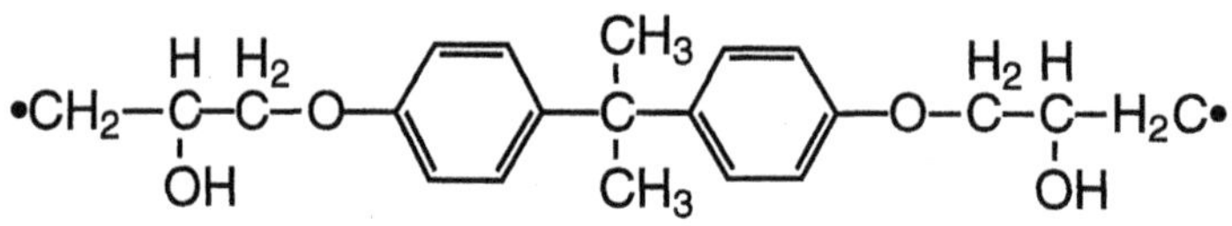

Figure 10.6 Chemical structures of the hole transport agent *p*-diethylaminobenzaldehyde-diphenyl hydrazone (DEH), the nonlinear optical chromophore 4-nitro-1, 2-phenylenediamine (NPDA) and the polymer derivative of diglycidylether of Bisphenol-A (Bis-A). Upon polymerization the hydrogens of the amines (denoted by a dot •) are polymerized in part to the end carbon groups of Bis-A (*Ducharme et al. 1991*).

$\phi\mu\tau$: As already mentioned in section 10.4.1 the photocarrier generation efficiency ϕ in polymers depends strongly on the applied electric field. One of the best efficiencies has been reported for TNF:PVK (up to 10^{-1} at 10^{6} V/cm but less than 10^{-5} at low fields, *Perlstein et al. 1982*). It seems probable that ϕ will not be larger in photorefractive polymeric compounds since doping levels of the photosensitizers must be kept low in order to obtain moderate absorption. For a typical material we estimate that ϕ will be less than 10^{-5} at low fields and smaller than 10^{-1} at high fields. For a photorefractive polymer system a carrier generation efficiency of 10^{-3} – 10^{-2} was observed at applied fields between 100 and 700 kV/cm (*Scott et al. 1992*).

As discussed in section 10.3.2, the carrier mobility μ can be expected to be low in glassy polymer systems. Best mobilities are around 10^{-5} cm^2/Vs at a very high fields and can be as low as 10^{-9} cm^2/Vs at low fields (see Table 10.1). For a photorefractive polymer system a hole mobility in the order of 10^{-8} – 10^{-7} cm^2/Vs was reported (*Scott et al. 1992*).

The value of $\phi\mu\tau_R$ is difficult to estimate since τ_R is not well known. It can be expected to be lower than in ordered compounds if we assume recombination times in the order of 1 μs.

Electro-optic response of polymers: Chapters 3 and 8 give an estimate for maximum electro-optic coefficients in stilbene and benzene single crystals, which we can use to obtain estimation for an optimum and a typical value of r in polymeric systems. If these molecules are suspended in a polymer host (as guests or side-chains) and ordered perfectly parallel, the electro-optic properties will still be somewhat smaller than in an optimized single crystal of the same chromophores. This effect is due to a smaller chromophore density and reduces the electro-optic coefficients by typically 50%.

The parameters obtained from these considerations are listed in Table 10.1.

10.5.2 Photorefractive single crystals

As discussed in section 10.4.2 the photoconductivity in a spectral range of low absorption is often very weak in organic single crystals. By appropriate doping with sensitizers photoconductivity can be increased. Dopants can either be acceptors (donors) or dye molecules. Both doping approaches seem reasonable since the exact mechanisms of photocarrier generation are not yet completely understood.

First photorefractive effects in an organic single crystal were found in electro-optic single crystals of 2-cyclooctylamino-5-nitropyridine (COANP) doped with the strong electron acceptor 7,7,8,8-tetracyanoquinodimethane (TCNQ) (*Sutter et al. 1990a*, Figure 10.7).

Recently, photorefractive behavior was also observed in nominally pure organic single crystals of 4′-nitrobenzylidene-3-acetamino-4-methoxyaniline (MNBA, *Sutter et al. 1993*, Figure 10.7).

More experimental details on COANP:TCNQ and MNBA will be given in sections 10.6 and 10.7.

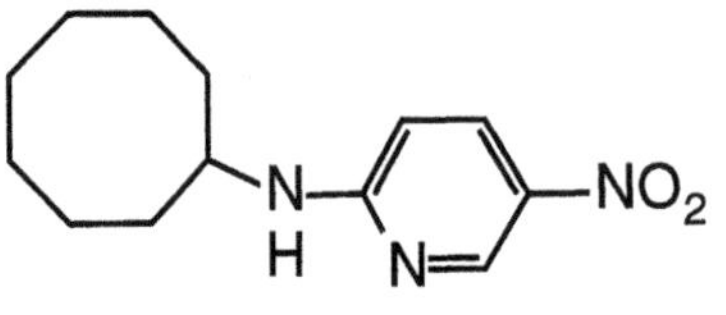

COANP

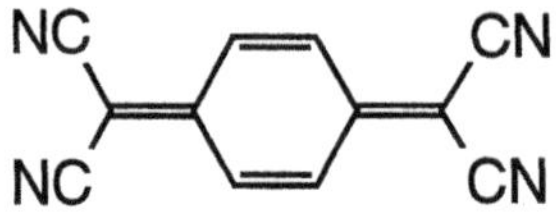

TCNQ

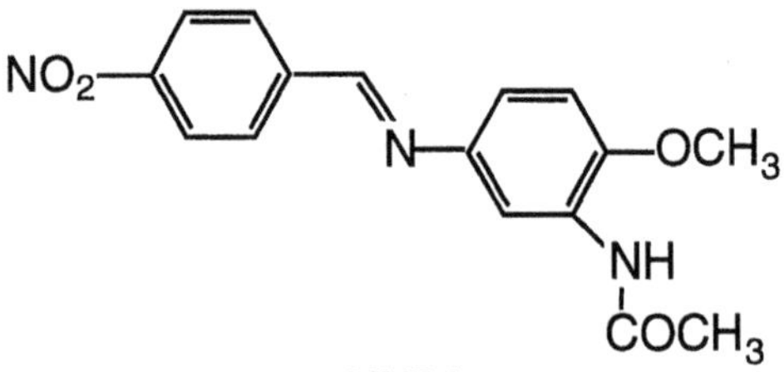

MNBA

Figure 10.7 2-cyclooctylamino-5-nitropyridine (COANP) doped with 7,7,8,8-tetracyanoquinodimethane (TCNQ) was the first known photorefractive organic material. 4′-nitrobenzylidene-3-acetamino-4-methoxyaniline (MNBA) is a nominally pure photorefractive single crystal.

In the following we discuss the basic limiting parameters (photoconductivity, charge transport, and electro-optic properties) of a photorefractive single crystal.

$\phi\mu\tau$: The maximum photocarrier generation efficiency ϕ is difficult to estimate since few experimental data on photosensitization of crystalline material are available. As discussed in section 10.4.2, it will probably be in the order of 10^{-4} to 10^{-2} for the best compounds. In contrast to inorganic compounds, geminate recombination may play a more important role in organic compounds and reduce the quantum efficiency (see section 10.4.1). The influence of geminate recombination on the photorefractive response is discussed by Twarowksi *(1989)*.

The mobility μ of electrons or holes in very pure organic crystals at room temperature is around 0.1 to 1 cm^2/Vs. However, as already mentioned, even low impurity concentrations can decrease μ by two orders of magnitude. For a doped organic crystal we therefore expect carrier mobilities between 0.001 and 1 cm^2/Vs. At lower temperatures the mobility can be expected to increase in high quality crystals.

The recombination time τ_R depends on crystalline quality and dopant concentrations as well as on the processes that are resolved in the corresponding measurements. Typical values are in the μs range.

From the properties discussed above, we see that $\phi\mu\tau_R \approx (0.01 \text{ to } 1{,}000) \times 10^{-12} cm^2/V$, which is slightly lower than but comparable to inorganic oxides.

Table 10.1 Overview on typical properties of some material classes at room temperature.

Material	Dielectric constant $\varepsilon[-]$	Refractive index n	Unclamped electro-optic coefficient r [pm/V]	Maximum electro-optic figure of merit n^3r[pm/V]	Trap density [10^{16} cm^{-3}]	Mobility μ_e [cm^2/Vs]	Quantum yields[1] $\phi[-]$	Recombination time τ_R	$\phi\mu\tau_R$ [10^{-12} cm^2/V]	Photo-conductivity[1] at 1 W/cm^2 [Ω^{-1} cm^{-1}]	Dielectric response time τ_{Di}[s][16]	Limiting beam coupling gain Γ[19] at $\Lambda \cong 2\ \mu m$ [cm^{-1}]	References, notes
Organic	3–7	2–2.4	<1200	<16000		$10^{-3}-1$	$<10^{-2}$	~μs(?)	<1000	$<4\times10^{-10}$	>0.001	160	limits[18]
single	5	2.2	200	2000		10^{-2}	10^{-4}	~μs (?)	1	10^{-13}	4	20	typical estimations[18]
crystals													
anthracene[2]			≪1[7]		0.04–1500	0.4–2	$10^{-4}-10^{-2}$ [3]						*Kao 1981, Pope 82*
MNBA[4]		ca.2[7]	220	ca.1760			[5]					18	*Tsunekawa 1990a,b*
DMSM[6]		ca.2[7]	432	ca.3460			[5]					35	*Yoshimura 1989*
MNA[12]		2.1	68	630			[5]					6	*Lipscomb 1981*
COANP		1.7	13	63								0.7	see text
benzene[13]		2.2	209	2225								22	see text
stilbene[14]		2.4	1148	15870								160	see text
Polymers	3-7	1.6–2.4	<550	7600	high	$\le10^{-5}$	$\le10^{-1}$ [22]	~μs (?)	1	$<4\times10^{-9}$	>0.01	75	limits[18]
(saturated	typ. 5	typ. 2	100	800		10^{-8}	10^{-5}	~μs (?)	0.000001	4×10^{-16}	1000	8	typical estimations[18,26]
backbone,						10^{-7}	10^{-2}	~μs (?)	0.001	4×10^{-12}	0.1	>8	typical estimations[18,22]
unordered)													
TPA:PKA[8]			0 (centric)	0			10^{-5} [9]						*Mort 1980*
TNF:PVK[10]			0 (centric)	0			10^{-3}						*Perlstein 1982*
DVC:MMA[11]			18	105		$10^{-9}-10^{-6}$	[5]					1	*Singer 1988*
bisA-													*Scott 1992*
NAT:DEH[24]	3			2.1–6.9		$10^{-8}-10^{-7}$	10^{-2} [22]			10^{-13} [22,23]	0.16[22,23]	2.2[22,23]	*Ducharme 1993*
PVK:F-DE-													*Silence 1993*
ANST:TNF[25]				3.1						3×10^{-12} [22,23]	0.22[22,23]	8.6[22,23]	*Donckers 1993*
Polymers (polydiacetylenes, crystalline)	see organic single crystals	see organic single crystals	no efficient electro-optic compounds known	dto	see organic single crystals	see organic single crystals	see organic single crystals	see organic single crystals	see organic single crystals	see organic single crystals	see organic single crystals	see r	
Ferroelectrics and oxides							0.01–1						
$LiNbO_3$	29,84	2.2	31	360		0.8			0.03–0.6	$10^{-14}-2.10^{-13}$	10–700	3.5	*Günter 88, Somm.88*
$KNbO_3$	55,1000	2.2	64,380	700,4000[20]	0.02–5	0.5		~1 μs	4.9–19000	$10^{-12}-10^{-8}$	$5\times10^{-4}-100$	7,30[21]	*Gün.88, Biag.90, Medr.88*
$BaTiO_3$	168,4300	2.4	80,1640	1100,22000[20]	0.5–8.7	0.5 ($\mu_h = 0.15$)			10–1000	$10^{-12}-10^{-10}$ [17]	0.1–400	11,170[21]	*Gün.88, Mich.78*
$Bi_{12}SiO_{20}$	56	2.5	5	80		0.03		0.05–10 μs	100000	4×10^{-8} [17]	6×10^{-5}	0.8	*Gün.88, Jona.88, Lesa.86*

Table 10.1 Continued.

Material	Dielectric constant ε[–]	Refractive index n	Unclamped electro-optic coefficient r [pm/V]	Maximum electro-optic figure of merit n^3r[pm/V]	Trap density [10^{16} cm^{-3}]	Mobility μ_e [cm^2/Vs]	Quantum yields[1] ϕ[–]	Recombination time τ_R	$\phi\mu\tau_R$ [10^{-12} cm^2/V]	Photo-conductivity[1] at 1 W/cm^2 [Ω^{-1} cm^{-1}]	Dielectric response time τ_{Di}[s][16]	Limiting beam coupling gain Γ[19] at $\Lambda \cong 2$ μm [cm^{-1}]	References, notes
Semiconductors													
GaAs	13.6	3.5	1.2	51.5		8500			100000	4×10^{-8} [17]	3×10^{-5}	0.5	*Günter 88, Glass 88*
CdTe	9.4	2.8	6.8	152		1050						1.5	*Günter 88, Glass 88*
InP	12.6	3.3	1.45	52		4600		100 ps–20 ns[15]	100000	4×10^{-8} [17]	3×10^{-5}	0.5	*Günter 88, Glass 88, Hamm. 84*

1) Best results at wavelengths in the visible spectrum.
2) Nominally pure crystals
3) For UV wavelengths
4) 4′-Nitrobenzylidene-3-acetamino-4-methoxyaniline
5) No photoconductivity experiments available
6) 4′-dimethylamino-N-methyl-4-stilbazolium methylsulfate
7) Estimations of the author
8) Polycarbonate doped with triphenylamine
9) At low fields
10) Polyvinylcarbazole doped with trinitrofluorenone
11) Dicyanovinylazo dye dissolved in poly(methylmethacrylate)
12) 2-methyl-4-nitroaniline
13) Hypothetical optimized benzene structure as obtained from semiempirical model, see text
14) Hypothetical optimized stilbene structure as obtained from semiempirical model, see text
15) depending on dopant level, *Hammod 1984*
16) at 1 W/cm^2, experimental values or as calculated from photoconductivity
17) calculated from $\phi\mu\tau$
18) theoretical limits and reasonable values for optimized compounds. See text.
19) Experimental values or calculated from n^3r using equation (10.8) with $\lambda = 500$nm and $\cos\theta = 0.5$. $E_g = E_d = 812$ V/cm
20) Non-diagonal element
21) theoretical values from n^3r
22) with applied electric fields
23) at large absorption $\alpha > 1$ cm^{-1}
24) bisphenol A 4,4′-nitroaminotolane doped with *p*-diethylamino-benzaldehyde-diphenyl hydrazone
25) poly(N-vinylcarbazole) doped with 3-fluoro-4-N,N diethylamino-β-nitrostyrene and 2,4,7 trinitro-9-fluorenone
26) with no applied field

Electro-optic coefficients: In Chapters 3 and 8 an estimation of the maximum electro-optic coefficients for benzene and stilbene derivatives using a semi-empirical model was given. Table 10.1 gives values for two of these hypothetical materials: an optimized benzene compound with an absorption edge around 500 nm and an optimized stilbene compound with an absorption edge at 580 nm.

10.5.3 Overview, comparison with inorganic compounds

It must be noted that most of the experimental results and estimations given in Table 10.1 are for conditions without applied electric fields (i.e. $E_o = 0$). Only some of the values for the polymers are given for $E_o \neq 0$. In these compounds, the carrier generation efficiency without applied field is very small and a large field (in the order of several 100 kV/cm) must be applied for observing the photorefractive effect. Such high fields will also increase the space charge field and affect the beam coupling gain Γ. This effect was, however, neglected in the comparison of Table 10.1.

From the parameters given in Table 10.1 we can calculate the limiting values for the dielectric response time τ_{di}, the beam coupling gain Γ, and the response time — gain coefficient Γ/τ (see Eqs. 10.16 to 10.18)).

For our analysis we will use the following typical parameters: average illumination intensity $I_o = 1$ W/cm^2, wavelength $\lambda = 515$ nm, grating spacing $\Lambda = 2$ µm, absorption constant $\alpha = 1$ cm^{-1}, and dielectric constant $\varepsilon = 3$.

The limits for τ, Γ and Γ/τ are summarized in Table 10.2.

Table 10.2 Typical limiting parameters for polymers and molecular single crystals.

compound class	ϕ	n^3r[pm/V]	τ_{di}[s]	Γ[cm^{-1}]	Γ/τ[cm^{-1}s^{-1}]
polymers (large applied electric field)	10^{-1}	7600	0.01	75	$\leq$ 725000 (c)
	10^{-1}	800	0.01	8	$\leq$ 76000 (d)
organic single crystals (no applied field)	0.01	16000	0.001	160	$\leq$ 152000 (a)
	0.01	2000	4	20	$\leq$ 19000 (b)

(a), (b), (c) and (d) denote the corresponding lines in Figure 10.8.

This comparison shows that the fundamental limit for Γ/τ is higher for polymers than for single crystals, while the individual limit for τ_{di} is lower. The reason for this discrepancy can be attributed to the fact the low carrier mobility μ in polymers does not affect the theoretical limit for Γ/τ according to Eq. (10.18), which neglects transport phenomena, but μ does affect the limit for τ_{di} of Eq. (10.16).

Figure 10.8 shows calculated values and limits of the beam coupling gain versus the dielectric response time for some organic materials. For comparison, experimental results obtained on inorganic compounds are added.

It must be noted that the estimated values for the response times given in Table 10.1 and Figure 10.8 are for holographic recording at an absorption coefficient of 1 cm^{-1}. Larger values of the absorption can lead to faster response times. To simplify a comparison of the materials, the values shown in Table 10.2 and Figure 10.8 are shown for conditions without applied electric field — only the measurements for polymers as well as the estimations for τ and Γ/τ for polymers

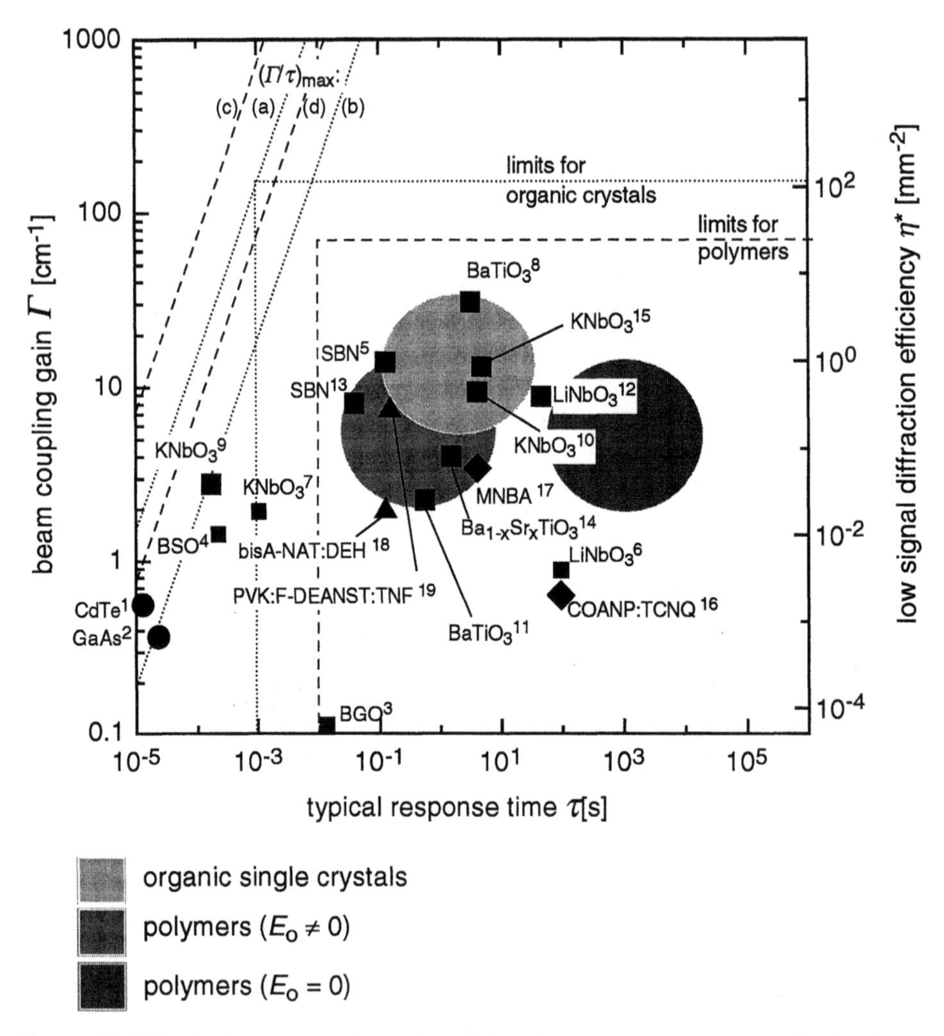

Figure 10.8 Typical response times (or dielectric response times as calculated from photoconduction) at 1 W/cm^2 and beam coupling gain Γ (or diffraction efficiency η) at Λ = 2 μm. The areas for the organic crystals and polymers are centered around the typical values for hypothetical optimized compounds found in Table 10.1 The dotted and dashed lines represent theoretical limits for single crystals and polymers, respectively. References for the experimental values for inorganics and oxides, denoted by squares and dots, are given in Table 10.3. Limiting values for Γ/τ are given for molecular single crystals (line (a) for n^3r = 16000 pm/V and (b) for n^3r = 2000 pm/V) and polymers (line (c) for n^3r = 7600 pm/V and (d) for n^3r = 800 pm/V, cf. Table 10.2). The low signal diffraction efficiency η^* is defined by $\eta^* = \eta/d^2$, where η is the total diffraction efficiency ($\ll 1$) at crystal thickness d.

are given for large applied fields because the photorefractive properties of these compounds cannot be observed without an applied field.

The following table lists the values and references of the experimental points in Figure 10.8.

Table 10.3 Experimental parameters (beam coupling gain, response time) of the materials shown in Figure 10.8. No applied field.

#	Material	$\Gamma[\mathrm{cm}^{-1}]$ a)	$\tau[\mathrm{s}]$ b)	Reference
1	CdTe:V	0.7 (1 μm)	2×10^{-7} (2–5 W/cm^2)	*Bylsma et al. 1987*
2	GaAs	0.4 (0.8 μm)	0.00002 (4 W/cm^2)	*Klein 1984*
3	BGO	0.1	0.01 (100 mW/cm^2)	*Powell et al. 1986*
4	BSO	1	0.25 ms	*Huignard et al. 1981, 1989*
5	SBN:Ce	12 (2 μm)	0.1	*Rakuljic et al. 1986*
6	$LiNbO_3$:Fe	ca. 1	200 (0.6 W/cm^2)	*Staebler et al. 1972*
7	$KNbO_3$ (nominally pure, reduced)	1 (1.2 μm)	0.002 (0.5 W/cm^2)	*Krumins et al. 1979*
8	$BaTiO_3$	20	2	*Rak et al. 1984*
9	$KNbO_3$ (nominally pure, reduced)	ca. 3	0.0001	*Voit et al. 1987, Amrhein 1991*
10	$KNbO_3$ (reduced)	10 (0.9 μm)	ca. 1–10	*Voit et al. 1988, Amrhein 1991*
11	$BaTiO_3$	2.2 (0.8 μm)	1 (2.5 W/cm^2)	*Rytz et al. 1988*
12	$LiNbO_3$	10 (0.4 μm)	20	*Kukhtarev et al. 1979b*
13	SBN:Cr	8 (5 μm)	0.07	*Vazquez et al. 1991*
14	$Ba_{1-x}Sr_xTiO_3$ (BST)	5 (0.5 μm)	0.8	*Zhuang et al. 1990*
15	$KNbO_3$:Fe (unreduced)	18 (2.4 μm)	1–10 (?) c)	*Zhang et al. 1990b*
16	COANP:TCNQ	0.8 (1.2 μm)	200–800 (100 mW/cm^2)	*Sutter et al. 1990b*
17	MNBA	2–4 (1.5 μm)	5–50 (100 mW/cm^2)	*Sutter et al. 1993*
18	bisA-NAT:DEH	2.2 (1.6 μm)	0.16 (2 W/cm^2)	*Silence et al. 1993*
19	PVK:F-DEANST:TNF	8.6 (1.6 μm)	0.22	*Donckers et al. 1993*

a) At $\Lambda = 2\mu$m,unless otherwise stated
b) At $I_o = 1$ W/cm^2, unless otherwise stated
c) Estimated from other experiments on $KNbO_3$:Fe

The photorefractive sensitivities S_n and S_η are defined in Eqs. (10.10) and (10.13). S_n gives the change of the refractive index per absorbed unit energy, S_η the change of diffraction efficiency per absorbed unit energy. We can give an upper estimate for these sensitivities if we assume an optimized transport length $L_{eff} = L_D/2$ (Eq. (10.11). Table 10.4 gives S_n and S_η for organic materials as calculated form the parameters in Table 10.1, together with reference values for inorganic compounds (*Günter 1982*).

Note that S_n and S_η in Table 10.4 depend linearly on the recombination time τ_R, the order of magnitude of which is not well known for organic compounds. The limit of S_n in organic single crystals seems rather high (1 cm^3/J), but it should be kept in mind that this value was calculated using limiting model parameters (n^3r = 16,000 pm/V, mobility $\mu = 1$ cm^2/Vs, recombination time $\tau_R = 1$ μs, quantum yield ϕ = 0.01). It will be difficult to reach all these values simultaneously in a real material.

Table 10.4 The photorefractive sensitivities S_n and S_η and the photorefractive figure of merit $n^3 r/\varepsilon$.

Material	$n^3 r/\varepsilon$[pm/V]	S_n[cm³/J]	S_η[cm²/J]
organic single crystals			
limits	5300	1	6×10^4 1)
typical	670	10^{-4}	6
unordered saturated polymers (with applied electric field)			
limits	2500	10^{-2}	600
typical	270	10^{-5}	0.6
inorganic single crystals			
$KNbO_3$	13 2)	0.0125	
$BaTio_3$	6.5 2)	—	10^2–10^4
$Bi_{12}SiO_{20}$	1.4 2)	0.071	1400

1) Calculated values from Eq. (10.13) with $\pi/\lambda\cos\theta = 6 \times 10^4$ cm^{-1}
2) r_{33}

The electro-optic figures of merit $n^3 r_{eff}$ or $n^3 r_{eff}/\varepsilon$ of polymers can be expected to become comparable to those of conventional non-semiconducting photorefractive materials and considerably larger than the values found in semiconductors. However, the low quantum efficiency ϕ and the low mobility μ make it necessary to apply large electric fields (around 100 V/μm). In practice such high fields can only be applied to thin films but not to bulk devices.

Therefore, organic single crystals seem more suitable for classical photorefractive applications, in particular for volume holography, while photorefractive polymers are better suited for film-applications. Presently, the electro-optic figures of merit of the best organic single crystals are already comparable to those of good ferroelectric materials and our estimations show that materials with even higher electro-optic coefficients may become available in the near future. If they can be made sufficiently photoconductive, their gains can be expected to reach values comparable to good inorganic compounds. The charge carrier mobility is slightly smaller than that of the inorganic oxides, but this is partially compensated by a lower dielectric constant. Typical response times (at illumination intensities of 1 W/cm^2 and in the absence of applied fields) will be above some milliseconds, more probably around seconds.

Presently available experimental results on photorefractive single crystals will be discussed in the following sections.

10.6 PHOTOREFRACTIVE PROPERTIES OF COANP:TCNQ

2-cyclooctylamino-5-nitropyridine (COANP) doped with the molecular acceptor 7,7,8,8-tetracyanoquinodimethane (TCNQ) (Figure 10.9) was the first known organic photorefractive single crystal (*Sutter et al. 1990a,b*).

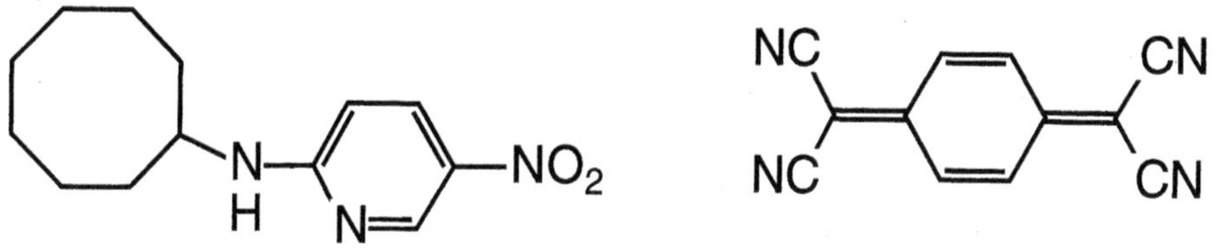

COANP TCNQ

Figure 10.9 2-cyclooctylamino-5-nitropyridine (COANP) and 7,7,8,8-tetracyano-quinodimethane (TCNQ).

COANP is a well known nonlinear optical compound and has been investigated thoroughly (*Günter et al. 1987, Bosshard et al. 1989*). Pure optical quality single crystals of COANP have been grown from supercooled melts and solutions (*Hulliger et al. 1990b*). The crystals have point group mm2 with four molecules per unit cell. The highest nonlinear optical coefficient is d_{32} = 43 pm/V at a fundamental wavelength of 1,064 nm. Its electro-optic coefficients are shown in Table 10.5 (*Bosshard et al. 1993*).

7,7,8,8-Tetracyanoquinodimethane (TCNQ) has been studied for its ability to form a stable radical (*Acker et al. 1962, Melby et al. 1962, Hertler et al. 1962*). Its electrophilic nature is ascribed to the electron-poor π-orbital system of the quinone. It is known to form solid complexes with many compounds (*Acker et al. 1962, Schlenk 1909*), most famous of which is the charge transfer complex with tetrathiafulvalene (TTF). TTF:TCNQ was the first organic compound to show metallic conductivity at room temperature (*Ferrais 1973*). Besides complex formation, TCNQ is also known to undergo addition and substitution reactions with a large number of molecules (*Bespalov et al. 1975*).

The electronic states of TCNQ can be distinguished by their absorption or electron-spin-resonance spectra. In its radical state, $TCNQ^-$ is characterized by the ESR signal of its unpaired electron (*Chesnut 1961*). $TCNQ^-$ can undergo dimerization to form $TNCQ_2^{-2}$, especially in polar solvents. This dimer does not reveal ESR resonance, because its two excess electrons are paired. However, it can be easily recognized by its optical absorption spectrum: while the $TCNQ^-$ monomer exhibits strong absorption resonances at 740 nm and 840 nm, the dimer $TNCQ_2^{-2}$ is characterized by an absorption peak at 640 nm (*Boyd et al. 1965b*).

Table 10.5 The electro-optic coefficients of 2-cyclooctylamino-5-nitropyridine.

Wavelength	r_{23}[pm/V]	r_{33}[pm/V]
515	26 ± 4	28 ± 5
633	13 ± 2	15 ± 2

Uncharged TCNQ does not reveal any absorption peaks in the spectral range from 600 nm to 1,000 nm.

Doped crystals of COANP:TCNQ used for optical experiments were grown from an undercooled melt of COANP containing small amounts of TCNQ. Details of this procedure are given in Chapter 4.

The doped crystals showed a strong green or brown color originating from an absorption peak at 650 nm (attributed to the dimer $TNCQ_2^{-2}$) and an additional absorption band near the absorption edge at 500 nm (*Sutter et al. 1991b*).

The COANP:TCNQ crystals were found to be photoconductive at wavelengths between 480 and 650 nm. Typical photoconductivities were around 0.3 to 1.5 × $10^{-14}\ \Omega^{-1}\text{cm}^{-1}$ at 100 mW/cm^2, depending on dopant levels and wavelengths. The steady state dark conductivity was estimated to be much smaller than $10^{-14}\ \Omega^{-1}\text{cm}^{-1}$.

Figure 10.10 shows a typical measurement of the photoconductivity of COANP:TCNQ at 515 nm.

The photorefractive properties of COANP:TCNQ were investigated by optical Bragg diffraction and beam coupling methods (*Sutter et al. 1990b*). These investigations showed that the photorefractive gratings in COANP:TCNQ were pure phase gratings with a phase shift of 90° in respect to the illumination grating. The majority carriers are holes.

The response time τ was found to be in the order of 100 to 1,000 seconds at 100 mW/cm^2, depending strongly on the grating spacing Λ. Figure 10.11 shows the dependence of τ on the square of the grating vector $K = 2\pi/\Lambda$.

The dependence of τ on Λ is described by Eq. (10.7). It is determined by the material parameters L_D (diffusion length), L_S (Debye screening length) and τ_{di} (dielectric response time).

Figure 10.11 shows that the points for $\tau(K^2)$ lie nearly on a straight line. From this, we conclude that the screening length L_S is small and $\tau(K^2)$ is completely determined by the diffusion length L_D and the dielectric response time, i.e.

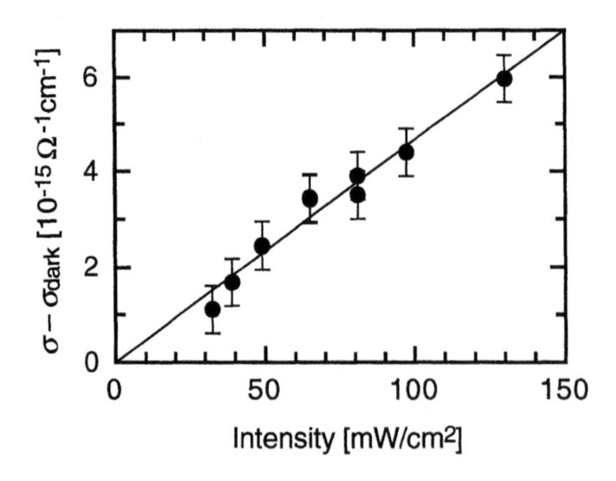

Figure 10.10 The photoconductivity of COANP:TCNQ as a function of illumination intensity at 515 nm, polarization || b.

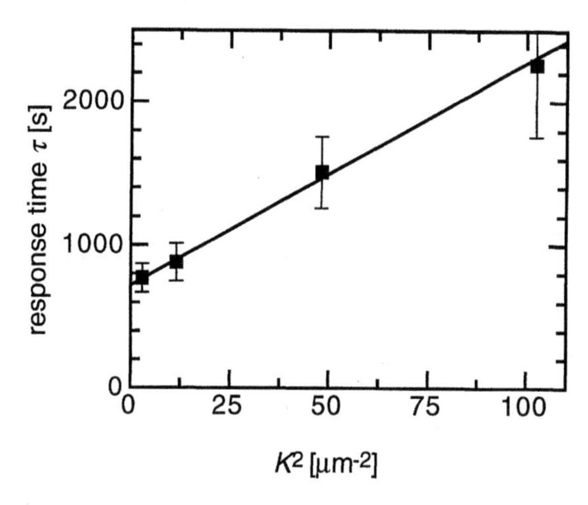

Figure 10.11 Response time versus the square of the grating vector. A straight line is fitted through the points (for parameters see text). Experimental parameters: I_1 = 79 mW/cm^2, I_2 = $I_1/10$. $K \parallel$ c, polarization $\parallel$ b, wavelength 515 nm.

$$\tau = (1 + K^2 L_D^2)\,\tau_{di} \tag{10.19}$$

Linear regression for our data yielded the following parameters:

$\tau_{di} = 718 \pm 145$ s, $L_D^2 \tau_{di} = 15.6 \pm 7\ \mu m^2 s$, $L_D = 0.15 \pm 0.07\ \mu$ m, $\mu\ \tau_{Re} = 0.87 \times 10^{-12} m^2/V$

Combining the value of $\mu\tau_{Re}$ with the result for $\phi\mu\tau_{Re}$ obtained from photoconductivity measurements, the quantum efficiency ϕ can be estimated to be in the order of 10^{-6} to 10^{-5}.

The photorefractive gratings in all available samples of COANP:TCNQ were found to be virtually stable in darkness. Decay rates were estimated to be smaller than 20% per week. This suggests a dark decay time τ_{dark} that is much larger than 10^6 seconds.

The dependence of the space charge field E_1 on the grating spacing Λ is shown in Figure 10.12.

The data in Figure 10.12 was fitted with an equation derived from Eq. (10.4), which is corrected by an empirical reduction factor R:

$$E_1 = Rm \frac{E_q\,(iE_d - E_o)}{(E_q + E_d)\,iE_o} \tag{10.20}$$

From the experimental data the following parameters were obtained: $R = 0.50 \pm 0.07$ and $N^* = (8.5 \pm 2.0)\ 10^{13}\ cm^{-3}$.

A discussion of the microscopic parameters of COANP:TCNQ is given in the conclusions, section 10.8.

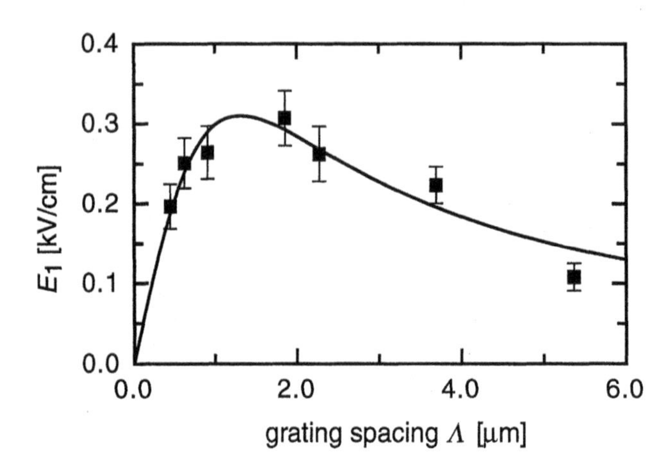

Figure 10.12 The dependence of the space charge field on the grating spacing (wavelength 515 nm, polarization || b).

10.7 PHOTOREFRACTIVE PROPERTIES OF MNBA

4′-nitrobenzylidene-3-acetamino-4-methoxyaniline (MNBA, Figure 10.13) is a known strongly nonlinear optical organic single crystal (*Tsunekawa et al. 1990a, b*). Its electro-optic and nonlinear optical properties are well known (*Knöpfle et al. 1994*). MNBA belongs to point group *m* and has a nearly optimized structure for electro-optic applications with all molecular charge transfer axes lying nearly parallel. Its largest electro-optic coefficient is r_{11} = 50 *pm*/V (at a wavelength of 532 nm). The short wavelength absorption edge of MNBA lies at approximately 510 to 530 nm.

A first observation of the photorefractive effect in MNBA was recently reported (*Sutter et al. 1993*). In these investigations nominally pure crystals of MNBA were used. The wavelength of the writing beams was 515 nm with a polarization along the main charge-transfer axis. Therefore, the writing beams were strongly absorbed (absorption constant 60 cm^{-1}).

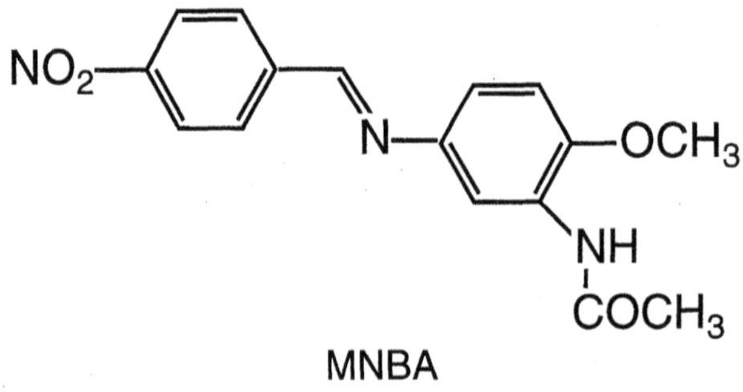

NO_2 N OCH_3 NH $COCH_3$

MNBA

Figure 10.13 4′-nitrobenzylidene-3-acetamino-4-methoxyaniline (MNBA).

The response time τ of the photorefractive effect in MNBA was found to be in the order of 10 to 100 seconds for writing intensities around 100 mW/cm^2. As in COANP:TCNQ, τ depends strongly on the grating spacing Λ, with a drastic increase in τ for small values of Λ, indicating a large diffusion and a small screening length. Figure 10.14 shows the response time τ versus the square of the grating vector K^2.

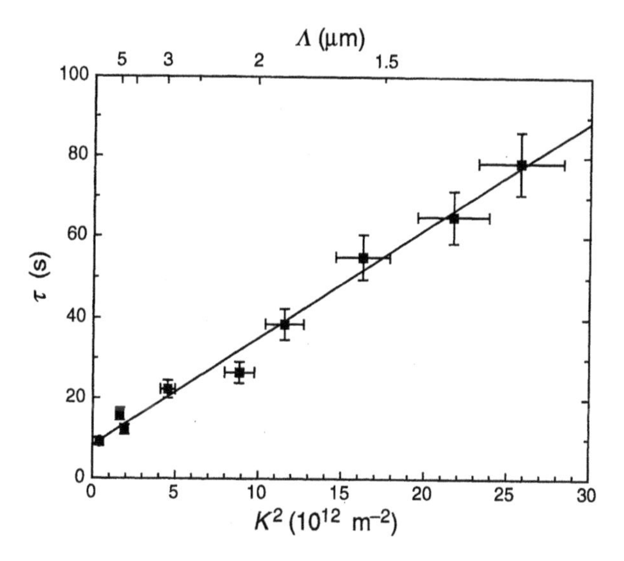

Figure 10.14 Response time of the photorefractive effect in MNBA as a function of the square of the grating vector $K^2 = (2\pi/\Lambda)^2$. Writing intensity 2×65 mW/cm^2 (*Sutter et al. 1993*).

Fitting Eq. (10.19) to these data yielded $\tau_{di} = 8.3$ s and $L_D = 0.57$ μm. From the dielectric response time and Eq. (10.15) we can calculate $\phi\mu\tau_{Re} = 1.32 \times 10^{-18}$ m^2/V. From $L_D = (\mu\tau_{Re}k_B T/e)^{1/2}$ we obtained $\mu\tau_{Re} = 1.28 \times 10^{-11}$ m^2/V. From this the quantum efficiency can be estimated: $\phi = 1.0 \times 10^{-7}$. This indicates a very inefficient mechanism of carrier generation. The comparatively fast response times as compared to COANP:TCNQ originate mainly from the strong absorption of the writing beams.

Figure 10.15 shows the beam coupling gain as a function of the grating spacing.

The experimental data in Figure 10.15 are well described by Eqs. (10.8, 10.20) with the values $N^* = 6.3 \times 10^{13}$ cm^{-3} and $R = 0.62$. As in COANP:TCNQ, the effective trap density N^* is very low.

A discussion of these experimental parameters is given in the next section.

10.8 CONCLUSIONS

Table 10.6 summarizes some relevant material parameters of the photorefractive effect measured in organic and inorganic compounds.

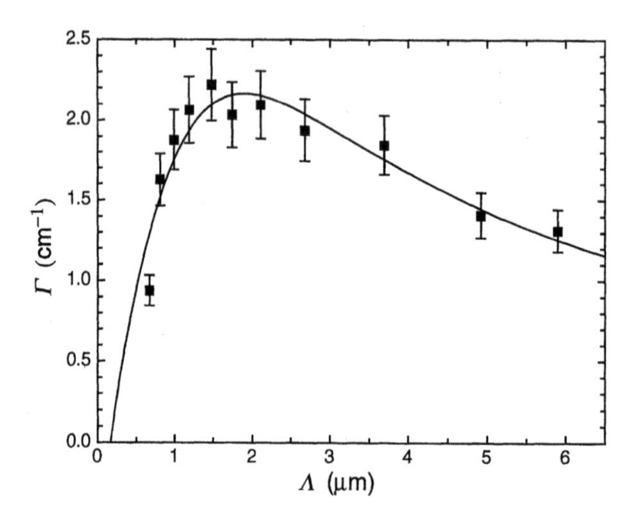

Figure 10.15 Beam coupling gain as a function of the grating Λ in MNBA.

Table 10.6 Photorefractive parameters of some materials.

material	$\mu\tau_R[m^2/V]$	$\phi[-]$	$N^*[cm^{-3}]$	Ref.
COANP:TCNQ	$(0.8–8) \times 10^{-13}$	$10^{-6}–10^{-5}$	8×10^{13}	*Sutter et al. 1991b*
MNBA	1.3×10^{-11}	10^{-7}	6×10^{13}	*Sutter et al. 1993*
Inorganic ferroelectrics [a)]	$10^{-13}–10^{-8}$	0.01–1	$10^{15}–10^{18}$	Table 10.1
Semiconductors [a)]	10^{-6}	0.001–1		Table 10.1

a) Typical values

The most striking characteristic of MNBA and COANP:TCNQ is their low quantum efficiency and the corresponding low photoconductivity and slow response. It is a consequence of an inefficient mechanism of carrier generation. Better photosensitizers could probably increase the quantum efficiency. Also arrangements where very high fields are applied (such as in the experiments of *Ducharme et al. 1991* on polymers) should increase the carrier generation efficiency. It should also be pointed out that the dielectric response time is proportional to the dielectric constant (see Eq. (10.16)), which is very low in organics. Therefore, organic compounds with quantum efficiencies in the order of 10^{-4} could already show reasonably fast effects.

The effective trap density N^* is very low. In the donor-acceptor band transport model, we have $N^* = N_A(N_D - N_A)/N_D$ implying that $N_D - N_A$ is very low (a low value of N_A seems improbable, considering the only moderate quality of the investigated samples). This corresponds to a low density of available charge carriers $(N_D - N_A)$.

The low trap density leads to a low limiting field $E_q = N^* \, q/K\varepsilon\varepsilon_0$. However, the generally low dielectric constants in organics compensate this effect at least partially, since typical values for N^*/ε do not differ strongly between the known organic single crystals and inorganic oxides.

Consulting Figure 10.8 we see that MNBA and COANP:TCNQ are not optimized materials. Their response time is one or two orders of magnitude slower than what we expect from a well optimized compound.

The maximum photorefractive gain found so far in an organic single crystal (about 2 cm^{-1} in MNBA) is also one order of magnitude below the value that can be expected for an optimized organic compound. This is mainly due to the fact that the electro-optic coefficients of MNBA or COANP:TCNQ are not very high. An investigation of presently known better electro-optic compounds could therefore lead to the discovery of stronger photorefractive systems.

The investigation of the photorefractive effect in organic compounds stands at its very beginning. First studies show that it is possible to find the effect in single crystals as well as in polymers. Even though performance of inorganic photorefractive materials is still superior to the presently known organic compounds, our discussion shows that an improvement of the strength as well as the speed of organics should be possible. Also, photorefractive investigations provide the means to determine charge transport parameters (such as mobility, lifetime, or trap density) that are of great interest to related fields of research.

11. REFERENCES

Acker D. S. and Hertler W. R., J. Am. Chem. Soc. **84**, 3370 (1962)
Adamson, A. W., *Physical Chemistry of Surfaces*, John Wiley and Sons, New York (1990)
Akaba R., Tokumaru K. and Kobayashi T., Bull. Chem. Soc Jpn. **53**, 1993 (1980)
Akaba R., Sakugari H. and Tokumaru K., Bull. Chem. Soc. Jpn. **58**, 1186 (1985)
Aktsipetrov O. A., Akhmediev N. N., Baranova I. M., Mishina E. D. and Novak V. R., Sov. Tech. Phys. Lett. **11** (5), 249 (1985)
Albrecht O., Gruler H. and Sackmann E., J. Physique **39**, 301 (1978)
Allen A., Hann R. A., Gupta S., Gordon P. F., Bothwell B., Ledoux I., Vidakovic P., Zyss J., Robin P., Chastaing E. and Dubois J., SPIE Proc. **682**, SPIE-The International Society for Optical Engineering, Bellingham, Washington (1986), p. 97
Allen S., McLean T. D., Gordon P. F., Bothwell B. D., Hursthouse M. B. and Karaulov S. A., J. Appl. Phys. **64** (5), 2583 (1988)
Allen S. in *Organic materials for non-linear optics,* editors R. A. Hann and D. Bloor, Royal Soc. of Chemistry, London (1989), pp. 137–150
Amos A. T. and Burrows B. L., in *Advances in Quantum Chemistry,* Vol. 7, editor P.-O. Löwdin Academic Press, New York (1973), pp. 289–313
Amrhein P., personal communication (1991)
Andelmann D., Brochard F. and Joanny J., J. Chem. Phys. **86**, 3673 (1987)
Andreazza P., Lefauchaux F., Robert M. C., Josse D. and Zyss J., J. Appl. Phys. **68**, 8–13 (1990a)
Andreazza P., *Croissance en Gel de Cristaux Organiques: POM et NPP. Caracterisation par Topographie aux Rayons X et par l'Etude des Propriétés Optiques non-Lineaires* Ph.D. thesis, University P. and M. Curie, Paris VI, France (1990b)
Arend H., Peret R., Wüest H. and Kerkoc P., J. Cryst. Growth **74**, 321 (1986)
Armstrong J. A., Bloembergen N., Ducuing J. and Pershan P. S., Phys. Rev. **127**, 1918 (1962)
Asai N., Tamada H., Fujiwara I. and Seto J., J. Appl. Phys. **72** (10), 4521 (1992)
Ashwell G. J., Dawnay E. J. C., Kuczynski A. P., Szablewski M., Sandy I. M., Bryce M. R., Grainger A. M. and Hasan M., J. Chem. Soc. Faraday Trans. **86**, 1117 (1990)
Ashwell G. J., Dawnay E. J. C., Kuczynski A. P. and Martin P. J., SPIE Proc **1361** SPIE-The International Society for Optical Engineering, Bellingham, Washington (1991), p. 589
Ashwell G. J., Hargreaves R. C., Baldwin C. E., Bahra G. S. and Brown C. R., Nature **357**, 393 (1992a)
Ashwell G. J., personal communication (1992b)
Atkins P. W., *Molecular Quantum Mechanics,* 2nd edn., Oxford University Press (1983)
Aveyard R., Binks B. P., Fletcher P. D. I. and Ye X., Thin Solid Films **210/211**, 36 (1992)
Azzam R. M. A. and Bashava N. M., *Ellipsometry and Polarized Light,* North-Holland, Amsterdam (1977)
Badan J., Hierle R., Perigaud A. and Vidakovic P. V., in *Nonlinear optical properties of organic molecules and crystals,* editors D. S. Chemla and J. Zyss., Academic Press Inc., Orlando (1987), pp. 297–356
Bailey R. T., Cruickshank F. R., Guthrie S. M. G., McArdle B. J., Morrison H., Pugh D., Shepherd E. A., Sherwood J. N., Yoon C. S., Kashyap R., Nayar B. K. and White K. I., Opt. Commun. **65**, 229 (1988)
Bailey R. T., Bourhill G. H., Cruickshank F. R., Pugh D., Sherwood J. N., Simpson G. S. and Varma K. B. R., J. Appl. Phys. **71** (4), 2012 (1992)

Bareman J. P., Cardini G. and Klein M. L., Phys. Rev. Lett. **60**, 2152 (1988)
Barraud A., Rosilo C. and Ruaudel-Teixier, J. Colloid Interface Sci.**62**, 509 (1977)
Barraud A. and Leloup J., Fr. Patent 83, 03, 578 (1983)
Barraud A. and Palacin S. (eds), *Langmuir-Blodgett Films 5*, Proceedings of the Fifth International Conference on Langmuir-Blodgett Films, Thin Solid Films **210/211**, Elsevier, Amsterdam (1992)
Bartolini R. A., Bloom A. and Escher J. S., Appl. Phys. Lett. **28** (9), 506 (1976)
Barzoukas M., Josse D., Fremaux J., Nicoud J. F. and Morley J. O., J. Opt. Soc. Am. B **4** (6), 977 (1987)
Barzoukas M., Josse D., Zyss J., Gordon P. and Morley J. O., Chem. Phys. **139**, 359 (1989)
Baumert J. C., Ph.D. thesis Nr. 7802, ETH Zürich (1985)
Baumert J. C., Twieg R. J., Bjorklund G. C., Logan J. A. and Dirk C. W., Appl. Phys. Lett. **51**, 1484 (1987)
Bechthold P. S., Ph.D. thesis, Universität Köln (1976)
Bergman J. G. and Crane G. R., J. Chem. Phys. **60** (6), 2470 (1974)
Bespalov B. P. and Titov V. V., Russian Chem. Rev. **44** (12), 1091 (1975) or Uspekhi Khimii **44**, 2249 (1975)
Biaggio I., Zgonik M. and Günter P., Opt. Commun. **77** (4), 312 (1990)
Bibo A. M. and Peterson I. R., Thin Solid Films **178**, 81 (1989)
Bierlein J. D., Cheng L. K., Wang Y. and Tam W., Appl. Phys. Lett. **56** (5), 423 (1990)
Blodgett, K. B., J. Am. Chem. Soc. **57**, 1007 (1935)
Blythe A. R., *Electrical properties of polymers*, Cambridge University Press (1979)
Boettcher C. J., *Theory of electric polarization*, Elsevier, Amsterdam (1952)
Born M. and Wolf E., *Principles of Optics*, 6th edn., Pergamon Press, Oxford (1980a), p. 94
Born M. and Wolf E., *Principles of Optics*, 6th edn., Pergamon Press, Oxford (1980b), pp. 90–98
Born M., *Optik*, 2nd edn., Springer-Verlag, Berlin (1985), p. 45
Bornside D. E., Macosko C. W. and Scriven L. E., J. Appl. Phys. **66** (11), 5185 (1989)
Borsenberger P. M., Pautmeier L. and Bässler H., J. Chem. Phys. **94** (8), 5447 (1991)
Bosshard Ch., Sutter K., Günter P. and Chapuis G., J. Opt. Soc. Am. B **6** (4), 721 (1989)
Bosshard Ch., Küpfer M., Günter P., Pasquir C., Zahir S. and Seifert M., Appl. Phys. Lett. **56**, 1204 (1990)
Bosshard Ch., Doctoral thesis at the Swiss Federal Inst. of Techn., Zürich, Switzerland, Diss. No. 9407 (1991a)
Bosshard Ch., Flörsheimer M., Küpfer M. and Günter P., Opt. Commun. **85** (2, 3), 247 (1991b)
Bosshard Ch. and Hulliger J., unpublished work (1992a)
Bosshard Ch., Knöpfle G., Prêtre P. and Günter P., J. Appl. Phys. **71** (4), 1594 (1992b)
Bosshard Ch., Küpfer M., Flörsheimer M. and Günter P., Thin Solid Films **210/211**, 153 (1992c)
Bosshard Ch., Küpfer M., Flörsheimer M., Borer Th., Günter P., Tang Q. and Zahir S., Thin Solid Films **210/211**, 198 (1992d)
Bosshard Ch., Küpfer M., Flörsheimer M. and Günter P., Mol. Cryst. Liq. Sci. Technol.-Sec. B: Nonlinear Optics **3**, 215 (1992e)
Bosshard Ch., Sutter K., Schlesser R. and Günter P., J. Opt. Soc. Am. B **10** (5), 867 (1993)
Bourdiev L., Ronsin O. and Chatenay D., Science **259**, 798 (1993)
Bourhill G., Bredas J.-L., Cheng L.-T., Friedli A., Gorman C. B., Marder S. R., Meyers F., Perry J. W., Pierce B. M., Skindhoj J. and Tiemann B. G., in 1993 Technical Digest Series Volume 17, Optical Society of America, Washington, DC 20036–1023, pp. 46–49

Boyd G. D., Ashkin A., Dziedzic J. M. and Kleinman D. A., Phys. Rev. B **137**, 1305 (1965a)
Boyd R. H. and Phillips W. D., J. Chem. Phys. **43** (9), 2927 (1965b)
Boyd G. D. and Kleinman D. A., J. Appl. Phys. **39**, 3597 (1968)
Boyd G. D., Kaspar H. and McFee J. H., IEEE J. Quantum Electron. **QE-7** (12), 563 (1971)
Boyd G. D., Moshrefzadeh R. S. and Ender D. A., SPIE Proc. **971**, SPIE-The International Society for Optical Engineering, Bellingham, Washington (1988), p. 230
Březina B., and Hulliger J., Cryst. Res. Technol. **26**, 155 (1991)
Broussoux D., Chastaing E., Esselin S., Le Barny P., Robin P., Bourbin Y., Pocholle J. P. and Raffy J., Revue Technique Thomson-CSF20–21, 151 (1989)
Bubeck Ch., Laschewsky A., Lupo D., Neher D., Ottenbreit P., Paulus W., Prass W., Ringsdorf H. and Wegner G., Adv. Mater. **3**, 55 (1991)
Bueche F., *Physical Properties of Polymers*, Interscience, New York (1962)
Bylsma R. B., Bridenbaugh P. M., Olson D. H. and Glass A. M., Appl. Phys. Lett. **51**, 889 (1987)
Caseri W., Sauer T. and Wegner G., Makromol. Chem. Rapid Commun. **9**, 651 (1988)
Chemla D. S., Kupecek P., Schwartz C., Schwab C. and Goltzene A., IEEE J. Quantum Electron. **7**, 126 (1971a)
Chemla D. S. and Kupecek P., Rev. de Phys. Appl. **6**, 31 (1971b)
Chemla D. S., Oudar J. L. and Jerphagnon J., Phys. Rev. B **12**, 4534, (1975)
Chemla D. S., Oudar, J. L. and Zyss J., Echo Rech. **103**, 3 (1981)
Cheng L. T., Tam W., Meredith G.R., Rikken G. L. J. A. and Mejer E. W., SPIE Proc. **1147**, editor G. Khanarian, SPIE-The International Society for Optical Engineering, Bellingham, Washington (1989), pp. 61–72
Cheng L.T., Tam W., Stevenson S.H., Meredith, G.R., Rikken G. and Marder S., J. Phys. Chem. **95**, 10631 (1991)
Chesnut D. B. and Philips W. D., J. Chem. Phys. **35** (3), 1002 (1961)
Chikuma K. and Umegaki S., J. Opt. Soc. Am. B. **7** (5), 768 (1990)
Choy M. M. and Byer R. L., Phys. Rev. B **14** (4), 1693 (1976)
Clays K. and Persoons A., Phys. Rev. Lett. **66** (23), 2980 (1991)
Clays K. and Persoons A., Rev. Sci. Instrum. **63** (6), 3285 (1992)
Clays K., Armstrong N. J., Ezenyilimba M. C. and Penner T. L., Chem. Mater. (5), 1032 (1993a)
Clays K., Armstrong N. J. and Penner T. L., J. Opt. Soc. Am. B **10** (5), 886 (1993b)
Cohen M. D., Hirshberg Y. and Schmidt G. M. J., in *Hydrogen Bonding*, editor D. Hadzi, Pergamon Press, Oxford (1957), pp. 293 ff
Cross G. H., Girling I. R., Peterson I. R., Cade N. A. and Earls J. D., J. Opt. Soc. Am. B **4**, 962 (1987)
Cross G. H., Peterson I. R., Girling I. R., Cade N. A., Goodwin M. J., Carr N., Sethi R. S., Marsden R., Gray G. W., Lacey D., McRoberts A. M., Scrowston R. M. and Toyne K. J., Thin Solid Films **156**, 39 (1988)
Cummins P. G., Dunmur D. A., Munn R. W. and Newham R. J., Acta Cryst. A **32**, 847 (1976)
Daniel M. F., Dolphin J. C., Grant A. J., Kerr K. E. N. and Smith G. W., Thin Solid Films **133**, 235 (1985a)
Daniel M. F., and Hart J. T. T., J. Mol. Electron. **1**, 97 (1985b)
Decher G., Tieke B., Bosshard Ch. and Günter P., J. Chem. Soc., Chem. Commun. 933 (1988)
Decher G., Tieke B., Bosshard Ch. and Günter P., Ferroelectrics **91**, 193 (1989a)
Decher G., Klinkhammer F., Peterson I. R. and Steits R., Thin Solid Films **178**, 445 (1989b)

Deshpande L. V., Joshi M. B. and Mishra R. B., J. Opt. Soc. Am. **70**, 1160 (1980)
Dietrich A., Möhwald H., Rettig W. and Brezesinski G., Langmuir **7**, 539 (1991)
Donckers M. C. J. M., Silence S. M., Walsh C. A., Hache F., Burland D. M., Moerner W. E. and Twieg R. J., Opt. Lett. **18** (13), 1044 (1993)
Ducharme S., Feinberg J. and Neurgaonkar R. R., IEEE J. Quantum Electron. **QE-23**, 2116 (1987)
Ducharme S., Risk W. P., Moerner W. E., Lee V. Y., Twieg R. J. and Bjorklund G. C., Appl. Phys. Lett. **57** (6), 537 (1990)
Ducharme S., Scott J. C., Twieg R. J. and Moerner W. E., Phys. Rev. Lett. **66** (14), 1846 (1991)
Ducharme S., Jone B., Takacs J. M. and Zhang L., Opt. Lett. **18** (2), 152 (1993)
Duda G., Schouten A. J., Arndt T., Lieser G., Schmidt G. F., Bubeck C. and Wegner G., Thin Solid Films **159**, 221 (1988a)
Duda G. and Wegner G., Makromol. Chem., Rapid Commun. **9**, 495 (1988b)
Dulcic A. and Sauteret C., J. Chem Phys. **69** (8), 3453 (1978a)
Dulcic A. and Flytzanis C., Optics Commun. **25**, 402 (1978b)
Dutta P., Peng J. B., Lin B., Ketterson J. B., Prakash M., Georgopoulos P. and Ehrlich S., Phys. Rev. Lett. **58**, 2228 (1987)
Eckhardt R. C., Masuda H., Fan Y. X. and Byer R. L., IEEE J. Quantum Electron. **26** (5), 922 (1990)
Ehringhaus A., Zeitschrift f. Kristallographie **102**, 85 (1951)
Eich M., Looser H., Yoon D. Y., Twieg R., Bjorklund G. and Baumert J. C., J. Opt. Soc. Am. B. **6** (8), 1590 (1989a)
Eich M., Reck B., Yoon D. Y., Wilson C. G. and Bjorklund G. C., J. Appl. Phys. **66** (7), 3241 (1989b)
Eich M., Sen A., Looser H., Bjorklund G. C., Swalen J. D., Twieg R. and Yoon D. Y., J. Appl. Phys. **66** (6), 2559 (1989c)
Eichler H. J., Günter P. and D. Pohl, *Laser induced dynamic gratings,* Springer, Heidelberg (1986)
Embs F. W., Wegner G., Neher D., Albouy P., Miller R. D., Wilson C. G. and Schrepp W., Macmolecules **24**, 5068 (1991a)
Embs F., Funhoff D., Laschewsky A., Licht U., Ohst H., Prass W., Ringsdorf H., Wegner G. and Wehrmann R., Adv. Mater. **3**, 25 (1991b)
Emin D., in *Handbook of conducting polymers,* vol. 2 editor T. A. Skotheim, Dekker, New York (1986), pp. 915–938
Ezumi K., Nakai H., Sakata S., Hishikida K., Shiro M. and Kubota T., Chem. Lett. **1393** (1974)
Fainman Y., Klancnik E. and Lee S. H., Optical Engineering **25** (2), 228 (1986)
Falk U., Hickel W., Lupo D., Prass W. and Scheunemann U., in *Nonlinear Optics,* editor S. Miyata, Elsevier Science Publishers, Amsterdam (1992) pp. 271–282
Feast W. J., J. Mater. Sci., **25** (9) 3796–805 (1990a)
Feast W. J. and Friend R. H., Phil. Trans. R. Soc. (London) A **303**, 117–25 (1990b)
Ferrais J., J. Am. Chem. Soc. **95** (3), 948 (1973)
Fischer A. and Sackmann E., J. Physique **45**, 517 (1984)
Fischer A., Ph.D. thesis, Technical University of Munich (1985)
Flörsheimer M. and Möhwald H., Thin Solid Films **159**, 115 (1988)
Flörsheimer M., Ph.D. thesis, Technical University of Munich (1989a)
Flörsheimer M. and Möhwald H., Chem. Phys. Lipids **49**, 231 (1989b)
Flörsheimer M. and Möhwald H., Thin Solid Films **189**, 379 (1990)
Flörsheimer M. and Möhwald H., Colloids and Surfaces **55**, 173 (1991)

Flörsheimer M., Küpfer M., Bosshard Ch. and Günter P., in *Nonlinear Optics*, editor S. Miyata, Elsevier Science Publishers, Amsterdam (1992a) pp. 255–269
Flörsheimer M., Küpfer M., Bosshard Ch., Looser H. and Günter P., Adv. Mater. **4**, 795 (1992b)
Flörsheimer M., Looser H., Küpfer M. and Günter P., Thin Solid Films **224**, 1001 (1993a)
Flörsheimer M., Sutter K. and Günter P., in *Organic Materials for Nonlinear Optics III*, editors G. J. Ashwell and D. Bloor, Royal Society of Chemistry, Cambridge, (1993b) pp. 307–313
Flörsheimer M., Steinfort A.J. and Günter P., Thin Solid Films, **247**, 190 (1994a)
Flörsheimer M., Jundt D., Looser H., Sutter K. and Günter P., J. Phys. Chem. **98**, 6399 (1994b)
Fluck D., Amrhein P. and Günter P., J. Opt Soc.Am B **8** (10), 2196 (1991)
Fluck D., Günter P., Fleuster M. and Buchal C., Advanced Solid-State Lasers and Compact Blue-Green Lasers Technical Digest, **Vol. 2**, 473–475 (1993)
Fousek J., Ferroelectrics **20**, 11 (1978)
Fox S. J., in *Photoconductivity in Polymers: An Interdisciplinary Approach*, editors A. V. Patsis and D. A. Seanor, Technomic, Westport, USA (1976), pp. 253–277
Fuith A. and Warhanek H., J. Cryst. Growth **69**, 96 (1984)
Fukuda T. and Sano T., J. Jap. Assoc. Cryst. Growth **16**, 26 (1989) (Special issue: organic crystals, see other contributions)
Gahm J., Zeiss Mitteilungen **2**, 389 (1962) and **3**, 3 (1963)
Garnaes J., Schwartz D. K., Viswanathan R. and Zasadzinski J. A. N., Nature **357**, 54 (1992)
Garoff S., Deckman H. W., Dunsmuir J. H. and Alvarez M. S., J. Physique **47**, 701 (1986)
Girling I. R. and Milverton D. R. J., Thin Solid Films **115**, 85 (1984)
Girling I. R., Kolinsky P. V., Cade N. A., Earls J. D. and Peterson I. R., Opt. Commun. **55**, 289 (1985)
Girton D. G., Kwiatokowski S. L., Lipscomb G. F. and Lytel R. S., Appl. Phys. Lett. **58** (16), 1730 (1991)
Glass A. M. and Strait J., in *Photorefractive Materials and Their Applications I*, editors P. Günter and J.-P. Huignard, Springer, Berlin (1988), pp. 237–262
Gotoh T., Tsunekawa T. and Egawa K., Toray Industries, patent: JP 63/319/297 (1988)
Gotoh T., J. Jap. Assoc. Cryst. Growth **17**, 95 (1990)
Gotoh T., Fukuda S. and Yamashiki T., in *Nonlinear Optics*, editor S. Miyata, Elsevier Science Publishers, Amsterdam (1992), pp. 219–224
Grammatica S. and Mort J., Appl. Phys. Lett. **38** (6), 445 (1981)
Guggenheim E. A., Trans. Faraday Soc. **45**, 714 (1949)
Guha S., Frazier C. C. and Wenpeng Chen, SPIE Proc. **971**, SPIE-The International Society for Optical Engineering, Bellingham, Washington (1988), pp. 89–96
Günter P., in *Proceedings of Electro-optics/Laser International '76*, editor H. G. Jerrard, IPC Science and Technology Press Ltd., Surrey, England (1976) pp. 121–130
Günter P., Ferroelectrics **24**, 35 (1980)
Günter P., Phys. Reports **93**, 199 (1982)
Günter P., Bosshard Ch., Sutter K., Arend H., Chapuis G., Twieg R. J. and Dobrowolski D., Appl. Phys. Lett. **50** (9), 486 (1987)
Günter P. and Huignard J.-P., in *Photorefractive Materials and Their Applications I*, editors P. Günter and J.-P. Huignard, Springer, Berlin (1988), pp. 7–74
Günter P. and Zgonik M., Opt. Lett. **16** (23), 1825 (1991)
Haas D., Yoon H. and Man H. T., SPIE Proc. **1147** editor G. Khanarian, SPIE — The International Society for Optical Engineering, Bellingham, Washington (1989) p. 222

Hall T. J., Jaura R., Connors L. M. and Foote P. D., Progress Quantum Electron. **10**, 77 (1985)
Hammond R. B., Paulter N. G., Wagner R. S. and Springer T. E., Appl. Phys. Lett. **44**, 620 (1984)
Hampsch H. L., Torkelson J. M., Bethke S. J. and Grubb S. G., J. Appl. Phys. **67** (2), 1037 (1990)
Harada A., Okazaki Y., Kamiyama K. and Umegaki S., Appl. Phys. Lett. **59** (13), 1535 (1991)
Hariharan R., *Optical Interferometry*, Academic Sydney, Australia (1985), pp. 33–56
Hartshorne N. H. and Stuart A., *Crystals and the Polarizing Microscope*, Edward Arnold Ltd., London (1970)
Hayden L. M., Sauter G.F., Ore F. R., Pasillas P. L., Hoover J. M., Lindsay G. A. and Henry R. A., J. Appl. Phys. **68** (2), 456 (1990)
Hecht E., *Optics* Addison-Wesley World Student Series Edition, Reading, UK (1987a), pp. 163–166
Hecht E., *Optics* Addison-Wesley World Student Series Edition, Reading, UK (1987b), pp. 297–299
Heckl W. M., Lösche M., Cadenhead D. A. and Möhwald H., Eur. Biophys. J. **14**, 11 (1986a)
Heckl W. M. and Möhwald H., Ber. Bunsenges. Phys. Chem. **90**, 1159 (1986b)
Heflin J. R., Wong K. Y., Zamani-Khamiri O. and Garito A. F., Phys. Rev. B **38** (2), 1573 (1988)
Hellwarth R. W., J. Opt. Soc. Am. **67**, 1 (1977)
Helm Ch., Laxhuber L., Lösche M. and Möhwald H., Colloid and Polymer Sci. **264**, 46 (1986)
Helm C. A., Möhwald H., Kjaer K. and Als-Nielsen J., Biophys. J. **52**, 381 (1987)
Helm Ch. and Möhwald H., J. Phys. Chem. **92**, 1262 (1988)
Helm C. A., Tippmann-Krayer P., Möhwald H., Als-Nielsen J. and Kjaer K., Biophys. J. **60**, 1457 (1991)
Hénon S. and Meunier J., Rev. Sci Instrum. **62**, 936 (1991)
Hermann J. P. and Ducuing J., J. Appl. Phys. **45**, 5100 (1974)
Hertler W. R., Hartzler H. D., Acker D. S. and Benson R. E., J. Am. Chem. Soc. **84**, 3387 (1962)
Hewig G. H. and Jain K., Opt. Commun. **47** (5), 347 (1983)
Hickel W., Duda G., Jurich M., Kröhl T., Rochford K., Stegeman G. I., Swalen J. D., Wegner G. and Knoll W., Langmuir **6**, 1403 (1990)
Hickel W., Bauer J., Lupo D., Menzel B., Falk U. and Scheunemann U. in *Organic Materials for Nonlinear Optics III*, editors G. J. Ashwell and D. Bloor, Royal Society of Chemistry, Cambridge, UK (1993) pp. 80–87
Hill J. R., Pantelis P., Dunn P. L. and Davies G. J., SPIE Proc. **1147**, editor G. Khanarian, SPIE-The International Society for Optical Engineering, Bellingham, Washington (1989) p. 165
Hiroshi I., Kazushi H., Hidehiko T. and Keisuke S., Denshi Joho Tsushin Gakkai Ronbunshi, J70-C, 224–228 (1987)
Hoesterey D. C. and Letson G. M., J. Phys. Chem. Solids **24**, 1609 (1963)
Holcroft B., Petty M. C., Roberts G. G., and Russel G. J., Thin Solids Films **134**, 83 (1985)
Hönig D. and Möbius D., Thin Solid Films **210/211**, 64 (1992)
Hoppe W., Lohmann W., Markl H. and Ziegler H., Biophysik, Heidelberg (1982)
Hosomi T., Suzuki T., Yamamoto H., Watanabe T., Sato H. and Miyata S., SPIE Proc. **1147**, editor G. Khanarian, SPIE-The International Society for Optical Engineering, Bellingham, Washington (1989), pp. 124–128

Huang B., Su G. and He Y., J. Cryst. Growth **102**, 762–764 (1990)
Huang J., Lewis A. and Rasing Th., J. Phys. Chem. **92**, 1756 (1988)
Huignard J. P. and Marrakchi A., Opt. Commun. **38** (4), 249 (1981)
Huignard J. P. and Günter P., in *Photofractive materials and thier applications II,* editors P. Günter and J. P. Huignard, Springer, Berlin (1989), pp. 205–274
Huijts R. A. and Hesselink G. L. J., Chem. Phys. Lett. **156** (2, 3), 209 (1989)
Hulliger J., Wang W. S. and Ehrensperger M., J. Cryst. Growth **100**, 640–642 (1990a)
Hulliger J., Březina B. and Ehrensperger M., J. Cryst. Growth **106**, 605–610 (1990b)
Hulliger J., Schumacher Y., Sutter K., Březina B. and Ammann H., Mat. Res. Bull. **26**, 887 (1991)
Hulliger J., Sutter K., Schumacher Y., Březina B. and Ivanshin V. A., J. Cryst. Growth **128**, 886 (1993)
Hulliger J., Angew. Chem. **106**, 151 (1994), Int. Ed. **33**, 143 (1994)
Hurst M. and Munn R. W., J. Mol. Electron. **2**, 139 (1986)
Hutchings M. G., Gordon P. F. and Morley J. O., Inst. Phys. Conf. Ser. **103**, 97 (1989)
Ito H. and Inaba H., Opt. Lett. **2** (6), 139 (1978)
Jacquemain D., Leveiller F., Weinbach S. P., Lahav M., Leiserowitz L., Kjaer K. and Als-Nielsen J., J. Am. Chem. Soc. **113**, 7684 (1991)
Jacquemain D., Wolf S. G., Leveiller F., Deutsch M., Kjaer K., Als-Nielsen J., Lahav M. and Leiserowitz L., Angew. Chem. Int. Ed. Engl. **31**, 130 (1992)
Jain K., Crowley J. I., Hewig G. H., Cheng Y. Y. and Twieg R. J., Opt. Laser Technol. **297** (1981)
Jen A. K.-Y., Rao V. P., Wong K. Y. and Drost K. J., J. Chem. Soc., Chem. Commun. 90 (1993)
Jenekhe S. A., Lo S. K. and Flom S. R., Appl. Phys. Lett. **54** (25), 2524 (1989)
Jenekhe S. A., Chen W. C., Lo S. and Flom S. R., Appl. Phys. Lett. **57** (2), 126 (1990)
Jerphagnon J. and Kurtz S. K., J. Appl. Phys. **41**, 1667 (1970)
Jonathan J. M. C., Roosen G. and Roussignol Ph., Opt. Lett. **13**, 224 (1988)
Jungbauer D., Reck B., Twieg R., Yoon D. Y., Willson C. G. and Swalen J. D., Appl. Phys. Lett. **56** (26), 2610 (1990)
Kajikawa K., Takezoe H. and Fukuda A., Jap. J. Appl. Phys. **30**, L1525 (1991)
Kanetake T., Ishikawa K., Hasegawa T. Koda T., Takeda K., Hasegawa M., Kubodera K. and Kobayashi H., Appl. Phys. Lett. **54** (23), 2287 (1989)
Kao K. C. and Hwang W., *Electric Transport in Solids,* Pergamon Press, Oxford (1981)
Karl N., High Purity Organic Molecular Crystals, in *Crystals (Growth, Properties, and Applications),* vol. 4, Springer-Verlag, Berlin (1980)
Karl N., in *Materials for Non-linear and Electro-optics* editor M. H. Lyons, Inst. of Physics, Conference Series Number 103, IOP, Bristol, 1989, pp. 107–118
Katz H. E., Dirk C. W., Schilling M. L., Singer K. D. and Sohn J. E., Material Research Society **Vol. 109**, Pittsburgh, Pennsylvania (1988)
Kawakami T. and Sonoda N., Appl. Phys. Lett. **62** (19), 2167 (1993)
Keller D. J., McConnell H. M. and Moy V. T., J. Phys. Chem. **90**, 2311 (1986)
Keller D. J., Korb J. P. and McConnell H. M., J. Phys. Chem. **91**, 6417 (1987)
Kenn R. M., Böhm C., Bibo A. M., Peterson I. R., Möhwald H., Als-Nielsen J. and Kjaer K., J. of Phys. Chem. **Vol. 95**, 2093 (1991)
Kerkoc P., Zgonik M., Sutter K., Bosshard Ch. and Günter P., Appl. Phys. Lett. **54** (21), 2062 (1989)
Kerkoc P. and Hulliger J., J. Cryst. Growth **99**, 1023–1027 (1990)
Kerkoc P., Doctoral thesis at the Swiss Federal Inst. of Techn., Zurich, Switzerland, Diss. No. 9413 (1991)

Khanarian G., Norwood R. A., Haas D., Feuer B. and Karim D., Appl. Phys. Lett. **57** (10), 977 (1990)
Kiguchi M., Kato M. Okunaka M. and Taniguchi Y., Appl. Phys. Lett. **60**, 1933 (1992)
Kinoshita T., Nonaka Y., Nihei E., Koike Y. and Sasaki K., in *Nonlinear Optics*, editor S. Miyata, Elsevier Science Publishers, Amsterdam (1992), pp. 479–485
Kitaigorodski A. I., Mixed Crystals, Solid State Sciences **33**, Springer (1984)
Kjaer K., Als-Nielsen J., Helm C. A., Laxhuber L. A. and Möhwald H., Phys. Rev. Lett. **58**, 2224 (1987)
Kjaer K., Als-Nielsen J., Helm C. A., Tippman-Krayer P. and Möhwald H., J. Phys. Chem. **93**, 3200 (1989)
Klein M. B., Opt. Lett. **9** (8), 350 (1984)
Kleinman D. A., Phys. Rev. **126**, 1977 (1962)
Klinkhammer F., diploma thesis, University of Mainz (1989)
Knöpfle G., Bosshard Ch., Schlesser R. and Günter P., IEEE J. Quantum Electron. **30**,1303 (1994)
Kobayashi T., Isoda S., Maeda T. and Hosino A., Cryst. Prop. Prep. **32–34**, 731 (1991)
Kondo S. (Hitachi Ltd.), patent: Jpn. Kokai Tokyo Koho JP 0328, 190 [9128, 190] (Cl. C30B23/08), Appl. 89/1607/748] (1989a)
Kondo T., Morita R., Ogasawara N., Umegaki S. and Ito R., Jap. J. Appl. Phys. **28** (9), 1622–1628 (1989b)
Kondo T., Ogasawara N., Umegaki S. and R. Ito, Mol. Cryst. Liq. Cryst. **182A**, 83 (1990)
Krumins A. and Günter P., Appl. Phys. **19**, 153 (1979)
Kubota Y. and Yoshimura T., Appl. Phys. Lett. **53**, 2579 (1988)
Kuhn H., Möbius D. and Bücher, H., in *Physical Methods of Chemistry, part III B*, editors A. Weissberger and B. W. Rossiter, Wiley-Interscience, New York (1972), p. 577
Kuktarev N. V. and Odulov S. G., ETP Lett **30**, 4 (1979a)
Kukhtarev N. V., Markov V. B., Odulov S. G., Soskin M. S. and Vinetskii V. L., Ferroelectrics **22**, 961 (1979b)
Kuroda S.-I., Ikegami K., Saito K., Saito M. and Sugi M., Thin Solid Films **178**, 555 (1989)
Kurthen C. and Nitsch W., Adv. Mater. **3**, 445 (1991)
Kurtz S. K. and Perry T. T., J. Appl. Phys. **19**, 3768 (1968)
Kurtz S. K., *Materials for nonlinear optics, Laser Handbook Vol. 1.* editors F. T. Arecchi and E. O. Schulz-DuBois, North Holland Publishing Company (1972), pp. 923–974
Kurtz S. K., private communication (1990)
Landolt-Börnstein (vol. 8, group III, 1972)
Landolt-Börnstein, *Elastische, piezoelektrische, pyroelektrische, piezooptische Konstanten und nichtlineare dielektrische Suszeptibilitäten von Kristallen, Band III/11*, editors K.-H. Hellwege and A. M. Hellwege, Springer-Verlag, Berlin (1979a)
Landolt-Börnstein , *Elastische, piezoelektrische, pyroelektrische, piezooptische Konstanten und nichtlineare dielektrische Suszeptibilitäten von Kristallen, Band III/11*, editor K.-H. Hellwege and A. M. Hellwege, Springer-Verlag, Berlin (1979b), for the nonlinear optical susceptibility d_{11} of quartz we used the arithmetic average of the different values reported there.
Langmuir I., J. Am. Chem. Soc. **39**, 1848(1917)
Langmuir I., J. Chem. Phys. **1**, 756 (1933)
Laschewski A., Paulus W., Ringsdorf H., Lupo D., Ottenbreit P., Prass W., Bubeck C., Neher D. and Wegner G., Makromol. Chem., Macromol. Symp. **46**, 205 (1991)
Lawless W. N. and Vries R. C., J. Opt. Soc. Am. **54**, 1225 (1964)
Lawrence B., M. Sc. thesis, Dept. of Electrical Engineering, MIT (1992)

Leblanc Jr. O. H. in *Physics and chemistry of the organic solid state*, Vol. III, editors D. Fox, M. M. Labes, and A. Weissberger A., Wiley and Sons, New York (1967), pp. 133–192

Leblanc R.M. and Salesse C. (eds.), Proceedings of the Sixth International Conference on Organized Molecular Films, Thin Solid Films **242–244**, Elsevier, Lausanne (1994)

Ledoux I., Josse D., Fremaux P., Piel J.-P., Post G., Zyss J., McLean T., Hann R. A., Gordon P. F. and Allen S., Thin Solid Films **160**, 217 (1988)

Ledoux I., Lepers C., Perigaud A., Badan J. and Zyss J., Opt. Commun. **80** (2), 149 (1990)

Ledoux I., Zyss J., Jutand A. and Amatore C., Chem Phys. **150**, 177 (1991)

Lesaux G., Lauanay J. C. and Brun A., Opt. Commun. **57** (3), 166 (1986)

Leslie T. M., De Martino R. N., Choe E. W., Khanarian G., Hass D., Nelson G., Stamatoff J. B., Stuetz D. E., Teng C. C. and Yoon Y. N., Mol. Cryst. Liq. Cryst. **153**, 451 (1987)

Levine B. F. and Bethea C. G., J. Chem. Phys. **63** (6), 2666 (1975)

Levine B. F. and Bethea C. G., J. Chem. Phys. **65** (6), (1976)

Levine B. F., Bethea C. G., Thurmond C. D., Lynch R. T. and Bernstein J. L., J. Appl. Phys. **50** (4), 2523 (1979)

Levy Y. Dumont M., Chastaing E., Robin P., Chollet P. A., Gadret G. and Kajzar F., Mol. Cryst. Liq. Cryst. Sci. Technol.-Sec. B: Nonlinear Optics **4**, 1 (1993)

Li D. Q., Ratner M. A. and Marks T. J., Am. Chem. Soc. **110**, 1707 (1988)

Li L., Lee J. Y., Yang Y., Kumar J. and Tripathy S. K., Appl. Phys. B **53**, 279 (1991)

Lin J. T., Gong-Fan Huang, Ming-Yi Hwang, and Chong S. W., SPIE Proc. **1220**, SPIE-The Internatoinal Society for Optical Engineering, Bellingham, Washington (1990a), pp. 18–23

Lin B., Shih M. C. and Bohanon T. M., Phys. Rev. Lett. **65**, 191 (1990b)

Lipscomb G. F., Garito A. F. and Narang R. S., J. Chem. Phys. **75**, 1509 (1981)

Looser H., (1990), unpublished

Lösche M., Sackmann E. and Möhwald H., Ber. Bunsenges. Phys. Chem. **87**, 848 (1983)

Loulergue J. C., Lévy Y., Dumont M., Robin P. and Pocholle J. P., in *Advanced Optoelectronic Technology*, SPIE Proc. **864**, SPIE-The International Society for Optical Engineering, Bellingham, Washington (1987), p. 14

Lytel R., Lipscomb G. F., Stiller M., Thackara J. I. and Ticknor A. J., *Nonlinear Optical Effects in Organic Polymers*, NATO ASI Series E Vol. 162, Kluver Academic Publisher (1989a) p. 277

Lytel R., Lipscomb G. F., Stiller M., Thackara J. I. and Ticknor A. J., *Proc. of Conf. on Materials for Non-linear Optics, Oxford, 1988* editor R. A. Hann and D. Bloor (1989b), pp. 382–389

Maker P. D., Terhune R. W., Nisenoff M. and Savage C. M., Phys. Rev. Lett. **8**, 21 (1962)

Mathy A., Mathauer K., Wegner G. and Bubeck C., Thin Solid Films **215** 98 (1992)

Matsumoto M., Sekiguchi T., Tanaka H., Tanaka M., Nakamura T., Tachabana H., Manda E., Kawabata Y. and Sugi M., J. Phys. Chem. **3** 5877 (1989)

McConnell H. M. and Moy V. T., J. Phys. Chem. **92**, 4520 (1988)

Medrano C., Voit E., Amrhein P. and Günter P., J. Appl. Phys. **64** (9), 4668 (1988)

Melby L. R., Harder R. J., Hertler W. R., Mahler W., Benson R. E. and Mochel W. E., J. Am. Chem. Soc. **84**, 3374 (1962)

Meredith G. R. in *Nonlinear Optical Properties of Organic and Polymeric Materials*, editor D. J. Williams, American Chemical Society, Washington, DC (1983), pp. 109–133

Meredith G. R., in *Nonlinear optics: materials and devices*, editor C. Flytzanis, J. L. Oudar, Erice, Sicily, Springer Proceedings in Physics 7, (1986), pp. 116–127

Michel-Calendini F. M. and Boyeaux J. P., Solid State Commun. **28**, 257 (1978)

Miller A., Knoll W. and Möhwald H., Phys. Rev. Lett. **56**, 2633 (1986)

Miller A. and Möhwald H., J. Chem. Phys. **86**, 4258 (1987)

Minari N., Ikegami K., Kuroda S., Saito K., Saito M. and Sugi M., Solid State Commun. **65**, 1259 (1988)

Minari N., Ikegami K., Kuroda S., Saito K., Saito M. and Sugi M., Journal of the Physical Society of Japan **Vol. 58**, 222 (1989)

Mittler-Neher S., Neher D., Stegeman G. I., Embs F. W. and Wegner G., Chem. Phys., **161** 289 (1992)

Möhlmann G. R., Horsthuis W. H. G., McDonach A., Copeland M. J., Duchet C., Fabre P. , Diemeer M. B. J., Trommel E. S., Suyten F. M. M., Van Tomme E., Baquero P., and Van Daele P., SPIE Proc. **1337**, SPIE-The International Society for Optical Engineering, Bellingham, Washington (1990), p. 215

Möhwald H., in *The Physics and Fabrication of Microstructures and Microdevices*, editors M. J. Kelly and C. Weisbuch, Springer-Verlag, Berlin, Springer Series in Physics, 13, (1986) p. 166

Möhwald H., Thin Solid Films **159**, 1 (1988a)

Möhwald H., Kirstein S., Haas H. and Flörsheimer M., J. Chim. Phys. **85**, 1009 (1988b)

Morita R., Ogasawara N., Umegaki S. and Ito R., Jap. J. Appl. Phys. **26** (10), L1711 (1987)

Morley J. D., Dochtery V. J. and Pugh D., J. Chem. Soc. Perkin Trans. **2**, 1361 (1987)

Morley J. O., in *Nonlinear Optics of Organics and Semiconductors*, editor T. Kobayashi, Springer-Verlag, Berlin, Heidelberg (1989) pp. 86–97

Mort J., Advances in Physics **29** (2), 367 (1980)

Mort J. and Pfister G., in *Electronic properties of polymers*, edited by J. Mort and G. Pfister, J. Wiley and Sons, New York (1982), pp. 215–266

Mortazavi M. A., Knoesen A., Kowel S. T., Higgins B. G. and Dienes A., J. Opt. Soc. Am. B **6** (4), 733 (1989)

Moy, V. T., Keller, D., Gaub, H. E. and McConnell, H. M., J. Phys. Chem. **90**, 3198 (1986)

Mullen R. A. in *Photorefractive Materials and Their Applications: Fundamental Phenomena*, editors P. Günter and J.-P. Huignard, Springer-Verlag, Berlin (1988), pp. 167–194

Muller P. and Gallet F., J. Phys. Chem. **95**, 3257 (1991)

Nahata A., Horn K. A. and Yardley J. T., IEEE J. Quantum Electron. **26** (9), 1521 (1990)

Nakanishi H., J. Jap. Assoc. Cryst. Growth **16**, 17 (1989)

Nayer B. K., in *Nonlinear optical properties of organic and polymeric materials*, editor D. J. Williams, American Chemical Society, Washington, DC (1983), pp. 153–166

Nelson, D. R. and Halperin, B. I., Phys. Rev. B **19**, 2457 (1979)

Nelson, D. R., Rubinstein, M. and Saepen, F., Phil. Mag. A **46**, 105 (1982)

Nelson, D. R., Phys. Rev. B **27**, 2902 (1983)

Noguchi K., Mitomi O., Kawano K. and Yanagibashi M., IEEE Photon. Technol. Lett. **5** (1), 52 (1993)

Nicoud J. F. and Twieg R. J., in *Nonlinear optical properties of organic molecules and crystals*, editors D. S. Chemla and J. Zyss, Academic Press Inc., Orlando (1987), pp. 227–296

Nitsch W., Kremnitz W. and Schweyer G., Ber. Bunsenges. Phys. Chem. **91**, 218 (1987)

Norwood R. A. and Khanarian G., Electron. Lett. **26**, 2105 (1990)

Nye J. F., *Physical Properties of Crystals*, Clarendon Press, Oxford (1967)

Okada S., Masaki A., Matsuda H., Nakanishi H., Koike T., Ohmi T. and Yoshikawa N., SPIE Proc. **1337**, editor G. Khanarian, SPIE-The International Society for Optical Engineering, Bellingham, Washington, (1990)

Ollenik R. and Nitsch W., Ber. Bunsenges. Phys. Chem. **85**, 900 (1981)

Ono S., Okano Y. and Fukuda T., *Preprints International Workshop on Crystal Growth of Organic Materials*, Tokyo, Japan (1989)

Orczyk M. E., Zieba J., and Prasad P. N., SPIE Proc. **2025**, SPIE-The International Society for Optical Engineering, Bellingham, Washington (1993), in print
Orthmann E. and Wegner G., Angew. Chem. Int. Ed. Engl. **25**, 1106 (1986)
Ottelenghi M. and McClure D. S., J. Chem. Phys. **46** (12), 4613 (1967)
Oudar J. L. and Chemla D. S., Opt. Commun. **13**, 10 (1975)
Oudar J. L., J. Chem. Phys. **67** (2), 446 (1977a)
Oudar J. L. and Hierle R., J. Appl. Phys. **48** (7), 2699 (1977b)
Oudar J. L. and Chemla D. S., Chem. Phys. **66** (6), 2664 (1977c)
Oudar J. L. and Zyss J., Phys. Rev. A **26**, 2076 (1982)
Paley M. S., Harris J. M., Looser, H. Baumert J. C., Bjorklund G. C., Jundt D. and Twieg R. J., J. Org. Chem. **54**, 3774 (1989)
Pauling L., *The Nature of the Chemical Bond*, Cornell Univ. Press, Ithaca, New York (1960)
Penn B. G., Shields A. W. and Frazier O. D., NASA Tech. Memo., NASA-TM-100341, NAS 1.15:100341 (1988); avail.: Sci. Tech. Aerosp. Rep. 27, abstr. no. N89-10656 (1989)
Penner T. L., Armstrong N. J., Willand C. S., Schildkraut J. S. and Robello D. R., in *Nonlinear Optical Properties of Organic Materials IV*, SPIE Proc. **1560**, SPIE-The International Society for Optical Engineering, Bellingham, Washington (1991), p. 377
Perigaud A. and Nicolau Y. F., J. Cryst. Growth **79**, 752–757 (1986)
Perlstein J. H. and Borsenberger P. M., in *Extended linear chain compounds*, Vol. 2., editor J. S. Miller, Plenum Press, New York (1982), pp. 339–385
Perry J. W., Marder S. R., Perry K. J., Sleva E. T., Yakymyshyn C. P., Stewart K. R. and Boden E. P., SPIE Proc. **1560** editor K. Singer, SPIE-The International Society for Optical Engineering, Bellingham, Washington (1991), pp. 302–309
Peterson I. R., J. Mol. Electron. **2**, 95 (1986)
Peterson I. R., J. Mol. Electron. **3**, 103 (1987a)
Peterson I. R., Earls J. D., Girling, I. R. and Russel G. J., Mol. Cryst. Liq. Cryst. **147**, 141 (1987b)
Peterson I. R., Russell G. C., Earls J. D. and Girling I. R., Thin Solid Films **161**, 325 (1988a)
Peterson I. R., Earls J. D., Girling I. R. and Barnes W. L., J. Phys. D: Appl. Phys. **21**, 773 (1988b)
Peterson I. R., J. Chim. Phys. **85**, 997 (1988c)
Peterson, I. R., J. Phys. D: Appl. Phys. **23**, 379 (1990)
Philipps M. C., Progr. Surf. Membr. Sci. **5**, 139 (1972)
Pierce B. M., SPIE Proc. **1560** editor K. Singer, SPIE-The International Society for Optical Engineering, Bellingham, Washington (1991)
Pitt C. W. and Walpita L. M., Electron. Lett. **12**, 479–481 (1976)
Pitt C. W. and Walpita L. M., Thin Solid Films **68** 101 (1980)
Pope M. and Swenberg C. E., *Electronic Processes in Organic Crystals*, Clarendon Press, Oxford (1982)
Pope M., Mol. Cryst. Liq. Cryst. **171**, 89 (1989)
Powell M. A. and Petts C. R., Opt. Lett. **11** (1), 36 (1986)
Puterman M., Fort T. and Lando J. B., J. Colloid Interface Sci. **47**, 105 (1974)
Rak D., Ledoux I. and Huignard J. P., Opt. Commun. **49** (4), 302 (1984)
Rakuljic G. A., Yariv A. and Ratnakar R., Optical Engineering **25** (11), 1212 (1986)
Rao V. P., Jen, A. K.-Y., Wong, K. Y. and Drost K. J., Tetrahedron Lett. **34** (11), 1747 (1993a)
Rao V. P., Jen, A. K.-Y., Wong, K.Y. and Drost K. J., J. Chem. Soc., Chem. Commun. 1118 (1993b)
Rayleigh (Lord Rayleigh), Phil. Mag. **48**, 337 (1890)
Regener R. and Sohler W., J. Opt. Soc. Am. B **4** (2), 267 (1988)

Reichardt Ch., *Solvents and solvent effects in organic chemistry*, Verlag Chemie, Weinheim (1988)
Reiter R., Motschmann H., Orendi H., Nemetz A. and Knoll W., Langmuir **8**, 1784 (1992)
Richardson T., Roberts G. G., Polywka M. E. C. and Davies S. G., Thin Solid Films **179**, 405 (1989)
Riegler H., Rev. Sci. Instrum. **59**, 2220 (1988)
Rikken G. L. J. A., Seppen C. J. E., Nijhuis S. and Meijer E. W., Appl. Phys. Lett. **58** (5), 435 (1991)
Rikken G. L. J. A., Seppen C. J. E., Staring E. G. J. and Venhuizen A. H. J., Appl. Phys. Lett. **62**, 2483 (1993)
Risk W. P., Kozlovsky W. J., Lenth W., Lau S. D., Bona G. L., Jaeckel H. and Webb D. J., *Advanced Solid-State Lasers and Compact Blue-Green Lasers Technical Digest*, **Vol. 2**, (1993), pp. 489–491
Roberts G. editor, *Langmuir-Blodgett Films*, Plenum Press, New York (1990)
Rustagi K. C. and Ducuing J., Opt. Commun. **10**, 258 (1974)
Rytz D., Klein M. B., Mullen R. A., Schwartz R. N., Valley G. C. and Wechsler B. A., Appl. Phys. **52** (32), 1759 (1988)
Sackmann E., Fischer A. and Frey W., in *Springer Proceedings in Physics* **21** (Physics of Amphiphilic Layers, editors J. Meunier, D. Langevin, and N. Boccara, 25 (1987)
Sagawa M., Kagawa H., Kakuta A. and Kaji M., Appl. Phys. Lett. **63** (14), 1877 (1993)
Saito K., Ikegama K., Kuroda S., Sioto M. and Sugi M., Thin Solid Films **179**, 369 (1989)
Sasaki K., Kinoshita T. and Karasawa N., Appl. Phys. Lett. **45** (3), 333 (1984)
Sauer T., Arndt T., Batchelder D. N., Kalachev A. A. and Wegner G., Thin Solid Films **187**, 357 (1990)
Scheffen-Lauenroth Th., Klapper H. and Becker R. A., J. Cryst. Growth **55**, 557–570 (1981)
Schein L. B. and Brown D. W., Mol. Cryst. Liq. Cryst. **87**, 1 (1982)
Schildkraut J. S., Penner T. L., Willand C. S. and Ulman A., Optics Letters **13**, 134 (1988)
Schildkraut J. S., Appl. Phys. Lett. **58** (4), 340 (1991)
Schlenk W., Ann. **368**, 277 (1909)
Schlossman M. L., Schwartz D. K., Pershan P. S., Kawamoto E. H., Kellogg G. J. and Lee S., Phys. Rev. Lett. **66**, 1599 (1991)
Schoenes, J., *Festkörperoptik, Skriptum zur Vorlesung*, ETH Zürich (1984/85)
Schwartz D. K., Garnaes J., Viswanathan R. and Zasadzinski J. A. N., Science **Vol. 257**, 508 (1992)
Scott R. A. M., Laridjani S. R. and Morantz D. J., J. Cryst. Growth **22**, 53 (1974)
Scott J. C., Pautmeier L. Th. and Moerner W. E., J. Opt. Soc. Am. B **9** (11), 2059 (1992)
Shen Y. R., *The Principles of Nonlinear Optics*, John Wiley and Sons, Inc., New York (1984)
Sherwood J. N., in *Organic Materials for Nonlinear Optics*, editors R. A. Hann and D. Bloor, Royal Society of Chemistry, Special Publication No. 69 (1989)
Shi Y., Psaltis D., Marrakchi A. and Tanuay Jr. A. R., Appl. Opt. **22**, 3665 (1983)
Shumate M. S., Appl. Opt. **5** (2) 327 (1966)
Shuto Y., Takara H., Amano M. and Kaino T., Jap. J. Appl. Phys. **28** (12), 2508 (1989)
Shuto Y., Amano M. and Kaino T., IEEE Trans. Photon. Techn. Lett. **3** (11), 1003 (1991)
Sigelle M. and Hierle R., J. Appl. Phys. **52**, 4199 (1981)
Silence S. M., Walsh C. A., Scott J. C., Matray T. J., Twieg R. J., Hache F., Bjoirklund G. C. and Moerner W. E., Optics Lett. **17** (16), 1107 (1992a)
Silence S. M., Walsh C. A., Scott J. C. and Moerner W. E., Appl. Phys. Lett. **61** (25), 2967 (1992b)

Silence S. M., Hache F., Donckers M., Walsh C. A., Burland D. M., Bjorklund G. C., Twieg R. H. and Moerner W. E., personal communication, to be published in SPIE proceedings (1993)

Silinsh E. A., *Organic Molecular Crystals*, Springer-Verlag, Berlin (1980)

Sinclair M., Moses D., Akagi K. and Heeger A. J., Material Research Society **Vol. 109**, Pittsburgh, Pennsylvania (1988), p. 205

Singer K. D. and Garito A. F., J. Chem. Phys. **75**, 3572 (1981)

Singer K. D., Sohn J. E. and Lalama S. L., Appl. Phys. Lett. **49**, 248 (1986)

Singer K. D., Kuzyk M. G. and Sohn J. E., J. Opt. Soc. Am. B **4** (6), 968 (1987)

Singer K. D., Kuzyk M. G., Holland W. R., Sohn J. E., Lalama S.J. Camizzoli R. B., Katz H. E. and Schilling M. L., Appl. Phys. Lett. **53** (19), 1800 (1988)

Singer K. D., Sohn J. E., King I. A., Gordon H. M., Katz H. E. and Dirk C. W., J. Opt. Soc. Am. B **6** (7), 1339 (1989)

Sloan G. J. and McGhie A. R., *Techniques of Melt Crystallization*, John Wiley, New York (1988)

Smith D. P. E., Hörber, J. K. H., Binnig G. and Nejoh H., Nature **344**, 641 (1990)

Sohn J. E., Singer K. D., Kuzyk M. G., Holland W. R., Katz H. E., Dirk C. W., Schilling M. L. and Comizzoli R. B., *Nonlinear Optical Effects in Organic Polymers*, NATO ASI Series E **Vol. 162**, Kluwer Academic Publishers, (1989), p. 291

Solymar L. in *Electro-optic and photorefractive materials*, editor P. Günter, Springer-Verlag, Berlin, (1987), pp. 229–265

Sommerfeldt R., Holtmann L., Krätzig E. and Grabmeier B. C., Phys. Stat. Sol. (a) **106**, 89 (1988)

Sorita T., Miyake S., Fujioka H. and Nakajima H., Jap. J. Appl. Phys. **30**, 131 (1991)

Spratte, K. and Riegler H., Makromol. Chem., Macromol. Symp. **46**, 113 (1991)

Staebler D. L. and Amodei J. J., J. Appl. Phys. **43** (3), 1042 (1972)

Stähelin M., Burland D. M. and Rice J. E., Chem. Phys. Lett. **191** (3, 4), 245 (1992)

Stegeman G. I. and Seaton C. T., J. Appl. Phys. **58** (12), R5 (1985)

Stegeman G. I., Seaton C. T., Hetherington W. M., Boardman A. D. and Egan P. in *Nonlinear optics: Materials and devices* (Springer Proc. in Phys. 7) editors C. Flytzantis and J. L. Oudar (1986) pp. 31–64

Stegeman G. I., Seaton C. T. and Zanoni R., Thin Solid Films **152**, 231 (1987)

Stegeman G. I. and Stolen R. H., J. Opt. Soc. Am. B **6** (4), 652 (1989)

Stegeman G. I., Otomo A., Bosshard Ch. and Flipse R., *Proceedings of the 2nd International Conference on Frontiers of Polymers and Advanced Materials*, editor P. Prasad, Plenum, N. Y. (1993) in press

Steinhoff R., Chi L. E., Makrowski G. and Möbius D., J. Opt. Soc. Am. B **6**, 843 (1989)

Stevenson J. L., J. Phys. D **6**, L13 (1973)

Stevenson S. H. and Meredith G. R., SPIE Proc. **682**, SPIE-The International Society for Optical Engineering, Bellingham, Washington (1987), p. 147

Stolka P. and Pai D. M., Adv. in Polymer Science **29**, 1 (1978)

Sugi M., Saito M., Fukui T. and Iizima S., Thin Solid Films **99**, 17 (1983)

Sugi M., J. Mol. Electron. **1**, 3 (1985a)

Sugi M., Sakai K., Saito M., Kawabata Y. and Iizima S., Thin Solid Films **132**, 69 (1985b)

Sugi M., Minari N., Ikegami K., Kuroda S., Saito K. and Saito M., Thin Solid Films **178**, 157 (1989)

Sugihara O., Kai S., Uwotoko K., Kinoshita T. and Sasaki K., Appl. Phys. Lett. **68**, 4990–4992 (1990)

Sugihara O., Toda T., Ogura T., Kinoshita T. and Sasaki K., Opt. Lett. **16** (10), 702 (1991a)

Sugihara O., Kinoshita T., Okabe M., Kunioka S., Nonaka Y. and Sasaki K., Appl. Opt. **30** (21) 2957 (1991b)

Sutter K., Bosshard C., Ehrensperger M., Günter P. and Twieg R. J., IEEE J. Quantum Electron. **24** (12), 2362 (1988a)

Sutter K., Bosshard Ch., Wang W. S., Surmely G. and Günter P., Appl. Phys. Lett. **53** (19), 1779 (1988b)

Sutter K., Bosshard Ch., Baraldi L. and Günter P. in *Materials for Non-linear and Electro-optics* (Inst. of Physics, Conference Series Number 103, IOP, Bristol), editor M. H. Lyons (1989) pp. 127–132

Sutter K., Hulliger J. and Günter P., Solid State Commun. **74**, 867 (1990a)

Sutter K. and Günter P., J. Opt. Soc. Am. **B7** (12), 2274 (1990b)

Sutter K., Knöpfle G., Saupper N., Hulliger J. and Günter P., J. Opt. Soc. Am. B **8** (7), 1483 (1991a)

Sutter K., Doctoral thesis at the Swiss Federal Inst. of Techn., Zürich, Switzerland, Diss. No. 9671 (1991b)

Sutter K., Hulliger J., Schlesser R. and Günter P., Opt. Lett. **18** (10), 778 (1993)

Suzuki H. and Hiratsuka H., SPIE Proc **971**, SPIE-The International Society for Optical Engineering, Bellingham, Washington (1988), pp. 97–106

Tabe Y., Ikegami K., Kuroda S., Saito K., Saito M. and Sugi M., Appl. Phys. Lett. **57**, 1191 (1990)

Tam W., Guerin B., Calabrese J. C. and Stevenson S. H., Chem. Phys. Lett.**154** (2), 93 (1989)

Tamura K., Padias A. B., Hall Jr. H. K. and Peyghambarian N., Appl. Phys. Lett **60** (15), 1803 (1992)

Taniuchi T. and Yamamoto K., *Digest of Conference on Lasers and Electro-Optics*, paper WR3, Optical Society of America, Washington D. C. (1986), p. 230

Teng C. C. and Garito A. F., Phys. Rev. B **28** (12), 6766 (1983)

Teng C. C. and Man H. T., Appl. Phys. Lett. **56** (18), 1734 (1990)

Teng C. C., Appl. Phys. Lett. **60** (13), 1538 (1992)

Tien P. K., Ulrich R. and Martin R. J., Appl. Phys. Lett. **17** (10), 447 (1970)

Tippmann-Krayer, P., Kenn, R. M. and Möhwald, H., Thin Solid Films **210/211**, 577–582 (1992)

Tipson R. S., Crystallization and Recrystallization, in *Technique of Organic Chemistry*, **vol. III**, part I: separation and purification, editor A. Weissberger, Wiley, New York, (1966)

Todorov T., Nikolova L. and Tomova N, Appl. Optics **23** (23), 4309 (1984)

Tohmon G., Ohya J., Yamamoto K. and Taniuchi T., IEEE Photon. Tech. Lett. **2** (9), 629 (1990)

Tomaru S., Matsumoto S., Kurihara T., Suzuki H., Ooba N. and Kaino T., Appl. Phys. Lett. **58** (23), 2583 (1991)

Tomita E., Sawara T. and Ataka T., Seiko Instruments and Electronics Ltd., patent: JP 63/201/1098 (1988)

Tomlinson W. J., Chandross E. A., Fork R. L., Pryde C. A. and Lamola A. A., Appl. Optics, **11** (3), 533 (1972)

Tomlinson W. J. and Chandross E. A., Adv. in Photochemistry **12**, 201 (1980)

Tredgold R. H., Young M. C. J., Hodge P. and Koshdel E., Thin Solid Films **151**, 441 (1987)

Tsunekawa T., Gotoh T. and Iwamoto M., Chem. Phys. Lett. **166** (4), 357 (1990a)

Tsunekawa T., Gotoh T., Mataki H., Kondoh T., Fukuda S. and Iwamoto M., SPIE Proc. **1337**, editor G. Khanarian, SPIE-The International Society for Optical Engineering, Bellingham, Washington 1990b, pp. 272ff

Twarowski A., J. Appl. Phys. **65** (7), 2833 (1989)
Twieg R., Azema A., Jain K. and Cheng Y. Y., Chem. Phys. Lett. **92** (2), 208 (1982)
Twieg R. J. and Jain K., in *Nonlinear optical properties of organic and polymeric materials*, editor D. J. Williams, American Chemical Society, Washington DC (1983), pp. 57–80
Twieg R. J. and Dirk C. W., J. Chem. Phys. **85**, 3537 (1986)
Uemiya U., Uenishi N., Shimizu Y., Yoneyama T. and Nakatsu K., Mol. Cryst. Liq. Cryst. **182A**, 51–57 (1990)
Uemiya U., Uenishi N., Okamoto S., Chikuma K., Kumara K., Kondo T., Ito R. and Umegaki S., Appl. Opt. **31** (36), 7581 (1992)
Ulman A., Willand C. S., Koehler W., Robello D. R., Williams D. J., and Handley L., J. Am. Chem. Soc. **112**, 7083 (1990)
Ulman A., *An Introduction to Ultrathin Organic Films*, Academic Press, Boston (1991)
Valley G. C., and Lam J. F., in *Photorefractive Materials and Their Applications I: Fundamental Phenomena*, editors P. Günter and J.-P. Huignard, Springer-Verlag, Berlin (1988), pp. 76–98
van der Poel C. J., Bierlein J. D., Brown J. B. and Colak S., Appl. Phys. Lett. **57** (20), 2074 (1990)
Van der Vorst C. P. J. M. and Picken S. J., J. Opt. Soc. Am. B **7** (3), 320 (1990)
van der Sluis P. and Kroon J., J. Cryst. Growth **97**, 645 (1989)
Vanderlick T. K. and Möhwald H., J. Phys. Chem. **94**, 886 (1990)
Vanherzeele H., Meth J., Jenekhe S. A. and Roberts M. F., Appl. Phys. Lett. **58**, 663 (1991)
Vazquez R. A., Ewbank M. D. and Neurgaonkar R. R., Opt. Commun. **80** (3, 4), 253 (1991)
Verbiest T., Hendrickx E. and Persoons A., in SPIE Proc. **1775**, editor D. Williams, SPIE-The International Society for Optical Engineering, Bellingham, Washington (1993), pp. 206–212
Voges E., in *Electro-optic and photorefractive materials*, editor P. Günter Springer-Verlag, Berlin, Heidelberg, New York (1987), pp. 150–158
Voit E. and Günter P., Opt. Lett. **12**, 769 (1987)
Voit E., Doctoral thesis at the Swiss Federal Inst. of Techn., Zürich, Switzerland, Diss. No. 8555 (1988)
Wada T., Yamada A. and Sasabe H., *Preprints International Workshop on Crystal Growth of Organic Materials*, Tokyo, Japan, (1989), pp. 229–233
Wahlstrohm E., *Optical Crystallography*, John Wiley, New York (1969)
Walsh C. A. and Moerner W. E., J. Opt. Soc. Am. **9** (9), 1642 (1992)
Wang W. S., Sutter K., Bosshard Ch., Pan Z., Arend H., Günter P., Chapuis G. and Nicolo F., Jap. J. Appl. Phys. **27**, 1138–1141 (1988a)
Wang Y., Tam W., Stevenson S. H., Clement R. A. and Calabrese J., Chem. Phys. Lett. **148**, 136–141 (1988b)
Wang W. S., Hulliger J. and Arend H. Ferroelectrics **92**, 113–119 (1989)
Ward J. F., Rev. Mod. Phys. **37**, 1 (1965)
Weiss R. M. and McConnell H. M., Nature **310**, 5972 (1984)
Wemple S. H. and DiDomenico M., in *Applied Solid State Science*, **Vol 3**, Advances in materials and device research, editor R. Wolfe, Academic Press, New York (1972), pp. 263–381
Wilke K. Th. and Bohm J., *Kristallzüchtung*, Deutscher Verlag der Wissenschaften (1988)
Williams D. J., editor, *Nonlinear Optical Properties of Organic Molecules and Polymeric Materials*, ACS Symposium Series N°233, Washington DC (1983)
Williams M. L., Landel R. F. and Ferry J. D., J. Amer. Chem. Soc., **77**, 3701 (1955)
Wong K. Y. and Garito A. F., Phys. Rev. A **34**, 5051 (1986)

Wu J. W., Heflin J. R., Norwood R. A., Wong K. Y., Zamini-Khamiri O., Garito A. F., Kalyanaraman P. and Sounik P., J. Opt. Soc. Am. B **6**, 707 (1989)

Wu J. W., Valley J. F., Ermer S., Binkley E. S., Kenney J. T., Lipscomb G. F. and Lytel R., Appl. Phys. Lett. **58** (3), 225 (1991)

Yamada M., Nada N., Saitoh M. and Watanabe K., Appl. Phys. Lett. **62** (5), 435 (1993)

Yamamoto K., Mizuuchi K., Kitaoka Y. and Kato M., Appl. Phys. Lett. **61** (21), 2599 (1993a)

Yamamoto K., Mizuuchi K., Kitaoka Y. and Kato M., *Advanced Solid-State Lasers and Compact Blue-Green Lasers Technical Digest*, **Vol. 2**, (1993b), pp. 480–482

Yariv A., IEEE J. Quantum Electron. **QE-9** (9), 919 (1973)

Yariv A., *Quantum Electronics*, John Wiley and Sons, New York (1975), pp. 327–370

Yeh P., Appl. Opt. **26** (4), 602 (1987)

Yitzchaik S., Berkovic G. and Kronganz V., Opt. Lett. **15** (20), 1120 (1990)

Yoshimura T., J. Appl. Phys. **62** (5), 2028 (1987)

Yoshimura T., Fujitsu Ltd., patent: JP 63/220219, Styrylpyridinium cyanine dye thin-film crystal electrooptical device (DSMS) (1988)

Yoshimura T. and Kubota Y., in *Nonlinear Optics of Organics and Semiconductors*, editor T. Kobayashi, Springer Procs. in Physics, **Vol. 36**, Berlin (1989), pp. 222–226

Zgonik M., Schlesser R., Biaggio I., Voit E., Tscherry J. and Günter P., J. Appl. Phys. **74**, 1287 (1993)

Zhang G. J., Kinoshita T., Sasaki K., Goto Y. and Nakayama M., J. Jap. Assoc. Cryst. Growth **17**, 116 (1990a)

Zhang H. Y., He X. H., Chen E. and Liu Y., Appl. Phys. Lett. **57** (13), 1298 (1990b)

Zhuang J., Li G.-S., Gao X.-C., Guo X.-B., Huang Y.-H., Shi Z.-Z., Weng Y.-Y. and Lu J., Opt. Commun. **82** (1, 2), 69 (1990)

Zyss J., Chemla D. S. and Nicoud J. F., J. Chem. Phys. **74** (9), 4800 (1981)

Zyss J. and Oudar J. L., Phys. Rev. A **26**, 2028 (1982)

Zyss J., Nicoud J. F. and Coquillay M., J. Chem. Phys. **81** (9), 4160 (1984)

Zyss J., J. Mol. Electron. **1**, **25** (1985)

Zyss J. and Chemla D. S., in *Nonlinear optical properties of organic molecules and crystals*, editors D. S. Chemla and J. Zyss, Academic Press Inc., Orlando (1987), pp. 23–191

Zyss J., Nonlinear Opt. **1**, 3 (1991)

Zyss J., J. Chem. Phys. **98** (9), 6583 (1993)

Zysset B., Biaggio I. and Günter P., J. Opt. Soc. Am. B **9** (3), 379 (1992)

APPENDIX: CONVERSION BETWEEN SI AND ESU UNITS

In nonlinear optics the esu (cgs) system is still widely used. Therefore the conversion factors between SI and esu units for the most important parameters in this work are given.

electric field E	$E_{esu} = \frac{10^6}{c} \cdot E_{SI}$	$c = 2.99792 \cdot 10^{10}$ (velocity of light in $cm \cdot s^{-1}$
macroscopic polarization P	$P_{esu} = \frac{c}{10^5} \cdot P_{SI}$	
molecular dipole moment μ, p	$\mu_{esu} = 10c \cdot \mu_{SI}$	
linear optical susceptibility $\chi^{(1)}$	$\chi_{esu} = \frac{\chi_{SI}}{4\pi}$	
electro-optic coefficient	$r_{esu} = \frac{c}{10^6} \cdot r_{SI}$	
nonlinear optical susceptibility $\chi^{(2)}$, d	$d_{esu} = \frac{c}{4\pi \cdot 10^6} \cdot d_{SI}$	
electric-field induced nonlinearity $\Gamma = d/E\,\Gamma_{esu}$	$\Gamma_{esu} = \frac{c^2}{4\pi \cdot 10^{12}} \cdot \Gamma_{SI}$	
second-order polarizability	$\beta_{esu} = \frac{c}{4\pi} \cdot \beta_{SI}$	

INDEX

For Product Safety Concerns and Information please contact our EU
representative GPSR@taylorandfrancis.com
Taylor & Francis Verlag GmbH, Kaufingerstraße 24, 80331 München, Germany

www.ingramcontent.com/pod-product-compliance
Ingram Content Group UK Ltd.
Pitfield, Milton Keynes, MK11 3LW, UK
UKHW051834180425
457613UK00022B/1243